Closing the Gap Between ASIC & Custom

Tools and Techniques for
High-Performance ASIC Design

The cover was designed by Steven Chan. It shows the Soft-Output Viterbi Algorithm (SOVA) chip morphed with a custom 64-bit datapath. The SOVA chip picture is courtesy of Stephanie Ausberger, Rhett Davis, Borivoje Nikolic, Tina Smilkstein, and Engling Yeo. The SOVA chip was fabricated with STMicroelectronics. The 64-bit datapath is courtesy of Andrew Chang and William Dally. GSRC and MARCO logos were added.

Closing the Gap Between ASIC & Custom

Tools and Techniques for
High-Performance ASIC Design

David Chinnery
Kurt Keutzer
University of California
Berkeley

Boston/Dordrecht/London

Distributors for North, Central and South America:
Kluwer Academic Publishers
101 Philip Drive
Assinippi Park
Norwell, Massachusetts 02061 USA
Telephone (781) 871-6600
Fax (781) 681-9045
E-Mail: kluwer@wkap.com

Distributors for all other countries:
Kluwer Academic Publishers Group
Post Office Box 322
3300 AH Dordrecht, THE NETHERLANDS
Telephone 31 786 576 000
Fax 31 786 576 474
E-Mail: services@wkap.nl

 Electronic Services < http://www.wkap.nl>

Library of Congress Cataloging-in-Publication Data

A C.I.P. Catalogue record for this book is available
from the Library of Congress.

Copyright © 2002 by Kluwer Academic Publishers

All rights reserved. No part of this work may be reproduced, stored in a retrieval system, or transmitted in any form or by any means, electronic, mechanical, photocopying, microfilming, recording, or otherwise, without the written permission from the Publisher, with the exception of any material supplied specifically for the purpose of being entered and executed on a computer system, for exclusive use by the purchaser of the work.

Printed on acid-free paper.

Printed in the United States of America.

Contents

Preface — xi

List of trademarks — xv

1. **Introduction and Overview of the Book** — 1
   David Chinnery, Kurt Keutzer – UC Berkeley
   1. WHY ARE CUSTOM CIRCUITS *SO* MUCH FASTER? 1
   2. WHO SHOULD CARE? 1
   3. DEFINITIONS: ASIC, CUSTOM, ETC. 3
   4. THE 35,000 FOOT VIEW: WHY IS CUSTOM FASTER? 4
   5. MICROARCHITECTURE 9
   6. TIMING OVERHEAD: CLOCK TREE DESIGN AND
      REGISTERS 12
   7. LOGIC STYLE 15
   8. LOGIC DESIGN 17
   9. CELL DESIGN AND WIRE SIZING 18
   10. LAYOUT: FLOORPLANNING AND PLACEMENT TO
       MANAGE WIRES 20
   11. PROCESS VARIATION AND IMPROVEMENT 22
   12. SUMMARY AND CONCLUSIONS 26
   13. WHAT'S NOT IN THE BOOK 28
   14. ORGANIZATION OF THE REST OF THE BOOK 28

Contributing Factors

2. **Improving Performance through Microarchitecture** — 33
   David Chinnery, Kurt Keutzer – UC Berkeley
   1. EXAMPLES OF MICROARCHITECTURAL
      TECHNIQUES TO INCREASE SPEED 34
   2. MEMORY ACCESS TIME AND THE CLOCK PERIOD 44
   3. SPEEDUP FROM PIPELINING 45

3. **Reducing the Timing Overhead** — 57
   David Chinnery, Kurt Keutzer – UC Berkeley
   1. CHARACTERISTICS OF SYNCHRONOUS
      SEQUENTIAL LOGIC 58

    2. EXAMPLE WHERE LATCHES ARE FASTER 77
    3. OPTIMAL LATCH POSITIONS WITH TWO CLOCK
       PHASES 81
    4. EXAMPLE WHERE LATCHES ARE SLOWER 83
    5. PIPELINE DELAY WITH LATCHES VS. PIPELINE
       DELAY WITH FLIP-FLOPS 87
    6. CUSTOM VERSUS ASIC TIMING OVERHEAD 90

4. **High-Speed Logic, Circuits, Libraries and Layout**   **101**
   Andrew Chang, William J. Dally – Stanford University
   David Chinnery, Kurt Keutzer, Radu Zlatanovici – UC Berkeley
    1. INTRODUCTION 101
    2. TECHNOLOGY INDEPENDENT METRICS 102
    3. PERFORMANCE PENALTIES IN ASIC DESIGNS
       FROM LOGIC STYLE, LOGIC DESIGN, CELL DESIGN,
       AND LAYOUT 108
    4. COMPARISON OF ASIC AND CUSTOM CELL AREAS 129
    5. ENERGY TRADEOFFS BETWEEN ASIC CELLS AND
       CUSTOM CELLS 133
    6. FUTURE TRENDS 138
    7. SUMMARY 139

5. **Finding Peak Performance in a Process**   **145**
   David Chinnery, Kurt Keutzer – UC Berkeley
    1. PROCESS AND OPERATING CONDITIONS 146
    2. CHIP SPEED VARIATION DUE TO STATISTICAL
       PROCESS VARIATION 155
    3. CONTINUOUS PROCESS IMPROVEMENT 157
    4. SPEED DIFFERENCES DUE TO ALTERNATIVE
       PROCESS IMPLEMENTATIONS 159
    5. PROCESS TECHNOLOGY FOR ASICS 161
    6. POTENTIAL IMPROVEMENTS FOR ASICS 164

DESIGN TECHNIQUES

6. **Physical Prototyping Plans for High Performance**   **169**
   Michel Courtoy, Pinhong Chen, Xiaoping Tang, Chin-Chi Teng,
   Yuji Kukimoto – Silicon Perspective, a Cadence Company
    1. INTRODUCTION 169
    2. FLOORPLANNING 170
    3. PHYSICAL PROTOTYPING 172

4.	TECHNIQUES IN PHYSICAL PROTOTYPING	180
5.	CONCLUSIONS	185

7. Automatic Replacement of Flip-Flops by Latches in ASICs — 187
David Chinnery, Kurt Keutzer – UC Berkeley
Jagesh Sanghavi, Earl Killian, Kaushik Sheth – Tensilica

1.	INTRODUCTION	187
2.	THEORY	191
3.	ALGORITHM	199
4.	RESULTS	203
5.	CONCLUSION	207

8. Useful-Skew Clock Synthesis Boosts ASIC Performance — 209
Wayne Dai – UC Santa Cruz
David Staepelaere – Celestry Design Technologies

1.	INTRODUCTION	209
2.	IS CLOCK SKEW REALLY GLOBAL?	210
3.	PERMISSIBLE RANGE SKEW CONSTRAINTS	211
4.	WHY CLOCK SKEW MAY BE USEFUL	213
5.	USEFUL SKEW DESIGN METHODOLOGY	216
6.	USEFUL SKEW CASE STUDY	218
7.	CLOCK AND LOGIC CO-DESIGN	220
8.	SIMULTANEOUS CLOCK SKEW OPTIMIZATION AND GATE SIZING	220
9.	CONCLUSION	221

9. Faster and Lower Power Cell-Based Designs with Transistor-Level Cell Sizing — 225
Michel Côté, Philippe Hurat – Cadabra, a Numerical Technologies Company

1.	INTRODUCTION	225
2.	OPTIMIZED CELLS FOR BETTER POWER AND PERFORMANCE	226
3.	PPO FLOW	228
4.	PPO EXAMPLES	235
5.	FLOW CHALLENGES AND ADOPTION	238
6.	CONCLUSIONS	239

10. Design Optimization with Automated Flex-Cell Creation — 241
Debashis Bhattacharya, Vamsi Boppana – Zenasis Technologies

1.	FLEX-CELL BASED OPTIMIZATION – OVERVIEW	244
2.	MINIMIZING THE NUMBER OF NEW FLEX-CELLS CREATED	249

3. CELL LAYOUT SYNTHESIS IN FLEX-CELL BASED
   OPTIMIZATION 254
4. GREATER PERFORMANCE THROUGH BETTER
   CHARACTERIZATION 255
5. PHYSICAL DESIGN AND FLEX-CELL BASED
   OPTIMIZATION 259
6. CASE STUDIES WITH RESULTS 261
7. CONCLUSIONS 266

11. **Exploiting Structure and Managing Wires to Increase
    Density and Performance**   **269**
    Andrew Chang, William J. Dally – Stanford University
    1. INHERENT DESIGN STRUCTURE 269
    2. SUCCESSIVE CUSTOM TECHNIQUES FOR
       EXPLOITING STRUCTURE 271
    3. FUTURE DIRECTIONS 285
    4. SUMMARY 285

12. **Semi-Custom Methods in a High-Performance
    Microprocessor Design**   **289**
    Gregory A. Northrop – IBM
    1. INTRODUCTION 289
    2. CUSTOM PROCESSOR DESIGN 290
    3. SEMI-CUSTOM DEISGN FLOW 291
    4. DESIGN EXAMPLE – 24 BIT ADDER 297
    5. OVERALL IMPACT ON CHIP DESIGN 300

13. **Controlling Uncertainty in High Frequency Designs**   **305**
    Stephen E. Rich, Matthew J. Parker, Jim Schwartz – Intel
    1. INTRODUCTION 305
    2. FREQUENCY TERMINOLOGY 305
    3. UNCERTAINTY DEFINED 306
    4. WHY UNCERTAINTY REDUCES THE MAXIMUM
       POSSIBLE FREQUENCY 309
    5. PRACTICAL EXAMPLE OF TOOL UNCERTAINTY ... 312
    6. FOCUSED METHODOLOGY DEVELOPMENT 314
    7. METHODS FOR REMOVING PATHS FROM THE
       UNCERTAINTY WINDOW 315
    8. THE UNCERTAINTY LIFECYCLE 317
    9. CONCLUSION 321

14. **Increasing Circuit Performance through Statistical Design Techniques**   323
   Michael Orshansky – UC Berkeley
   1. PROCESS VARIABILITY AND ITS IMPACT ON TIMING 324
   2. INCREASING PERFORMANCE THROUGH PROBABILISTIC TIMING MODELING 329
   3. INCREASING PERFORMANCE THROUGH DESIGN FOR MANUFACTURABILITY TECHNIQUES 334
   4. ACCOUNTING FOR IMPACT OF GATE LENGTH VARIATION ON CIRCUIT PERFORMANCE: A CASE STUDY 338
   5. CONCLUSION 342

DESIGN EXAMPLES

15. **Achieving 550MHz in a Standard Cell ASIC Methodology**   345
    David Chinnery, Borivoje Nikolić, Kurt Keutzer – UC Berkeley
    1. INTRODUCTION 345
    2. A DESIGN BRIDGING THE SPEED GAP BETWEEN ASIC AND CUSTOM 346
    3. MICROARCHITECTURE: PIPELINING AND LOGIC DESIGN 348
    4. REGISTER DESIGN 352
    5. CLOCK TREE INSERTION AND CLOCK DISTRIBUTION 356
    6. CUSTOM LOGIC VERSUS SYNTHESIS 357
    7. REDUCING UNCERTAINTY 358
    8. SUMMARY AND CONCLUSIONS 358

16. **The iCORE™ 520MHz Synthesizable CPU Core**   361
    Nick Richardson, Lun Bin Huang, Razak Hossain, Julian Lewis, Tommy Zounes, Naresh Soni – STMicroelectronics
    1. INTRODUCTION 361
    2. OPTIMIZING THE MICROARCHITECTURE 363
    3. OPTIMIZING THE IMPLEMENTATION 375
    4. PHYSICAL DESIGN STRATEGY 378
    5. RESULTS 379
    6. CONCLUSIONS 380

17. **Creating Synthesizable ARM Processors with Near Custom Performance**   **383**
    David Flynn – ARM
    Michael Keating – Synopsys
    1. INTRODUCTION 383
    2. THE ARM7TDMI EMBEDDED PROCESSOR 384
    3. THE NEED FOR A SYNTHESIZABLE DESIGN....................391
    4. THE ARM7S PROJECT 392
    5. THE ARM9S PROJECT 400
    6. THE ARM9S DERIVATIVE PROCESSOR CORES 403
    7. NEXT GENERATION CORE DEVELOPMENTS...................406

Preface
by Kurt Keutzer

Those looking for a quick overview of the book should fast-forward to the Introduction in Chapter 1. What follows is a personal account of the creation of this book.

The challenge from Earl Killian, formerly an architect of the MIPS processors and at that time Chief Architect at Tensilica, was to explain the significant performance gap between ASICs and custom circuits designed in the same process generation. The relevance of the challenge was amplified shortly thereafter by Andy Bechtolsheim, founder of Sun Microsystems and ubiquitous investor in the EDA industry. At a dinner talk at the 1999 *International Symposium on Physical Design*, Andy stated that the greatest near-term opportunity in CAD was to develop tools to bring the performance of ASIC circuits closer to that of custom designs. There seemed to be some synchronicity that two individuals so different in concern and character would be pre-occupied with the same problem. Intrigued by Earl and Andy's comments, the game was afoot.

Earl Killian and other veterans of microprocessor design were helpful with clues as to the sources of the performance discrepancy: layout, circuit design, clocking methodology, and dynamic logic. I soon realized that I needed help in tracking down clues. Only at a wonderful institution like the University of California at Berkeley could I so easily commandeer an able-bodied graduate student like David Chinnery with a knowledge of architecture, circuits, computer-aided design and algorithms. David has grown from graduate research assistant to true collaborator over the course of this work, and today he truly "owns" the book.

The search for the performance gap between ASICs and custom circuits soon led us far beyond our provincial concerns of logic and circuit design. We found ourselves touring lands as distant as processor microarchitecture and as exotic as semiconductor process variation. We got a chance to share our initial discoveries at an invited session at the *Design Automation Conference* (DAC) in 2000. Whatever concerns we had that the topic was too esoteric to be of broad interest were quickly allayed. The DAC session, chaired by Bryan Ackland, was parked at the very end of the conference at a time when most conference participants are already on the plane back home. Nevertheless, the room was packed more tightly than any conference room in my memory. Attendees sat in the aisles and some were bold enough to sit cross-legged on the speakers' dais during the talks. Later one of the video operators complained that she was unable to get past the crowd wedged at the door in order to staff her video monitor. The public interest in the topic was very encouraging.

Every weary traveler in an unfamiliar land knows the joy of meeting a compatriot. Andrew Chang and William Dally were already well on their own way to forming conclusions about the relationship between ASIC and custom performance when they presented their views in the DAC 2000 session mentioned above. While the angle of our work was to show that ASIC techniques could be augmented to achieve nearly custom performance, the focus of their work was to show the superiority of custom design techniques over those of ASICs. Our discussions on these disparate views have now continued for over two years and Chapter 4 shows the resulting synthesis. Detailed arguments between Andrew Chang and David Chinnery examined our assumptions and questioned our conclusions, enabled us to arrive at a thorough and careful analysis of ASIC and custom performance.

In the following Fall a new faculty member at Berkeley, Borivoje Nikolić, found his way to my office. Reading our paper at DAC 2000 he noted the relationship between the design techniques he had used at Texas Instruments and those identified in the paper. With Borivoje's knowledge of circuit design in general, and a read channel design (described in Chapter 15) in particular, our work got a much stronger foundation. As a result we were able to better identify and illustrate the key design techniques necessary to improve performance in an ASIC methodology. The initial results of this collaboration appeared at DAC 2001 and we were again encouraged by the large audience for the work

In our investigative travels we encountered a few other designers whose work we are pleased to include in this book. Michael Keating's work on synthesizable versions of the ARM is an invaluable example. It is one of the few cases where a synthesized ASIC design could be compared side-by-side with a custom version. Discussions with Michael had a significant influence on our thinking.

If you can imagine the thrill of a detective stumbling onto an unexpected clue then you'll understand our enthusiasm when we spotted STMicroelectronics' work on the design of the 520MHz iCORE™ processor. As soon as we saw it we were anxious to include it in the book.

Along our investigations we encountered many others attempting to build tools for improving the performance of ASIC design. Their work speaks for itself in Chapters 6 through 14.

In addition to our co-authors, we'd like to acknowledge at least a few of those individuals with whom we've had relevant discussions. We've had animated discussions on the sources of performance in integrated circuit design with: Bryan Ackland, Matthew Adiletta, Shekhar Borkar, Patrick Bosshart, Dan Dobberpuhl, Abbas El Gamal, Douglas Galbraith, Mehdi Hatamian, Bill Huffman, Mark Horowitz, Arangzeb Khan, Ram Krishnamurthy, Jan Rabaey, Mark Ross, Paul Rodman, Takayasu Sakurai, Mark Vancura, Kees Vissers, Tony Waitz, Scott Weber, and Neil Weste. In retrospect, an unconscious seed of this work may have been planted at an ISSCC panel in 1992. In a panel chaired by Mark Horowitz in which Bosshart, Dobberpuhl, Hatamian, and I debated the respective merits of synthesis and custom methodologies.

Over the years we've also had innumerable discussions on the role of tools and technologies for high performance design. I'd like to acknowledge just a few of those individuals: Robert Brayton, Raul Camposano, Srinivas Devadas, Antun Domic, Jack Fishburn, Masahiro Fujita, Dwight Hill, Joe Hutt, Andrew Kahng, Desmond Kirpatrick, Martin Lefebvre, Don MacMillan, Sharad Malik, David Marple, Richard Rudell, Alex Saldanha, Alberto Sangiovanni-Vincentelli, Ken Scott, Carl Sechen, Farhana Sheikh, Greg Spirakis, Dennis Sylvester, Chandu Vishewariah, and Albert Wang.

One of the points of the book is to demonstrate the role that semiconductor processing variation plays in determining circuit performance. Thanks to a number of people for enlightening us on this topic, including: Jeff Bokor, Christopher Hamlin, Chenming Hu, T-J King, and Costas Spanos.

Helpful editorial work came from Matthew Guthaus, Chidamber Kulkarni, Andrew Mihal, Michael Orshansky, Farhana Sheikh, and Scott Weber. The cover was beautifully rendered by Steve Chan.

The home of this work is the Gigascale Silicon Research Center (GSRC) funded by the Microelectronics Advanced Research Consortium (MARCO). This book is in some regards a clandestine effort to realize the vision of Richard Newton, GSRC's first Director. Richard's vision was for *Custom Performance with ASIC productivity*. At some point we wisely realized it would be easier to try to realize this dream than to convince Richard that it was impossible.

In closing we'd like to especially to thank Earl Killian. Not only did he pose the question that first inspired our investigations but he has been the most insightful critic of our work throughout. Earl has always been willing to take the time to respond in depth on any technical question or to read in detail any idea. Earl's commitment to technical clarity and integrity has been a continuous inspiration. We must confess that up to our most recent email exchange our work has still not fully answered Earl's question to his satisfaction. Nevertheless, our relentless attempts to do so have vastly improved the quality of this book.

For emotional support we'd also like to thank the people close to us. David gratefully thanks his grandparents, Ronald and Alex Ireland, for their wonderful support and encouragement throughout the years. Kurt thanks Barbara Creech for her understanding and support during the writing of this book.

List of Trademarks

Throughout this book we make reference to various chips and software. The following trademarks are referred to in this book:

Advanced Micro Devices, Inc.: Athlon, K6

Cadence Design Systems, Inc.: CTGen, Pearl, Silicon Ensemble, Verilog

Compaq, Inc.: Alpha

International Business Machines, Inc.: PowerPC

Intel Corporation, Inc.: Pentium II, Pentium III, Pentium 4

Mentor Graphics, Inc.: Calibre

STMicroelectronics, Inc.: iCORE

Synopsys, Inc.: Design Compiler, DesignWare, Module Compiler, Physical Compiler, Power Compiler, PrimeTime, PrimePower,

Tensilica, Inc.: Xtensa

Chapter 1

Introduction and Overview of the Book

David Chinnery, Kurt Keutzer
Department of Electrical Engineering and Computer Sciences,
University of California at Berkeley

1. WHY ARE CUSTOM CIRCUITS *SO* MUCH FASTER?

This book begins with a fascinating question: Why are custom circuits *so much* faster than application-specific integrated circuits (ASICs)? Given the human effort put into designing circuits in a custom methodology it is not surprising that custom circuits are faster than their ASIC counterparts; it is surprising that they are routinely 3× to 8× faster when fabricated in the same processing generation. The first aim of this book is to explain this disparity in performance.

Our second aim is to understand practical ways in which the performance gap between ASICs and custom circuits can be bridged. For each factor in which custom circuits are routinely faster we examine tools and techniques that can help ASICs achieve higher performance.

2. WHO SHOULD CARE?

2.1 ASIC and ASSP Designers Seeking High Performance

The obvious target of this book is ASIC designers seeking higher performance ASIC designs; however, we quickly acknowledge not all ASIC designers are seeking higher performance. Many ASIC designs need only to be cheaper than FPGAs or faster than general-purpose processor solutions to be viable. For these designs final part cost, low non-recurring engineering cost, and time-to-market concerns dominate the desire for higher performance. Non-recurring engineering costs are growing for ASICs. Mask-set costs are now exceeding one million dollars. Both the number, and cost, of tools required to do ASIC design are rising. Simultaneously, in order to recoup their massive investment in fabrication facilities, semiconductor vendors for ASICs are "raising the bar" for incoming ASIC designs. Specifically, ASIC semiconductor vendors are raising the minimum volumes and expected revenues required to enter into contract for fabricating ASICs. These different factors are causing more ASIC design groups to rethink their approach. Some groups are migrating to using FPGA solutions. Some groups are migrating to application-specific standard parts (ASSPs) that can be configured or programmed for their target application. Those groups that retain their resolve to design ASICs have a few common characteristics: First, these groups aim to amortize increasing non-recurring engineering costs for ASIC designs across multiple applications. Thus they are no longer designing "point solution" ASICs but tending toward more sustainable IC *platforms* [27]. Secondly, to achieve the re-targetability of these platforms the ASICs typically have multiple on-chip processors that allow for programmability. Finally, given the effort and attention required to design a highly complex ASIC, design groups are demanding more performance out of their investment. In short, this book targets ASIC and ASSP designers seeking high-performance within an ASIC methodology, and we contend that this number is *increasing* over time.

2.2 Custom Designers Seeking Higher Productivity

An equally important audience for this book is custom designers seeking high performance ICs, but in a design methodology that uses less human resources, such as an ASIC design methodology. Without methodological improvements custom-design teams can grow as fast as Moore's Law to design the most complex custom ICs. Even the design teams of the most commercially successful microprocessor cannot afford to grow at that rate.

We hope to serve this audience in two ways. First, we account for the relative performance impact of different elements of a custom-design methodology. Design resources are not unlimited, and must be used judiciously; therefore, design effort should be applied where it offers the greatest benefit. We believe that our analysis should help to determine where limited design resources are best spent.

Secondly, specific tools targeted to improve the performance of ASICs can be applied to custom design. The custom designer has always lacked adequate tool support. Electronic Design Automation (EDA) companies have never successfully found a way to tie their revenues to the revenues of the devices they help design. Instead, EDA tool vendors get their revenues from licensing design tools for each designer (known as a "design seat"); it doesn't matter if the chip designed with an EDA tool sells in volumes of ten million parts or one, the revenue to the EDA company is the same. It has been estimated that there are more than ten times as many ASIC designers (50,000 – 100,000 worldwide) as custom designers (3,000 – 5,000 worldwide). As a result EDA tool vendors naturally "follow the seats" and therefore have focused on tools to support ASIC designers rather than custom designers.

As a particular example of ASIC tools that can be applied to high-performance custom designs, Chapter 9 describes a tool for automating continuous sizing of standard cells within an ASIC design methodology. Chapter 12 describes a tool that attacks the same problem but is primarily targeted for use within custom microprocessors.

2.3 EDA Tool Developers and Researchers

This book broadly surveys the factors of design and manufacturing that most impact performance. Nevertheless, there is room for further research of each of the topics mentioned in this book. Also, many of the topics of this book, such as continuous cell-sizing (Chapters 9 and 12) and process impact analysis (Chapter 14), currently have no commercial tool offering. We hope that EDA researchers and developers that agree that the market for performance-oriented tools is emerging will find this book to be a useful starting point for their investigation of the topic.

3. DEFINITIONS: ASIC, CUSTOM, ETC.

The term *application-specific integrated-circuit*, and its acronym *ASIC*, has a wide variety of associations. Strictly speaking it simply refers to an *integrated circuit* (IC) that has been designed for a particular *application*. This defines a portion of the semiconductor market, like memories, microprocessors or FPGAs tracked by the semiconductor. Two industries grew to support the development of these devices: The ASIC semiconductor-vendor industry (or simply the ASIC Vendor industry) established by companies such as LSI Logic, provided the service of fabricating ASICs designed by other independent design groups. Companies such as Cadence and Synopsys provided commercial tools for designing these ASICs. These tools were known as *ASIC design tools*. Another key element of the ASIC design process is *ASIC libraries*. ASIC libraries are carefully characterized descriptions of the primitive logic-level building blocks provided by the ASIC vendors. Initially these libraries targeted gate-array implementations, but in time the higher-performance standard-cell targets became more popular. ASIC vendors then offered complete flows consisting of ASIC tools, ASIC libraries, and a particular design methodology. These embodied an *ASIC methodology* and were known as *ASIC design kits*. Smith's book on ASICs [40] is a great one-stop reference on the subject. With this broader context, let us pause to note that the use of the term ASIC is misleading: it most often refers an IC produced through a standard cell ASIC methodology and fabricated by an ASIC vendor. That IC may in fact belong to the application-specific standard-part portion of the semiconductor market.

The term *custom integrated-circuit,* or *custom IC,* also has a variety of associations, but it principally means a circuit produced through a custom-design methodology. More generally custom IC is used synonymously with the semiconductor market segments of high-performance microprocessors and digital signal processors.

The reader may now wonder about our wisdom of focusing a book on discussing the relationship between two ambiguous terms! In this book when we examine the relationship between ASIC performance and custom performance, we mean that we are examining the relationship between custom circuits designed without any significant restriction in design methodology and ASIC circuits designed in a high productivity ASIC methodology. When we consider ways to bridge the performance gap between ASIC and custom, we mean that we are considering ways to augment and enhance an ASIC methodology to achieve closer to custom performance. In Section 4 we will do a quick examination of the factors that make custom circuits go so much faster than ASIC. In the succeeding sections of this chapter we then give some supporting details for these differences.

4. THE 35,000 FOOT VIEW: WHY IS CUSTOM FASTER?

Typically, speeds of good application-specific integrated circuits (ASICs) lag that of the fastest custom circuits in the same processing geometry by factors of six or more. When designing a custom processor, the designer has a full range of choices in design style. These include architecture and microarchitecture, logic design, floorplanning and physical placement, and choice of logic family. Additionally, circuits can be optimized by hand and transistors individually sized for speed, lower power, and lower area.

4.1 A Quick Comparison

To quantify the differences between ASIC and custom chip speeds, we first examine speeds of high performance designs and typical ASIC designs in 0.25um, 0.18um, and 0.13um technologies. When we refer to a technology, we are referring to fabrication processes with similar design rules and similar effective transistor channel lengths. Process technologies from different vendors vary in a number of ways: the channel length; the interconnect used (copper or aluminum); and other parts of the process.

Introduction and Overview of the Book

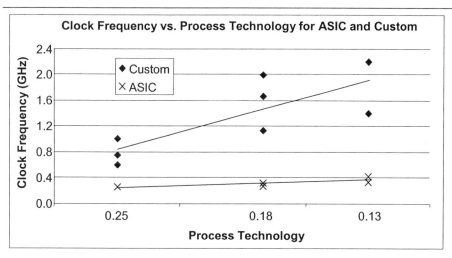

Figure 1. A graph of the clock frequency of high performance ASIC and custom processors in different process technology generations. The custom processors are about a factor of ×5 faster than the ASICs.

Figure 1 shows high performance custom and synthesizable ASIC processors in these technologies. The custom chips are a factor of 3× to 8× faster. Table 1 has the corresponding area and power for these designs.

All the custom designs, except the 1GHz IBM Power PC, are much larger designs than the synthesizable processors. They have out-of-order execution and can execute several instructions simultaneously (multiple instruction issue). The IBM Power PC and the ASICs are single issue, in-order processors. The IBM Power PC and Alpha 21264A are 64 bit machines; the rest are 32 bit.

Most of the custom designs use dynamic logic on critical paths and more pipeline stages to achieve higher speeds. The IBM Power PC has a RISC architecture and four short pipeline stages. To achieve high speed, the IBM Power PC has only a few levels of logic per pipeline stage [35].

The ASIC microprocessors are not necessarily representative of ASICs, because they bear more architectural similarity to custom microprocessors. They present a good mid-point between custom design and a typical ASIC design. Simply based on anecdotal information we postulate that average ASICs run about 20% to 50% slower than these high-performance ASIC CPUs. For example, 0.25um ASICs run at between 120MHz and 150MHz; and high speed network ASICs may run at up to 200MHz in 0.25um technology. Of course, one may find ASICs that operate at slower speeds, but in these devices we presume that performance was specifically *not* a criterion.

Custom	Frequency (GHz)	Technology (um)	Voltage (V)	Power (W)	Area (mm²)	Process Conditons	Operating Conditions
Pentium III (Katmai)	0.600	0.25	2.05	34.0	123.0		
Athlon (K7)	0.700	0.25	1.60	50.0	184.0		
Alpha 21264A	0.750	0.25	2.10	90.0	225.0		
IBM Power PC	1.000	0.25	1.80	6.3	9.8		
Pentium III (Coppermine)	1.130	0.18	1.75	38.0	95.0		
Athlon XP	1.733	0.18	1.75	70.0	128.0		
Pentium 4 (Willamette)	2.000	0.18	1.75	72.0	217.0		
Pentium III (Tualatin)	1.400	0.13	1.45	31.0	80.0		
Pentium 4 (Northwood)	2.200	0.13	1.50	55.0	146.0		
ASICs							
Tensilica Xtensa (Base)	0.250	0.25	2.50	0.20	1.0	typical	typical
Tensilica Xtensa (Base)	0.320	0.18	1.80	0.13	0.7	typical	typical
Lexra LX4380	0.266	0.18	1.80		1.8	typical	worst case
Lexra LX4380	0.420	0.13	1.20	0.05	0.8	typical	worst case
ARM1022E	0.325	0.13	1.20	0.23	7.9	worst case	worst case

Table 1. Custom and ASIC processors in 0.25um, 0.18um, and 0.13um. [2][3][4][12][17][18][33][34][37][45] Power for the LX4380 in 0.18um was not available. The custom speeds in 0.13um will continue to improve as this process technology matures. Worst case operating conditions are low supply voltage and high temperature (100°C or 125°C). Typical operating conditions are typical supply voltage and room temperature (25°C). Worst case process conditions are pessimistic to ensure high yield. Custom process frequencies are for the fastest speed sold in the process, with the exception of the IBM Power PC, which was not commercial.

Thus, at the outset, we can see that custom ICs operate 3× to 8× faster than ASICs in the same process. At first glance this gap seems staggering. If we put the speed improvement due to one process generation (e.g. 0.35um to 0.25um) as 1.5×, then this gap is equivalent to that of five process generations or nearly a decade of process improvement. In the following section we try to more precisely describe the factors that result in this significant speed differential.

FACTORS CONTRIBUTING TO SUPERIOR CUSTOM PERFORMANCE	vs. Poor ASIC	vs. Best Practice ASIC	Factor Affects
Microarchitecture: e.g. pipelining	×1.80	×1.30	# of stages, IPC
Timing overhead: clock tree design, registers, slack passing	×1.45	×1.10	$t_{timing\ overhead}$
High speed logic styles: e.g. dynamic logic	×1.40	×1.20	t_{comb}
Logic design	×1.30	×1.00	t_{comb}
Cell design and wire sizing, including transistor sizing	×1.45	×1.10	t_{comb}
Layout: floorplanning, placement, managing wires	×1.40	×1.00	overall performance
Exploiting process variation and accessibility	×2.00	×1.20	overall performance

Table 2. Maximum differences between custom and ASICs. A factor of ×1.00 indicates no difference. Some of these factors directly affect the overall delay; others affect only the combinational delay t_{comb} or timing overhead $t_{timing\ overhead}$ in a pipeline stage.

4.2 Adding Up the Numbers

Table 2 gives our overview of the contributions of various factors to the speed differential between ASICs and custom ICs. In the left hand column is the size of the factor if the comparable ASIC design ignores this factor altogether. However, in the case of microarchitecture we are comparing pipelined ASIC microprocessors to pipelined custom microprocessors. In the right hand column is the performance advantage that custom design retains even if the ASIC design uses best practices.

There are many limitations to such a comparison, but we felt that a rough accounting of the differences would be useful in focusing discussion on the most important factors. It is the thesis of this book that by using best tools and techniques, overall speed of ASICs can come within a factor of 3 of the highest performance custom circuits. This was the case for the Texas Instrument's SP4140 disk drive read channel, which achieved 550MHz in 0.21um technology. The fastest commercial custom processor released in the same year, 1999, was the 800MHz Pentium III in 0.18um technology. This is not an entirely fair comparison, as the SP4140 was not a processor, but it does show that it is possible to design high clock frequency ASICs – particularly if the datapath is only a few bits wide.

Moreover, each of the tools and techniques used to obtain this speed-up is described in subsequent chapters of this book. The remainder of this chapter goes into detail on the origins of the numbers used in Table 2 and then gives pointers into the topics in the remainder of the book. In each of the subsequent sections we will first try to justify the left-hand column of Table 2 by showing the sources of performance degradation. We will then describe the tools and techniques that can be employed to improve ASICs and how that can bring performance within the factor expressed from the right-hand column of Table 2.

4.3 How the Factors Combine to Cause ASICs to be Much Slower

It is tempting to multiply the factors in Table 2 to come up with a "worst possible" ASIC. However, the factors affecting primarily the combinational delay t_{comb} and the timing overhead $t_{timing\ overhead}$ are additive, affecting the delay per pipeline stage T_{stage}.

(1) $\quad T_{stage} = t_{comb} + t_{timing\ overhead}$

High speed logic styles, the design of the combinational logic (e.g. different styles for adders), and the design of the individual gates affect the delay of the combinational logic. Timing overhead can be reduced by careful clock tree design to minimize clock skew, and use of level-sensitive latches or high-speed pulse-triggered flip-flops.

Microarchitectural techniques, such as pipelining the logic into n pipeline stages, can reduce the delay per stage. Without pipelining, the delay would be

(2) $\quad T_{unpipelined} = nt_{comb} + t_{timing\ overhead}$

Poor layout and assuming worst case process variation exacerbate the impact of the other factors.

The numbers in Table 2 quantify the worst case impact of these factors on the overall speed of a design. However, poor timing overhead is only significant when the design is pipelined and the combinational delay in each pipeline stage has been reduced. Otherwise, the timing overhead is a very small factor compared to the large combinational delay.

To compare designs across process technology generations, we need a metric that is normalized with respect to the delay in the process. One normalized metric for the delay is the number of fanout-of-four (FO4) inverter delays (an inverter driving four times its input capacitance), in the given technology [19]. Chapter 2, Section 1.1, discusses the FO4 metric in further detail.

A very high-speed custom processor can achieve a clock period of about 10 FO4 delays per pipeline stage, with about 3 FO4 delays of timing overhead and 7 FO4 delays for the combinational delay. In comparison, ASIC processors have about 50 to 70 FO4 delays per pipeline stage, with about 15 to 20 FO4 delays of timing overhead (see Chapter 2 for more details). If these ASICs were not pipelined, their delay would be about 230

Introduction and Overview of the Book 9

to 310 FO4 delays. Thus an ASIC without pipelining might be a factor of 20× slower than custom.

A clock period of 250 FO4 delays corresponds to a clock frequency of about 100MHz in 0.13um (0.08um effective channel length). This is similar to the frequency of a front side bus. The Pentium 4 bus has a clock frequency of 100MHz, but it is quad-pumped (data is sent four times per clock cycle) to speed up the transfer of data [24].

5. MICROARCHITECTURE

In this section, microarchitecture refers to organization of functional units on an integrated circuit. In particular, this topic includes items such as: the number of functional units; the organization of the memory hierarchy; the input and output interfaces; the order of the functional units in the pipeline; the number of clock cycles these units have for computation; and logic for branch prediction and data forwarding. One of the most common microarchitectural improvements for performance is pipelining. Pipelines place additional registers in long chains of logic, reducing the levels of logic of the critical path, and allowing time stealing between pipeline stages with multi-phase clocking.

5.1 What's the Problem with ASIC Microarchitectures?

For pipelining to be of value, multiple tasks must be able to be initiated in parallel. Branch operations and other factors that limit instructions being performed in parallel will diminish performance. Many designs, such as bus interfaces, have a tight interaction with their environment in which each execution cycle depends on new primary inputs, and branches are common. In such cases, it is not clear how an ASIC may be reorganized to allow pipelining. Simply increasing the clock speed by adding registers would only increase latency due to the additional delay of the registers.

Pipelining in ASICs is limited by the larger timing overhead for the registers in the pipeline. If each stage has slow flip-flop registers, the clock frequency increases less for additional pipeline stages. Custom designs may also show superior logic-level design of regular structures such as adders, multipliers, and other datapath elements. This reduces the combinational delay of a pipeline stage. They achieve fewer levels of logic on the critical path with compact complex logic cells, and by combining logic with the registers. In a custom processor, changes can be made to balance the logic in pipeline stages *after placement*, ensuring that the delays in each stage are close. Whereas, an ASIC may have unbalanced pipeline stages, limiting the clock period to the delay of the pipeline stage with a longer critical path.

Additional processing speed can be achieved by issuing multiple instructions, but this requires complex hardware logic for speculative execution, unless the instructions have a high degree of parallelism. Pipeline stalls can occur when a stage is waiting for data to execute. To fully utilize more pipeline stages and multiple instruction streams requires forwarding logic to ensure that data is available. Better branch prediction logic is also needed, as a mispredicted branch will cause more logic to stall. There is a trade-off between more simultaneous processing and the penalties for branch misprediction and data hazards, which reduce the performance, and additional hardware and design cost [21]. The Alpha 21264 can issue up to six instructions per cycle, and has four integer execution units and two floating-point execution units [28], giving it significantly faster performance when instruction parallelism can be exploited.

It is difficult to estimate the precise performance improvement with microarchitectural changes. Large custom processors are multiple-issue, and can do out-of-order and speculative execution. ASIC microprocessors can do this too, but tend to have simpler implementations to reduce the design time.

The Alpha 21264 processor has seven pipeline stages, but it has out-of-order and speculative execution [28]. The Tensilica ASIC processors have a single-issue five stage pipeline [11] and the Lexra ASIC LX4380 processor has a single-issue seven stage pipeline [12]. Some ASIC designs may have no pipelining and as a result have significantly longer critical paths. The 1.0GHz IBM PowerPC chip also has a single-issue pipeline, with four stages [37].

There are about 13 FO4 delays in the critical path of both the Alpha 21264 [15] and the 1.0 GHz IBM PowerPC [35]. An ASIC typically has many more levels of logic on the critical path. Tensilica's synthesizable Xtensa processor is estimated to have about 62 FO4 delays for the Base configuration in 0.25um. The Lexra LX4380 ASIC with a MIPS instruction set has about 58 FO4 delays in 0.18um. The FO4 delays for each process technology were estimated from the effective channel length. For further details see Chapter 2, Section 3.3.

The overheads for pipelining are the register delays, larger impact of clock skew and clock jitter, and unbalanced pipeline stages. Estimating the pipelining overheads as about 30% for an ASIC design, the seven stage Lexra LX4380 processor is about $5.2\times$ faster. Estimating the clock and register overheads as about 20% for a custom design, the four stage IBM PowerPC processor is about 3.4 times faster. In general, we will refer to pipelining overhead and timing overhead synonymously, but obviously an unpipelined design doesn't have unbalanced pipeline stages.

These estimates don't consider the impact of pipeline hazards [21] on the instructions per cycle (IPC), which are discussed in more detail in Chapter 2,

Section 1.4.1. Assuming 30% timing overhead, the eight pipeline stages of the ASIC iCORE™ processor from STMicroelectronics (Chapter 16) give a 5.9× increase in the clock frequency. However, the IPC is about 0.7 (which has been optimized using branch prediction and forwarding), thus the actual performance increase by pipelining is only a factor of 4.1×.

If we compare ASIC and custom designs for the same application, then similar microarchitectural implementations are feasible and better comparisons between ASIC and custom designs are possible. The IPC of the super-pipelined Pentium 4 was 20% worse than the Pentium III [24], and the Pentium 4 was 1.6× faster [23], so the overall performance increase was only 28% going from about 10 pipeline stages to 20 pipeline stages.

Comparing a typical ASIC processor with 5 pipeline stages to the iCORE with 8 pipeline stages, the iCORE is about 1.4× faster because of more pipeline stages (see Chapter 2, Section 3.3.2, for details). Overall for custom and ASIC implementations of a chip targeting the same application (i.e. both designs can be pipelined), custom microarchitecture may contribute a factor of up to 1.8× compared to a more typical ASIC design with less attention to pipelining.

5.2 What Can ASICs Do About Microarchitecture?

If processing the data is interdependent, there is little that can be done to pipeline ASIC designs. On the other hand, if data can be processed in parallel it should be possible to pipeline circuitry performing the calculations or have parallel processing units. Pipelining and parallelism can increase the speed significantly, especially if large amounts of data can be processed in parallel.

The 520MHz iCORE discussed in detail in Chapter 16 has about 26 FO4 delays per pipeline stage, with 8 pipeline stages. It is the fastest ASIC processor that we are aware of in 0.18um technology (0.15um effective channel length [41]) and in many respects exhibits "best practice" for ASIC microarchitecture. The delay per pipeline stage would be 21 FO4 delays if custom clocking techniques were possible, which is comparable to the delay of the Pentium III. However, it seems unlikely that ASICs will be able to exploit pipelining in the same way that custom circuits do. ASICs are unable to have the same tight control of the combinational delay and the timing overheads. Thus we estimate a factor of 1.3× between best ASIC and custom implementations.

5.3 Microarchitecture in the Remainder of the Book

Chapter 2 provides a tutorial introduction to microarchitecture of ASICs. Each of our design examples in Chapters 15, 16, and 17, gives examples of developing a microarchitecture for a high-performance ASIC.

6. TIMING OVERHEAD: CLOCK TREE DESIGN AND REGISTERS

Pipelining ASICs is limited by the timing overhead for the registers in the pipeline. The timing overhead is the additional delay associated with pipeline registers and the arrival of the clock edge. The timing overhead consists of the delay through the registers; the setup time of the registers; the clock skew between arrival of the same clock edge at different points on the chip; and the jitter between the arrival of consecutive clock edges at the same point on the chip. Registers may be flip-flops that are triggered to store their input on a clock edge (e.g. D-type flip-flops), or latches that are level-sensitive and transparent for a portion of the clock cycle. When latches are transparent, the input propagates directly through to the latch output. When latches are opaque, the latch stores the last input. We also include the impact of unbalanced pipeline stages in the timing overhead factor.

The registers in the 600MHz Alpha 21264 took about 15% of the clock cycle, which is 250ps or 2 FO4 delays. The Alpha 21264 used high speed dynamic flip-flops, which are faster than the typical static CMOS D-type flip-flops used in ASICs. Standard cell D-type flip-flops have a delay of about 3 to 4 FO4 delays [30].

ASICs typically have about 4 FO4 delays of clock skew and jitter, which is about 10% of their clock period. Careful custom design techniques can reduce this to 1 FO4 delay. Comparing the absolute differences in clock skews, there is about a 10% increase in speed due to custom quality clock skew alone. The 600MHz Alpha 21264 has 75ps global clock skew in 0.35um, or about 5% of its clock period [15], whereas the Tensilica Xtensa Base processor has a budget of 200ps for clock skew and jitter in 0.13um – which is much worse, in what should be a faster technology!

Typical ASICs have a timing overhead of about 10 FO4 delays, which does not include the impact of pipeline stages being unbalanced. Whereas custom designs can keep the overhead down to as little as 3 FO4 delays. This is a difference of about 15% of the clock period of a typical ASIC, but it would amount to 35% of the clock period for a high speed ASIC such as the Texas Instruments SP4140 running at 550 MHz. It is essential for high speed ASICs to reduce the timing overhead.

Unbalanced pipeline stages also contribute to the pipelining overhead. Slack passing can compensate for unbalanced pipeline stages. Applying the information of Figure 4 to Figure 1 in Chapter 16, we determined that the critical sequential loop in the STMicroelectronics iCORE is IF1, IF2, ID1, ID2, OF1, back to IF1 through the branch target repair loop. This loop has an average delay of about 90% of the slowest pipeline stage (ID1), which has the worst stage delay and limits the clock period. Thus slack passing would give at most a 10% reduction in the clock period, or about 3 FO4 delays of the clock period of 26 FO4 delays.

Converting the Tensilica Xtensa flip-flops to latches improved the speed by up to 20% (see Chapter 7). Between 5% and 10% of this speed increase was from reducing the effect of setup time and clock skew on the clock period. The remainder is slack passing balancing pipeline stages. The slack passing in this latch-based design gives about a 10% improvement in clock speed.

Including the contribution of unbalanced pipeline stages in the timing overhead, custom designs may be a factor of 1.45× faster than ASICs due to the timing overhead.

6.1 What's the Problem with Timing Overhead in ASICs?

ASICs have larger clock skew and clock jitter, and slower registers than custom designs. Custom designs can also include some logic within the register to reduce the overhead. The clock skew in ASICs is worse because clock-trees that are automatically generated by synthesis and place and route tools are not as balanced as the clock trees produced manually. Custom clock trees can have pairs of inverters in the clock tree that are resized to balance the delay for the different load conditions post-layout. Some recent custom designs have used programmable delays to compensate for the impact of process variation on the clock skew.

Some of the high speed registers that have been used in custom designs are more subject to noise and races. ASIC designers try to avoid races and increase tolerance to noise, as they have far less control of variation accompanying additional layout iterations. ASIC designers will be cautious and design the circuitry to work for the range of possible conditions. Consequently, ASICs primarily use edge-triggered flip-flops, and standard cell library vendors have not supported high speed registers in the libraries. Tight control of the layout allows hold time violations and noise to be carefully avoided. Custom designs may run long wires with shielding wires, or use low-swing signaling to reduce the effects of noise.

Traditional ASIC design methodology does not allow slack passing and time borrowing to balance the delay of pipeline stages. Limited control of clock skew has prevented adjusting the arrival of the clock edge for cycle stealing. Level-sensitive latches haven't been used because there is a larger window for hold time violations. Multi-phase clocking schemes allow time borrowing in skew tolerant domino logic, with its precharge and evaluate phase [19]. However, ASICs cannot use domino logic for a variety of tool-flow related reasons. Thus for good performance, pipeline stages in ASICs have to be carefully balanced. Sometimes this is not possible, such as when accessing cache memory, which also requires logic for tag comparison and data alignment [20].

Timing overhead is most significant when a design has been highly pipelined to reduce the combinational delay in each pipeline stage. The iCORE with 8 pipeline stages is an example of such an ASIC.

6.2 What Can ASICs Do About Timing Overhead?

Current EDA tools can verify that hold times are not violated and insert delay elements to avoid races on short paths [36]. However, ASIC designers have not taken the next step of using this as an opportunity to use faster registers, now that races can be avoided.

Slack passing is possible with level-sensitive latches or cycle stealing, by careful scheduling of the arrival of the clock edges at different registers. Multi-phase clocking will not be a viable solution in the deep submicron, because of signal integrity issues and the increasing difficulty of distributing several clocks across the chip [36].

If the latch inputs arrive within the window while the latches are transparent, the setup time and clock skew have far less impact on the clock period. Some high speed pulsed flip-flops have about zero setup time,[30] but they are not transparent, so the impact of the clock skew is not reduced. As ASICs have large clock skew, latches have substantial benefits for reducing the clock period.

Also, level-sensitive latches reduce the impact of inaccuracy of wire load models and process variation. The clock period is not limited by the delay of the slowest pipeline stage, because of slack passing. Adjusting the clock skew after layout and extraction of parasitic capacitances can compensate for wire load model inaccuracies. However, changing the clock trees requires additional layout iterations.

The clock skew can be reduced by using better clock tree synthesis tools, (Chapter 8) or resorting to manual design. Additional techniques that are now commonly used in custom designs need to be supported by EDA tools. Clock tree synthesis tools need to allow schemes such as inverter pairs and programmable delays with phase detectors. Inverter pairs can be resized to balance the load and delays. This allows the clock tree to be rebalanced for layout variation without perturbing the layout excessively. Programmable delays are necessary to reduce the impact of process variation.

A voltage controlled oscillator adjusts the phase lock loop (PLL), which generates the clock from a multiple of a low frequency reference crystal

Introduction and Overview of the Book 15

oscillator. Reduction of jitter requires careful shielding of the PLL from voltage supply noise. RC filters or a voltage supply regulator can be used to reduce the impact of the supply noise.

Using latches allows slack passing and reduces the impact of clock skew and setup time. The timing overhead with latches for an ASIC can be as little as 5 FO4 delays (see Chapter 3, Table 2), reduced from up to 10 FO4 delays for a typical ASIC. Custom techniques further reduce the timing overhead to 3 FO4 delays. Latches and typical ASIC clocking techniques are 2 FO4 delays slower than custom techniques. A very fast ASIC may achieve a clock period of 20 FO4 delays (e.g. the SP4140), and 2 FO4 delays is 10% of this clock period. Thus a custom design may still be 1.1× faster than an ASIC design that uses latches to reduce the timing overhead.

6.3 Timing Overhead in the Remainder of the Book

Clock-related timing issues are tutorially reviewed in Chapter 3. Chapter 7 describes a prototype tool that automatically converts a gate net list with flip-flops to use latches – experimental results of 10% to 20% speed improvement are reported for a commercial synthesizable ASIC in a high-performance standard cell ASIC flow. Chapter 7 also discusses the timing overhead in the Xtensa microprocessor. Chapter 8 considers issues associated with clock-tree synthesis and use of carefully adjusted clock skew in clock-tree design for cycle stealing.

Chapter 15 discusses a very high speed ASIC design, where reduction of the timing overhead was essential. Clock trees were manually routed, and both latches and high speed pulsed flip-flops were used in some portions of the design.

7. LOGIC STYLE

Logic style connotes the circuit family in which gates are implemented. Dynamic logic can be used to speed up critical paths within the circuit by reducing gate delays. It is significantly faster than static CMOS logic and has smaller area, but requires careful design to ensure no glitching of input signals. Static CMOS logic has far less sensitivity to noise and consumes less power. As dynamic logic is faster, lowering the supply voltage can sometimes reduce the power consumption to less than static logic for the same speed constraint. Both the IBM PowerPC integer processor and the Alpha 21264 make use of dynamic logic for increased circuit speed [13][37].

7.1 What's the Problem with ASICs and Dynamic Logic?

Dynamic logic works in two phases. In the first "precharge" phase, charge is dynamically stored at a node in the gate. This charge may be discharged in the evaluate phase. Dynamic logic is particularly susceptible to noise. Any glitches on the inputs may cause a discharge of the charge stored, which should only occur when the logic function evaluates to false.

Dynamic logic functions used in the IBM 1.0 GHz design are 50% to 100% faster than static CMOS combinational logic with the same functionality [35][Kevin Nowka, personal communication]. This implies that sequential circuitry can be 1.4× faster, as dynamic logic reduces the combinational delay, but not the timing overhead (high speed registers can also be used in ASICs, which we discussed in Section 6).

Other high speed circuit styles such as differential cascode voltage switch logic (DCVSL) and complementary pass-transistor logic (CPL) are also faster than static CMOS logic. These logic styles are sometimes used in custom designs. Domino logic is 30% to 40% faster than DCVSL, which is faster than CPL. ASIC libraries are not available for these logic styles either.

7.2 Can Dynamic Logic be Used in an ASIC Methodology?

To be useful in an ASIC design methodology a logic family must be robust in a variety of circuit conditions, appropriate for typical ASIC applications, and supported by tools for static-timing analysis and manufacturing testing. Design of dynamic logic requires careful consideration of noise. The clock determines when pre-charging occurs, and inputs must not glitch during or after the pre-charge. These problems become more pronounced with deeper submicron technologies. These problems make dynamic much less robust that static CMOS. High speed dynamic logic has higher power consumption and requires careful design of the power and clock distribution. This also makes the use of dynamic logic less attractive for ASIC design. Finally, dynamic logic is not supported by ASIC static-timing and manufacturing test tools. For these reasons dynamic logic libraries are not available for ASIC design.

There has been some progress in dynamic logic circuit synthesis [46], but it has yet to produce commercially available libraries. It is used as an aid to in-house custom design. It seems unlikely that the methodological obstacles described above will be overcome, to enable dynamic logic synthesis for ASIC designs.

While it is unlikely that dynamic logic will be available for ASICs the gap between static CMOS and dynamic can be reduced. In some cases, custom designed static logic with pulsed inputs can achieve speeds within 20% of dynamic logic (see Section 3.2.2 of Chapter 4). The logic design and gates need to be carefully optimized for the different logic style. In particular instances, dynamic logic may only be 1.2× faster than highly optimized static logic.

7.3 Logic Style in the Remainder of the Book

Chapter 4 quantifies the performance improvement by using dynamic logic in more detail, and includes an example of a high speed 64 bit adder using static logic. Section 6.1 of Chapter 4 examines the limitations on future use of domino logic and discusses some alternative high-speed logic styles.

8. LOGIC DESIGN

While *logic style* connotes the circuit family in which gates are implemented, *logic design* describes the topology or interconnectivity of the gates. Logic design sits between the functional unit execution of the microarchitecture and the detailed circuit design of the most primitive cells of the IC. In structured or regular datapath logic, the *logic design* determines the choice of adder (e.g. carry look-ahead versus ripple-carry). The choice to pipeline the multiplier is a microarchitectural decision, but the use of a particular implementation of a Booth multiplier is a matter of logic design. A ripple carry adder computes the sum of two n-bit numbers in order n, whereas a radix-4 (4 bit generate and propagate stages) carry look-ahead adder computes the sum in order $\log_4 n$.

8.1 What's the Problem with ASICs and Logic Design?

For random logic both ASIC designers and custom designers are likely to use logic synthesis. For structured-datapath logic ASIC designers are typically not as aware of logic design alternatives as custom designers.

ASIC designers have to carefully specify the logic design in the RTL to get the desired implementation. High speed logic design often considers the actual layout of the design to minimize the load on each level of gates. A compact layout reduces wire lengths. For example, Wallace tree multipliers have a triangular layout – if not carefully constrained, layout tools will try and fit the gates to a rectangular region, which is sub-optimal. The poor layout penalizes irregularly shaped high-speed logic designs. This can force the ASIC designer to choose a "slower" implementation that is more amenable to layout.

Custom designs may also suffer poor logic design. In Chapter 12, Section 4.4, the wrong path in a 24 bit adder was optimized in its initial version. When this was fixed, the adder speed improved by 25%.

8.2 What can ASIC Designers do about Logic Design?

Fast datapath designs, such as carry look-ahead and carry-select adders and other regular elements, do exist in pre-designed libraries (e.g. Synopsys

DesignWare), but are not automatically invoked in register-transfer level logic synthesis of ASICs. Use of predefined macro cells for an ASIC can significantly improve the resulting design, by reducing the number of logic levels for implementing complex logic functions and reducing the area taken up by logic [6].

Contemporary logic synthesis does an adequate job of synthesizing random logic. Using pre-designed logic-level components is a natural solution for the logic-level design of more regular datapath elements such as adders, multipliers, and other arithmetic units. The Synopsys DesignWare intellectual property library contains a comprehensive library of synthesizable datapath components. For creating non-standard datapath elements the Synopsys Module Compiler has been successfully used in a variety of designs.

Researchers have examined techniques for synthesizing arithmetic circuits to faster implementations, such as logic with a carry-save design. This improved the speed by up to 30% [25]. We attribute a factor of up to $1.3\times$ between a typical ASIC and a high performance design using optimal logic design. The automated techniques available for improving logic design ensure that ASICs can achieve what would be possible by manual design of the logic. We believe that by using synthesis tools that target high speed arithmetic structures, ASIC logic design can come to parity with custom logic design. Successful use of Designware is detailed in Chapter 17. Retrospectively, a chapter detailing the use of Module Compiler [44] would have been well suited to this book.

8.3 Logic Design in the Remainder of the Book

Logic design is considered further in Chapter 4 of this book. Chapter 4, Section 3.2.1 looks at the particular example of a 64 bit adder. Section 3.4 of Chapter 4 looks at the relative performance impact of good logic design.

9. CELL DESIGN AND WIRE SIZING

In an ideal circuit, each gate is optimally crafted from transistors and each transistor is individually sized to meet the drive requirements of the capacitive load it faces, subject to timing constraints. Also wires may be widened to reduce the delays (proportional to the product of resistance and capacitance) by reducing the resistance. Additional buffers may be included to drive large capacitive loads that would be charged and discharged too slowly otherwise. Only in a custom design methodology can this ideal be realized. Any current ASIC methodology requires cell selection from a fixed library, where transistor sizes and drive strengths are determined by the choices in the library, and wire sizes are fixed.

9.1 What's the Problem with Cell Design for ASICs?

One element of the performance degradation of ASIC designs is the poverty of standard cell libraries. If in an ideal circuit each transistor is optimally sized, then clearly the limited number of discretely-sized cells in an ASIC library can only approximate custom designs. In addition, ASIC cells typically include design guard banding, such as buffering flip-flops, which introduces timing overhead. More fundamentally, the discrete transistor sizes of a library only approximate the continuous transistor sizing of a custom design. With sufficiently many drive strengths for each cell, the performance degradation of discrete sizes may only be 2% to 7% or less [14][16]. However, many ASICs still do not use good standard cell libraries with varied drive strengths. A cell library with only two drive strengths may be 25% slower than an ASIC library with a rich selection of drive strengths and buffer sizes, as well as dual polarities for functions (gates with and without negated output) [39]. A richer library also reduces circuit area [29]. Compounding custom design's advantage of continuous transistor sizing, development of macro-cells, and use of wire sizing, we estimate that custom designs can achieve a factor of 1.4× speed improvement over poor ASIC design.

9.2 What Can We do About Cell Design for ASICs?

ASIC designs should be using rich standard cell libraries with dual gate polarities and several drive strengths for each gate. Several tools are available for automating the creation of cells that are optimized for a design, and characterizing these cells for use in an ASIC design flow. This reduces the cell design time, which should enable high performance ASICs to take advantage of design specific cells where needed.

Crafting specific cells for a design, or generating an entire library optimized for the design, can give significant performance improvement. The STMicroelectronics iCORE gained a 20% speed improvement in comparison to a large standard cell library by using a library crafted specifically for the design (see Section 3.4 of Chapter 16). In this example creating the design-specific library required substantial design time. The iCORE design-specific library took five people about five months (25 man-months) to create. However, tools are available to automate most steps of the process. Using cells with skewed drive strength in a static CMOS 64-bit adder improved the speed by 25%.

There are alternatives to hand-crafting design-specific libraries. One promising prototype tool flow enables automatic creation of design-specific cells in order to optimize the design after layout. The advantage of a post-layout flow is that wiring capacitances are known. The prototype flow implemented by Cadabra was able to increase the speed of a bus controller by 13.5% (see Chapter 9, Section 4.2 for details). With these "liquid cells" late arriving signals can be routed closer to the gate output. Transistors can also be moved to maximize the adjacent drains and sources for diffusion sharing [22]. Other references support the notion that iterative transistor resizing and re-synthesis can improve speeds by 20% [10].

Tools for optimizing the widths of individual wires may be available in the future (e.g. [7]), but are not currently commercial available.

By creating some design specific cells and choosing the best library available for the process, ASICs can close the gap to custom cell design. We estimate that by using continuous (versus discrete) cell sizing and by using wire sizing custom designs can achieve speeds about 1.1× faster than the best ASIC designs.

9.3 Cells and Sizing in the Remainder of the Book

Chapter 4 gives a tutorial overview of issues in libraries. Cell generation and sizing is an active area of industrial research and as a result we have a number of contributions on this topic. Chapter 10 considers automatically finding macro-cells in logic. Chapter 9 looks at a commercial prototype tool for providing continuous cell sizing for an ASIC flow. Chapter 12 gives an internal IBM ASIC-like flow for the same problem, although this flow is targeted for custom designs.

10. LAYOUT: FLOORPLANNING AND PLACEMENT TO MANAGE WIRES

Wire delays associated with "global" wires between physical modules can be a dominant portion of the total path delay. The delay associated with wires depends on the length of the wire, the width and aspect ratios of the wire, and on driving the wire properly. Drivers need to be sized optimally and long wires need repeaters inserted. The primary factor in wire delay is wire length. Routing congestion and the position of cells in the layout affect wire length. If connected cells are placed far apart, there will be longer wires. Partitioning a design into modules keeps connected gates near each other. Floorplanning the chip and specifying global routing channels to limit congestion also help reduce wire lengths.

Careful design also considers the impact of noise and cross-coupling capacitance between wires. The Pentium 4 used a low-swing clock to reduce the crosstalk noise from the switching clock signal [31]. In addition, shielding can be used to reduce noise. Shielding is necessary for logic styles that are more subject to noise, such as dynamic logic. Shielding may require additional area as well as additional "wire planning" to determine how to route wires to limit cross-coupling. "Twisted pair" wiring can be used to substantially reduce inductive coupling noise [49].

10.1 What's the Problem with Layout in ASICs?

During pre-layout synthesis, the load a gate drives needs to be estimated. Traditionally, this has been done by creating a "wire-load model" that estimates the capacitive load as a function of block size and fanout. In smaller process geometries, wire load is becoming an increasing portion of the capacitive load. As a result, wire load model inaccuracy is causing more problems. Wire-loads may not properly take into account block size or inter-module routing [26].

Using an overly conservative wire load model increases the size of gates, and buffers are inserted to drive unrealistic loads. The area of the design increases, wire lengths increase, and each gate has to drive a larger load. This reduces the speed of the design after layout.

On the other hand, underestimating the wire loads leads to gates being too small and a lack of buffering for large loads. This is difficult to fix in layout, and also increases the post-layout clock period. The post-synthesis clock period will be under-estimated due to the small loads.

Failing to partition a design into small modules increases the number of wires with large capacitance (larger intra-module wires), and further reduces the accuracy of wire load models. We compared the impact of partitioning a $1cm^2$ design into small and large blocks of gates (emulating careful floorplanning). Small block sizes localize many critical paths to within the module, reducing routing congestion and use of scarce global routing resources [43]. Based on our simulations (with BACPAC [42]), careful floorplanning and placement to minimize wire lengths may increase circuit speed by up to 25%. Overall, we estimate that layout of a carefully partitioned design can increase the speed by a factor of 1.4× compared to an ASIC design that has large blocks of gates and uses inaccurate wire load models in synthesis.

10.2 What can ASICs do About Layout?

In our experience, an accurate wire load model can increase the speed of a design by 15%. However, even good wire load models do not accurately model the load faced by *each* gate, so some gates are undersized and some are oversized. Further improvements in performance require partitioning the design, careful floorplanning, and resynthesis of the design as the floorplan is generated.

Custom ICs are typically manually floorplanned. A number of tools are now reaching the ASIC market to facilitate chip-level floorplanning. These should diminish the gap between ASIC and custom designs due to layout. For example, adding good floorplanning, global routing, and physical resynthesis (e.g. using Physical Compiler) to a synthesis-layout flow with good wire load models can improve the speed by a further 15%.

In addition, tools with the capacity to identify similar structures that may be abutted or placed next to each other appropriately will reduce area, reducing wire lengths and increasing performance. A bit-slice may be laid out automatically then tiled, rather than the circuitry being placed without considering that it may be abutted [6]. Regular data paths can be best laid out by hand or tiling slices for abutment, but custom design is not as effective with irregular structures. Looking at an integrated circuit design as a whole, we estimate that ASIC designs should be able to achieve parity with custom designs with respect to floorplanning and detailed placement.

10.3 Layout in the Remainder of the Book

Chapter 4, Section 3.6, discusses the impact of wire load models, partitioning and layout tools in further detail. Floorplanning techniques for ASICs are described in Chapter 6. Chapter 11 gives an example where manual layout, such as bit-slicing tiling, improved the performance.

11. PROCESS VARIATION AND IMPROVEMENT

Traditionally, a typical ASIC designer perceives the actual semiconductor fabrication of chips in a highly simplified way: The semiconductor process is represented as being fully determined by a given technology generation, with the identical implementations of this process at different plants and at different times. The semiconductor process is also described through a set of worst-case numbers that abstract the complexity of the actual semiconductor manufacturing.

In reality, many different realizations of the same technology generation are possible. First, there are differences between the implementations of the same process technology at different foundries. Second, speed of the process is improved as the process matures. Also, in typical operating conditions, ASIC chips fabricated on a typical process may be 60% to 70% faster than the worst case speeds quoted by ASIC library estimates for worst case operating conditions (comparison in Chapter 5, Section 1.3). In addition, the fastest speeds produced in a plant may be 20% to 40% faster than the slowest clock frequencies [1][Costas Spanos, Chenming Hu and Michael Orshansky, personal communication], but without sufficiently high yield for use in ASIC manufacturing. We will discuss these different factors in more detail below.

Introduction and Overview of the Book 23

11.1 Performance Variation between Different Plants

In the same nominal process technology, the speed varies significantly for the same ASIC design synthesized to different libraries and different fabrication plant processes. Some semiconductor vendors have processes with copper interconnect and silicon-on-insulator technology, while other plants in the same technology generation may still be using aluminum interconnect and bulk CMOS. The dielectrics used also vary; lower-k dielectrics reduce capacitive coupling noise. Custom foundries have been aggressively scaling the effective channel length to reduce speed – for example, Intel has 0.06um channel length in 0.13um [47], whereas the ASIC vendor TSMC has 0.08um channel length [48].

Intel's P858.5 0.18um process had a ring oscillator delay of 10ps/stage, with an effective channel length of 0.10um. Examining ASIC processes with similar channel lengths: TSMC's 0.15um CL015HS process (0.11um L_{eff}) had a ring oscillator delay of 14ps/stage; UMC's 0.15um L150 MPU process (0.10um L_{eff}) had a ring oscillator delay of 14ps/stage. Thus a custom process may be up to 40% faster than a comparable ASIC process.

11.2 Continuous Process Improvement

For custom ICs, additional improvements to the process or the design are possible. In Intel's 0.25um P856 process, a shrink of 5% was achieved, giving a speed improvement of 18% [5]. There is room for making minor changes to increase the speed of a design, increasing the speed of a custom IC design over the process generation. Down-binning of chips with higher clock frequency to meet demand also spreads the range of frequencies seen in a process generation. Down-binning occurs when stores of slower chips are depleted, and it's evidenced by the ease of over-clocking many chips.

11.3 Speed Variation due to Process Variation

The semiconductor process cannot be perfectly controlled, which leads to statistical variation of many process variables. Several types of process variations can occur: batch-to-batch, wafer-to-wafer, die-to-die, and intra-die (intra-chip). These process variations cause the delays of wires and gates between chips and within a chip to vary, and chips are produced with a range of working speeds. Some chips with minor imperfections will only operate correctly at slower speeds. It appears that when Intel and AMD introduce a new architecture, the process variation is up to about 30% (see Chapter 5, Section 2 for more details). This variation decreases as the process matures.

11.4 What's the Problem with ASICs and Process Variation?

The ASIC design and manufacturing industry was organized around a "hand-off" point. ASIC foundries gave ASIC designers a golden toolset including timing tools. If the timing tools said that an ASIC design ran at a speed, say 100MHz, then the ASIC foundry guaranteed to produce parts at that speed. ASIC designers didn't have to worry about the details of ASIC foundries semiconductor processes and ASIC foundries didn't have to worry about the implementation details of ASIC designs. The cornerstone of the timing models that ASIC foundries gave designers were ASIC libraries that were well-characterized for timing. Even though these timing models hid the fact that delay of a gate was not a fixed number but a random variable with a certain probability distribution, the simplicity and reliability of the approach made it attractive.

This simplified approach to design that served designers so well for over a decade is now crippling the industry. First of all, practically speaking it is impossible for an ASIC foundry to provide a timing tool that predicts the timing of a circuit before layout. However well-characterized the cell library is, too much of the final circuit delay can reside in the wires.

Secondly, the increasing variation of processes (see Chapter 14, Section 1.1) and the rise of uncorrelated within-chip delay variation implies that traditional approach to worst-case timing analysis leaves significant performance potential of semiconductor processes unharvested. As discussed at length in Chapter 14, timing modeling for ASICs is based on two key assumptions: (1) intra-chip parameter variability within the chip is negligible compared to inter-chip variability, and (2) all types of digital circuits, e.g. all cells and blocks, behave statistically similarly (i.e. in a correlated manner) in response to parameter variations. Each of these assumptions is failing and the result is lost performance for the ASIC designer.

Custom design does not obviate the problems mentioned immediately above, and timing tools for custom designers do not necessarily do a better job of modeling process-related timing issues. What allows a custom designer to better exploit the performance potential of a process is that high-performance/high-volume custom designs are speed-binned after manufacturing. Speed-binning entails testing the parts at different speeds to determine the maximum possible operating frequency. This has two benefits. The first is the ability to harvest the fastest running parts in a process. The second benefit is that the process-profile offered by speed-binning gives both process engineers and circuit designers a very clear profile of the performance of the process.

If we compound the factors of foundry-to-foundry variance ($1.1\times$ to $1.4\times$) with ASICs need to stick to worst-case process numbers that reflect the variation within a process (1.2× to 1.3×) then it is easy to justify that some ASIC designs will operate at a deficit of 1.3× to 1.8× relative to custom designs fabricated at the best foundries. Continuous process improvements described in Section 11.2 could add an additional factor of 1.2× bringing the total deficit to 1.6× to 2.2×.

11.5 What can ASICs do about Process Variation?

There are certain advantages to the library-based approach used in ASIC designs. As a result of relying on pre-characterized cell libraries, ASICs are typically easy to migrate between technology generations. Thus synthesizable ASICs can be easily switched to use the best fabrication plants available for ASIC production. In contrast, custom designs cannot simply be mapped to a new gate library for the next technology generation. The custom designs must have transistors resized and circuits altered to account for design rules, voltage, current and power considerations not scaling linearly. For high volume custom designs where high speeds sell, it is profitable to have a design team making changes to take best advantage of a process.

Three sources of variation in performance were named in the sections above: variation within a plant; variation from foundry to foundry; and continuous process improvement. Regarding continuous process improvement, some ASIC foundries *do* provide new ASIC libraries throughout a technology generation as their technology matures. These foundries provide updated libraries for higher process speeds achieved by shorter effective transistor channel lengths and other changes. These libraries are close in speed to contemporary high-speed custom processes. For example, in the third quarter of 1998, IBM's CMOS7S 0.22um ASIC process and Intel's P856.5 0.25um custom process both had ring oscillator stage delays of 22ps. CMOS7S had six metal layers of copper interconnect; whereas P856.5 had five layers of aluminum interconnect [8]. These contemporary processes were comparable in speed, despite technology differences such as interconnect and gate length. Thus ASICs fabbed at the best foundries will enjoy the benefits of continuous process improvement.

Regarding performance variation from foundry to foundry: some very fast processes used in custom designs are also available for ASICs. One example is silicon-on-insulator (SOI), although it is more expensive to produce chips in this technology. Static logic is about 15% faster with SOI [2], and 5% faster using copper and a low-k dielectric [38].

Also the speed of ASICs is limited by the need to design circuits under the worst-case process conditions. As a result there is little ability for ASIC designs to harvest the fastest performance available in a process. ASIC foundries do not have the same level of process information gathered by custom foundries. ASIC foundries must contract with designers to deliver integrated circuits achieving a particular speed *before they've ever fabricated a single part*. On the other hand, high-volume custom designs get to market their fastest parts *after fabrication*.

Imagine trying to sign a contract with an ASIC foundry to produce a high-complexity design running at a moderately aggressive performance point, such as 650MHz in 0.18um. If worst-case timing numbers indicate speeds of 500MHz then it is likely that some parts coming off the line will operate at 650MHz, but how many? Profit margins for the ASIC foundry manufacturing will depend on the number of properly functioning die operating at 650MHz from each wafer. Evaluating this number requires accurately estimating the effect of process variation on this part, but this information is not available. Finally, guaranteeing the performance of the integrated circuit requires at-speed testing but current ASIC design methodologies do not produce manufacturing test vectors for *at-speed* testing. They do, of course, produce test vectors for stuck-at-fault testing. In short there are many obstacles that must be surmounted before ASIC designers can hope to get the same performance out of a process that custom designers are able to achieve. Until that time the best performance that ASICs can achieve in a process is almost certain to be limited by the distribution of the process variation. We estimate process variation as at least 20%, even in a mature process, and therefore estimate that ASICs will lose a factor of 1.2× performance to custom designs.

11.6 Process Variation in the Remainder of the Book

A tutorial overview of process related issues is given in Chapter 5. A very practical way of managing the impact of process variation on uncertainty is described by Intel in Chapter 13. A more speculative approach to exploiting processing is described in Chapter 14.

12. SUMMARY AND CONCLUSIONS

We first gave examples of performance of both custom ICs and ASICs. We then gave a top-level overview of what we feel accounts for the principal differences in performance between custom ICs and ASICs. We then examined in detail each of the factors of performance and attempted to justify the numeric percentage of our factors.

One key point in interpreting our results is that we have focused on *entire IC designs*. It is certainly true that if one restricts the focus to a single circuit element, such as a barrel-shifter for example, the influence of custom circuit design techniques may appear much more significant than we have indicated. However, when such elements are integrated into an entire path, such as in an ALU, their individual significance is naturally reduced.

Another important caveat is that because of space restrictions we have focused exclusively on speed differences between ASIC and custom ICs and not on area or power differences. Viewed from the standpoint of area our results and conclusions would be significantly different.

Based on our analysis we believe that the influence of the factors of floorplanning and circuit design, while significant, are relatively overstated in their importance in the performance gap between ASIC and custom ICs. From our analysis the two factors of equal or greater significance are pipelining and process variation. These factors account for a substantial portion of the performance difference between ASIC and custom. The use of dynamic-logic families is a third significant influence. Adding this factor to pipelining and process variation accounts for the majority of the performance difference ($\times 5$) between ASIC and custom design.

12.1 Conclusions About ASIC Design

Significant performance increases are available to designers using an ASIC methodology, but ASIC designers must become familiar with microarchitecture, physical design, clocking schemes, and sources of semiconductor process variation. To achieve these performance increases register-transfer level (RTL) design must mean more than synthesizing well structured code in a Hardware Description Language such as Verilog. The RTL designer must devote considerable attention to microarchitecture and exploit pipelining opportunities where they exist (Chapter 2 and design examples). To diminish the timing overhead associated with each pipeline stage latch-based design (Chapters 3 and 7) or constructive clock-skew (Chapter 8) needs to be understood and employed.

Achieving higher performance will require recovery from the physical-design allergy that infects so many ASIC designers today. High performance design require efficient floorplanning (Chapter 6) and attention to the cells that ultimately implement logical function (Chapters 4 and 9-12).

Finally, ASIC designers must look past the classic barrier of ASIC design "sign-off" and begin to understand and manage sources of process variation (Chapters 5 and 14) so that tool usage can be modified to better achieve design goals (Chapter 13).

The design examples shown in Chapters 15, 16, and 17 demonstrate that orchestration of many of the techniques described above may produce integrated circuits designed within an ASIC methodology that run $2\times$ to $3\times$ faster than most of their contemporary ASICs and may achieve clock rates within $2\times$ to $3\times$ of the fastest custom ICs.

12.2 Conclusions about Custom Design

It appears that by drawing attention to their ability to handcraft layout custom IC designers have been practicing a kind of magical misdirection. In full context it appears that the valuable work of a custom designer is much broader than detailed circuit design and includes a systematic orchestration of all the principles that we have named above: microarchitecture, floorplanning, wire-planning, cell-design, and process management. The systematic application of these principles to achieve system performance goals is the true characteristic of a high-performance custom IC designer.

13. WHAT'S NOT IN THE BOOK

This book focuses on sources of performance in integrated circuits and the tools and techniques by which performance can be achieved. Notably absent from this book are discussions of functional verification, manufacturing test, power dissipation (especially static power), or the impact of deep-submicron effects on integrated circuit design. These topics are important ones but are, unless mentioned, orthogonal issues to those that we discuss here. Where the techniques that we suggest here negatively impact one or more of these issues we have made every effort to point that out.

14. ORGANIZATION OF THE REST OF THE BOOK

To make it easy to randomly access information, this book is organized into four units. This introductory chapter forms the first unit. This chapter is intended to be a self-contained introduction to the entire topic. The second unit consists of Chapters 2 to 5. These four Chapters form a second independent tutorial overview of the subject. Following these are Chapters 6 to 14. These contributed chapters represent the state-of-the art tool approaches to improving ASIC performance. On the other hand their inclusion here should not be construed as a wholesale commercial endorsement, but as examples of groups bravely tackling some of the toughest problems in ASIC design. The sections above have provided contextual pointers to these chapters. The fourth unit of this book provides examples of ASIC designs that have realized very high performance utilizing an ASIC methodology.

Integrated circuit design and electronic-design automation are rapidly changing fields. We have resisted the temptation to keep soliciting new material and polishing existing chapters with the aim to deliver the book more quickly to the hands of its intended readers.

15. REFERENCES

[1] Allen, D. H., et al., "Custom circuit design as a driver of microprocessor performance," *IBM Journal of Research and Development*, vol. 44, no. 6, November 2000.

[2] Anderson, C, et al., "Physical Design of A Fourth-Generation POWER GHz Microprocessor," *IEEE International Solid-State Circuits Conference*, 2001.

[3] API Products: 21264A. March 2002. http://www.alpha-processor.com/products/21264A-processor.asp

[4] ARM, ARM1020E and ARM1022E - High-performance, low-power solutions for demanding SoC, 2002, http://www.arm.com/

[5] Brand, A., et al., "Intel's 0.25 Micron, 2.0 Volts Logic Process Technology," *Intel Technology Journal*, Q3 1998. http://developer.intel.com/technology/itj/q31998/pdf/p856.pdf.

[6] Chang, A., "VLSI Datapath Choices: Cell-Based Versus Full-Custom," SM Thesis, Massachusetts Institute of Technology, February 1998. ftp://cva.stanford.edu/pub/publications/achang_masterworks980427.pdf

[7] Chen, C., Chu, C., and Wong, D., "Fast and Exact Simultaneous Gate and Wire Sizing by Lagrangian Relaxation," *IEEE Transactions on Computer-Aided Design of Integrated Circuits and Systems*, vol. 18, no. 7, July 1999, pp. 1014-1025.

[8] Diefendoff, K., "The Race to Point One Eight," *Microprocessor Report*, vol. 12, no. 12, September 1998, pp. 10-22.

[9] Fishburn, J., and Dunlop, A., "TILOS: A Posynomial Programming Approach to Transistor Sizing," *Proceedings of the International Conference on Computer-Aided Design*, 1985, pp. 326-328.

[10] Gavrilov, S., et al., "Library-Less Synthesis for Static CMOS Combinational Logic Circuits," *Proceedings of the International Conference on Computer-Aided Design*, 1997, pp. 658-663.

[11] Gonzalez, R., "Configurable and Extensible Processors Change System Design," *Hot Chips 11*, 1999. ftp://www.hotchips.org/pub/hotc7to11cd/hc99/hc11_pdf/hc99.s4.3.Gonzalez.pdf

[12] Lexra, Lexra LX4380 Product Brief, 2002, http://www.lexra.com/LX4380_PB.pdf

[13] Gowan, M., Biro, L., and Jackson, D., "Power Considerations in the Design of the Alpha 21264 Microprocessor," *Proceedings of the Design Automation Conference*, 1998, pp. 726-731.

[14] Grodstein, J., et al., "A Delay Model for Logic Synthesis of Continuously-Sized Networks," *Proceedings of the International Conference on Computer-Aided Design*, 1995, pp. 458-462.

[15] Gronowski, P., et al., "High-Performance Microprocessor Design," *IEEE Journal of Solid-State Circuits*, vol. 33, no. 5, May 1998, pp. 676-686.

[16] Haddad, R., van Ginneken, L., and Shenoy, N., "Discrete Drive Selection for Continuous Sizing," *Proceedings of the International Conference on Computer Design*, 1997, pp. 110-115.

[17] Hare, C. 586/686 Processors Chart. http://users.erols.com/chare/586.htm

[18] Hare, C. 786 Processors Chart. http://users.erols.com/chare/786.htm

[19] Harris, D., and Horowitz, M., "Skew-Tolerant Domino Circuits," *IEEE Journal of Solid-State Circuits*, vol. 32, no. 11, November 1997, pp. 1702-1711.

[20] Hauck, C., and Cheng, C., "VLSI Implementation of a Portable 266MHz 32-Bit RISC Core," *Microprocessor Report*, November 2001.

[21] Hennessy, J., and Patterson, D. *Computer Architecture: A Quantitative Approach*, 2nd Ed. Morgan Kaufmann, San Francisco CA, 1996.

[22] Hill, D., "Sc2: A Hybrid Automatic Layout System," *Proceedings of the International Conference on Computer-Aided Design*, 1985, pp. 172-174.

[23] Hinton, G., et al., "A 0.18-um CMOS IA-32 Processor With a 4-GHz Integer Execution Unit," *IEEE Journal of Solid-State Circuits*, vol. 36, no. 11, November 2001, pp. 1617-1627.
[24] Intel, Inside the NetBurst Micro-Architecture of the Intel Pentium 4 Processor, Revision 1.0, 2000. http://developer.intel.com/pentium4/download/netburst.pdf
[25] Kim, T. and Um, J., "A Practical Approach to the Synthesis of Arithmetic Circuits using Carry-Save-Adders," *IEEE Transactions on Computer Aided Design of Integrated Circuits and Systems*, vol. 19, no. 5, May 2000, pp. 615-624,
[26] Kapadia, H., and Horowitz, M., "Using Partitioning to Help Convergence in the Standard-Cell Design Automation Methodology," *Proceedings of the 37th Design Automation Conference*, 1999, pp. 592-597.
[27] Keutzer, et al., "System-level Design: Orthogonalization of Concerns and Platform-Based Design," *IEEE Transactions on Computer-Aided Design*, vol. 19, no. 12, December 2000, pp. 1523-1543.
[28] Kessler, R., McLellan, E., and Webb, D., "The Alpha 21264 Microprocessor Architecture," *Proceedings of the International Conference on Computer Design*, 1998, pp. 90-95.
[29] Keutzer, K., Kolwicz, K., and Lega, M., "Impact of Library Size on the Quality of Automated Synthesis," *Proceedings of the International Conference on Computer-Aided Design*, 1987, pp. 120-123.
[30] Klass, F., et al., "A New Family of Semidynamic and Dynamic Flip-flops with Embedded Logic for High-Performance Processors," *IEEE Journal of Solid-State Circuits*, vol. 34, no. 5, May 1999, pp. 712-716.
[31] Kurd, N.A, et al., "A Multigigahertz Clocking Scheme for the Pentium® 4 Microprocessor," *IEEE Journal of Solid-State Circuits*, vol. 36, no. 11, November 2001, pp. 1647-1653.
[32] McDonald, C., "The Evolution of Intel's Copy Exactly! Technology Transfer Method," *Intel Technology Journal*, Q4 1998. http://developer.intel.com/technology/itj/q41998/pdf/copyexactly.pdf.
[33] MTEK Computer Consulting, AMD CPU Roster, January 2002. http://www.cpuscorecard.com/cpuprices/head_amd.htm
[34] MTEK Computer Consulting, Intel CPU Roster, January 2002. http://www.cpuscorecard.com/cpuprices/head_intel.htm
[35] Nowka, K., and Galambos, T., "Circuit Design Techniques for a Gigahertz Integer Microprocessor," *Proceedings of the International Conference on Computer Design*, 1998, pp. 11-16.
[36] Partovi, H., "Clocked storage elements," in *Design of High-Performance Microprocessor Circuits*, IEEE Press, Piscataway NJ, 2000, pp. 207-234.
[37] Posluszny, S., et al., "Design Methodology of a 1.0 GHz Microprocessor," *Proceedings of the International Conference on Computer Design*, 1998, pp. 17-23.
[38] Rohrer, N. et al., "A 480 MHz RISC Microprocessor in 0.12um Leff CMOS Technology with Copper Interconnects," *IEEE International Solid-State Circuits Conference*, 1998.
[39] Scott, K., and Keutzer, K., "Improving Cell Libraries for Synthesis," *Proceedings of the Custom Integrated Circuits Conference*, 1994, pp. 128-131.
[40] Smith, M., *Application-specific Integrated Circuits*, Addison-Wesley, Berkeley, CA, 1997.
[41] STMicroelectronics, "STMicroelectronics 0.25μ, 0.18μ & 0.12 CMOS," slides presented at the annual *Circuits Multi-Projets users meeting*, January 9, 2002. http://cmp.imag.fr/Forms/Slides2002/061_STM_Process.pdf
[42] Sylvester, D., Jiang, W., and Keutzer, K., BACPAC – Berkeley Advanced Chip Performance Calculator. 2000. http://www-device.eecs.berkeley.edu/~dennis/bacpac/

[43] Sylvester, D. and Keutzer, K., "Getting to the Bottom of Deep Sub-micron," *Proceedings of the International Conference on Computer Aided Design*, November 1998, pp. 203-211.

[44] Synopsys, Module Compiler: the next generation module compilation. 1999. http://www.synopsys.com/products/datapath/module_comp_ds.pdf

[45] Tensilica, Xtensa Microprocessor – Overview Handbook – A Summary of the Xtensa Microprocessor Databook. August 2001. http://www.tensilica.com/dl/handbook.pdf

[46] Thorp, T., Yee, G., and Sechen, C., "Domino Logic Synthesis Using Complex Static Gates," *Proceedings of the International Conference on Computer-Aided Design*, 1998, pp. 242-247.

[47] Thompson, S., et al., "An Enhanced 130 nm Generation Logic Technology Featuring 60 nm Transistors Optimized for High Performance and Low Power at 0.7 – 1.4 V," *Technical Digest of the International Electron Devices Meeting*, 2001.

[48] TSMC, TSMC – Technology and Manufacturing – 0.13 Micron. January 2002. http://www.tsmc.com/technology/cl013.html

[49] Zhong, G., Koh, C.-K., and Roy, K., "A Twisted-Bundle Layout Structure for Minimizing Inductive Coupling Noise," *Proceedings of the International Conference on Computer Aided Design*, 2000, pp. 406-411.

Chapter 2

Improving Performance through Microarchitecture
Pipelining, Parallelization, Retiming Registers, Loop Unrolling

David Chinnery, Kurt Keutzer
Department of Electrical Engineering and Computer Sciences,
University of California at Berkeley

In this book, microarchitecture refers to the high-level structural organization of the circuitry. This includes the number of functional units; the memory hierarchy; the input and output interfaces; the order of the functional units in the pipeline; the number of clock cycles these units have for computation; branch prediction and data forwarding logic.

Microarchitectural changes are the most significant way of reducing the clock period of a circuit and increasing the performance. A pipelined circuit can have substantially higher clock frequency, as the combinational delay of each pipeline stage is less. If the pipeline can be well utilized, each pipeline stage carries out computation in parallel. Thus pipelining can increase the circuit's *performance* (calculations per second). By allowing computation in parallel, duplicating logic increases the circuit *throughput* (calculations completed) per clock cycle, but increases the circuit area.

Section 1 gives examples of pipelining and retiming to balance the delay of pipeline stages. There are illustrations of simple parallelization, and loop unrolling to allow logic duplication. Brief comments about memory access and its impact on the clock frequency are in Section 2. Then Section 3 discusses the costs and reduction in clock period by pipelining.

Logic design encompasses circuit-level techniques for high-speed implementations of typical functional blocks. We restrict the focus of this chapter to microarchitecture and some discussion of pipeline balancing. Chapter 4 discusses logic design.

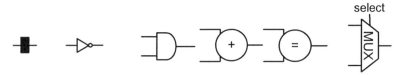

(a) Register (b) inverter gate (c) AND gate (d) Adder (e) Comparator (f) Multiplexer

Figure 1. Circuit symbols used in this chapter. Inputs are on the left side (and above for select on the multiplexer) and outputs are on the right.

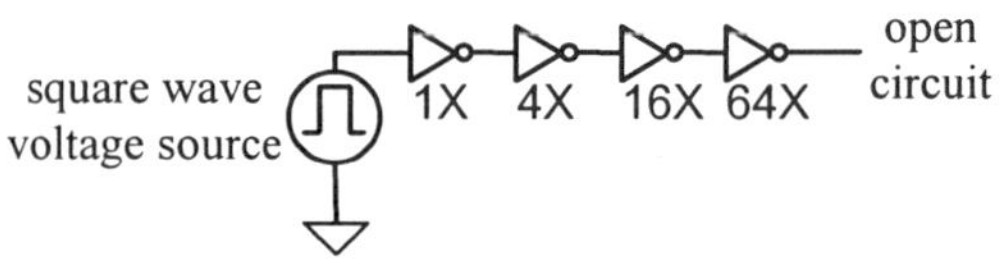

Figure 2. This illustrates a circuit to measure FO4 delays. The delay of the 4X drive strength inverter gives the FO4 delay. The other inverters are required to appropriately shape the input waveform to the 4X inverter, and reduce the switching time of the 16X inverter, which affect the delay of the 4X inverter [14].

1. EXAMPLES OF MICROARCHITECTURAL TECHNIQUES TO INCREASE SPEED

This section considers speeding up a variety of circuits by microarchitectural transformations, using the functional blocks shown in Figure 1. The examples assume nominal delays for these blocks, with delays measured in units of fanout-of-4 inverter (FO4) delays.

1.1 FO4 Delays

The fanout-of-4 inverter delay is the delay of an inverter driving a load capacitance that has four times the inverter's input capacitance [13]. This is shown in Figure 2. The FO4 metric is not substantially changed by process technology or operating conditions. In terms of FO4 delays, other fanout-of-4 gates have at most 30% range in delay over a wide variety of process and operating conditions, for both static logic and domino logic [14].

If it has not been simulated in SPICE or tested silicon, the FO4 delay can be calculated from the channel length. The rule of thumb for FO4 delay [18], based on the effective gate length L_{eff} is:

(1) $360 \times L_{eff}$ ps for typical operating and typical process conditions

(2) $500 \times L_{eff}$ ps for worst case operating and typical process conditions

where the effective gate length L_{eff} has units of micrometers. Typical process conditions give high yield, but are not overly pessimistic. Worst case operating conditions are lower supply voltage and higher temperature than typical operating conditions. Typical operating conditions for ASICs may assume a temperature of 25°C, which is optimistic for most applications. As other researchers have done [18], Equation (2) is used later in this chapter to estimate the FO4 delay in silicon for realistic operating conditions. Chapter 5 discusses process and operating conditions in more detail.

Typically, the effective gate length L_{eff} has been assumed to be about 0.7 of lambda (λ) for the technology, where λ is the base length of the technology (e.g. 0.18um process technology with L_{drawn} of 0.18um). As discussed in Chapter 5, many foundries are aggressively scaling channel length which significantly increases the speed.

Based on analysis in Table 5 of Chapter 5, typical process conditions are between 17% and 28% faster than worst case process conditions. Derating worst case process conditions by a factor of 1.2× gives

(3) $600 \times L_{eff}$ ps for worst case operating and worst case process

Equation (3) was used for estimating the FO4 delays of synthesized ASICs, which have been characterized for worst case operating and worst case process conditions. This allows analysis of the delay per pipeline stage, independent of the process technology, and independent of the process and operating conditions.

Note: these rules of thumb give approximate values for the FO4 delay in a technology. They may be inaccurate by as much as 50% compared to simulated or measured FO4 delays in silicon. These equations do not accurately account for operating conditions. Speed-binning and process improvements that do not affect the effective channel length are not accounted for. Accurate analysis with FO4 delays requires proper calibration of the metric: simulating or measuring the actual FO4 delays for the given process and operating conditions.

1.2 Nominal FO4 Delays for the Examples

The normalized delays are for 32-bit ASIC functional blocks (adder, comparator and multiplexer), and single bit inverter and AND gate. These delays do not represent accurate delays for elements in a real circuit.
- inverter gate delay $t_{inv} = 1$ FO4 delay
- AND gate delay $t_{AND} = 2$ FO4 delay

- adder delay $t_{add} = 10$ FO4 delays
- comparator delay $t_{comp} = 6$ FO4 delays
- multiplexer delay $t_{mux} = 4$ FO4 delays

To discuss the impact of the microarchitectural techniques, we need an equation relating the clock period to the delay of the critical path through combinational logic and the registers. In this chapter, we will assume that the registers are flip-flops.

1.3 Minimum Clock Period with Flip-Flops

Flip-flops have two important characteristics. The setup time t_{su} of a flip-flop is the length of time that the data must be stable before the arrival of the clock edge to the flip-flop. The clock-to-Q delay t_{CQ} of a flip-flop is the delay from the clock edge arriving at the flip-flop to the output changing. We assume simple positive edge triggered master/slave D-type flip-flops [30].

In addition, the arrival of the clock edges will not be ideal. There is some clock skew t_{sk} between the arrival times of the clock edge at different points on the chip. There is also clock jitter t_j, which is the variation between arrival times of consecutive clock edges at the same point on the chip.

Figure 3 shows the timing waveforms for a signal propagating along the critical path of a circuit. The nominal timing characteristics used are:
- Flip-flop clock-to-Q delay of $t_{CQ} = 4$ FO4 delays
- Flip-flop setup time $t_{su} = 2$ FO4 delays
- Clock skew of $t_{sk} = 3$ FO4 delays
- Clock jitter of $t_j = 1$ FO4 delay

The total delay along the critical path is the sum of the clock-to-Q delay from the clock edge arriving at *R1*, the maximum delay of any path through the combinational logic t_{comb} (the critical path), the setup time of the output flip-flop of the critical path, the clock skew and the clock jitter. This places a lower bound on the clock period, which may also be limited by other pipeline stages. The minimum clock period with flip-flops $T_{flip\text{-}flops}$ is [30]

(4) $\quad T_{flip-flops} = \max\{t_{CQ} + t_{comb} + t_{su} + t_{sk} + t_j\}$

The delay of a gate in the circuit will vary depending on its drive strength, its fanout load, and the arrival times and slews of its fanins. In this context, slews are the rise and fall times of the input signals. Thus the delay of flip-flops on each possible critical path would need to be calculated, rather than assuming the delays are equal. For example, the register storing input *b* may have slower clock-to-Q delay t_{CQ} than the register *R1* storing *a*.

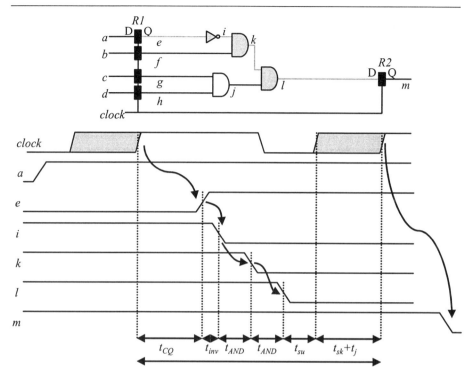

Figure 3. Waveform diagram of 0 to 1 transition on input *a* propagating along the critical path *aeiklm*. All other inputs *b*, *c*, and *d* are fixed at 1. The critical path is shaded in gray on the circuit. Clock skew and clock jitter between clock edge arrivals at registers *R1* and *R2* result in a range of possible arrival times for the clock edge at *R2*. The range of possible clock edge arrival times, relative to the arrival of the clock edge at the preceding register on the previous cycle, is shaded in gray on the *clock* edge waveform.

For simplicity, we assume equal clock-to-Q delay and setup times for all the registers, and equal clock jitter and clock skew at all registers, giving

(5) $\quad T_{flip-flops} = t_{CQ} + \max\{t_{comb}\} + t_{su} + t_{sk} + t_j$

We can group terms into the clocking overhead $t_{clocking}$, the clock skew and clock jitter, and the register overhead $t_{register}$, the setup and clock-to-Q delay:

(6) $\quad T_{flip-flops} = \max\{t_{comb}\} + t_{register} + t_{clocking}$

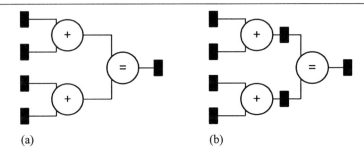

Figure 4. Diagram showing an add-compare operation (a) before and (b) after pipelining.

For the example shown in Figure 3, the clock period is

(7)
$$T_{flip-flops} = t_{CQ} + \max\{t_{inv} + t_{AND} + t_{AND}, t_{AND} + t_{AND}\} + t_{su} + t_{sk} + t_j$$
$$= 4 + 1 + 2 + 2 + 2 + 3 + 1$$
$$= 15 \text{ FO4 delays}$$

Now we can calculate the reduction in clock period by pipelining long critical paths.

1.4 Pipelining

If sequential logic produces an output every clock cycle, then reducing the clock period increases the calculations per second. Pipelining breaks up the critical path with registers for memory between clock cycles. This reduces the delay of each stage of the critical path, thus a higher clock speed is possible, and the circuit's speed is increased. The *latency* increases due to adding memory elements with additional delay, but the calculations per second increases because the clock period is reduced and each stage of the pipeline can be computing at the same time ideally.

Consider pipelining the add-compare operation shown in Figure 4, where the output of two adders goes to a comparator. From (4), the clock period before pipelining is

(8)
$$T_{flip-flops} = t_{add} + t_{comp} + t_{sk} + t_j + t_{CQ} + t_{su}$$
$$= 10 + 6 + 3 + 1 + 4 + 2$$
$$= 26 \text{ FO4 delays}$$

After pipelining with flip-flops, the minimum clock period is the maximum of the delays of any pipeline stage. Thus the clock period is

$$T_{flip-flops} = t_{CQ} + \max\{t_{add}, t_{comp}\} + t_{su} + t_{sk} + t_j$$
(9)
$$= 4 + \max\{10,6\} + 2 + 3 + 1$$
$$= 20 \text{ FO4 delays}$$

The 30% increase in speed may not translate to a 30% increase in performance, if the pipeline is not fully utilized. Note the pipeline stages are not well-balanced; if they were perfectly balanced, each stage would have the same combinational delay, and the clock period would be 18 FO4 delays.

The 30% decrease in the clock period comes at the cost of additional registers and wiring. The area increase is typically small relative to the overall size of the circuit. Sometimes the wiring congestion and increased number of registers can be prohibitive, especially if a large number of registers are required to store a partially completed calculation, such as when pipelining a multiplier.

The clock power consumption increases substantially in this example. Instead of the clock to 5 registers switching every 26 FO4 delays, the clock goes to 7 registers and switches every 20 FO4 delays. This gives an 82% increase in the power consumed by clocking the registers.

In a typical pipelined design, the clock tree may be responsible for 20% to 45% of the total chip power consumption [22]. Turning off the clock, *clock gating*, to modules that are not in use reduces the clock power [7].

1.4.1 Limitations to Pipelining

The most direct method of reducing the clock period and increasing throughput is by pipelining. However, pipelining may be inappropriate due to increased latency. For example, the latency for memory address calculation needs to be small, so the combinational delay must be reduced [28]. The pipeline may not be fully utilized, which reduces the instructions executed per clock cycle (IPC). It may not be possible to reduce the clock period further by pipelining, because other sequential logic may limit the clock period. Also, there is no advantage to performing calculations faster than inputs are read and outputs are written, so the clock period may be limited by I/O bandwidth.

Reduced pipeline utilization can be due to data hazards causing pipeline stalls, or branch misprediction [16]. Data hazards can be caused by instructions that are dependent on other instructions. One instruction may write to a memory location before another instruction has the opportunity to read the old data. To avoid data hazards, a dependent instruction in the pipeline must stall, waiting for the preceding instructions to finish. Branch misprediction causes the wrong sequence of instructions to be speculatively executed after a branch; when the correct branch is determined, the pipeline must be cleared of these incorrect operations. Forwarding logic and better branch prediction logic help compensate for the reduction in IPC, but there is additional area and power overhead as a result.

Hardware can improve pipeline utilization by data forwarding and better branch prediction. Compilation can reschedule instructions to reduce the hazards, and calculate branches earlier.

Duplicating sub-circuits or modules is an alternative to pipelining that does not suffer from increased latency or pipeline stalls. There may still be issues feeding inputs to and outputs from the logic to keep it fully utilized. Duplication entails a substantial area and power penalty.

1.5 Parallelization

Consider using a single adder to sum 8 inputs. This can be implemented with 7 add operations, as illustrated in Figure 5. Generally, pipelining an adder does not reduce the clock period, as high-speed adders are generally faster than other circuitry within the chip. Using a single adder to sum 8 inputs takes 7 cycles to perform the 7 add operations, giving a throughput of one 8-input sum per 7 clock cycles.

The circuit in Figure 5 performs the following operations each cycle:
- $i = a+b$, $j = c+d$, $k = e+f$, $l = g+h$, $m = i+j$, $n = k+l$, and $o = m+n$
- Denoting respective clock cycles with a superscript, the final output is $o^t = a^{t-3} + b^{t-3} + c^{t-3} + d^{t-3} + e^{t-3} + f^{t-3} + g^{t-3} + h^{t-3}$

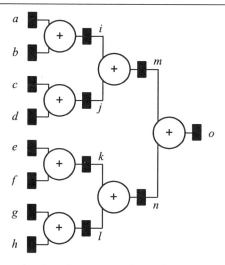

Figure 5. Adders computing the summation of 8 inputs.

The clock period for the Figure 5 circuit is

(10)
$$\begin{aligned} T_{flip-flops} &= t_{add} + t_{sk} + t_j + t_{CQ} + t_{su} \\ &= 10 + 3 + 1 + 4 + 2 \\ &= 20 \text{ FO4 delays} \end{aligned}$$

By implementing the sum of 8 inputs directly with 7 pipelined adders, the throughput is increased to calculating one 8-input sum per clock cycle. The latency for the summation is 3 clock cycles.

Compared to using a single adder, the seven adders have at least seven times the area and power consumption. The energy per calculation is the product of the power and the clock period divided by the throughput. Thus the energy per calculation of the sum of 8 inputs is about the same, as the throughput has also increased by a factor of seven.

The area and power cost of parallelizing the logic is substantial. Generally, computing the same operation k times in parallel increases the power and area of the replicated logic by more than a factor of k, as there is more wiring.

Sometimes because of recursive dependency of algorithms, it is not possible to simply duplicate logic. In such cases, the cycle dependency can be unrolled to allow logic duplication.

1.6 Loop Unrolling

Figure 6(a) shows the recursion relation of the Viterbi algorithm for a two-state Viterbi detector. The recursive add-compare-select calculations for the two-state Viterbi detector are

- $sm_1^n = \max\{sm_1^{n-1} + bm_{1,1}^{n-1}, sm_2^{n-1} + bm_{2,1}^{n-1}\}$
- $sm_2^n = \max\{sm_1^{n-1} + bm_{1,2}^{n-1}, sm_2^{n-1} + bm_{2,2}^{n-1}\}$

There are a number of approaches for speeding up circuitry for Viterbi detectors, which are discussed in the design of the Texas Instruments SP4140 disk drive read channel in Chapter 15. One simple approach to double the throughput is to unroll the recurrence relation, as shown in Figure 6(b), which doubles the throughput at the cost of doubling the area and power.

Loop unrolling can allow tight recurrence relations like the Viterbi algorithm to be unrolled into sequential stages of pipelined logic, increasing the throughput.

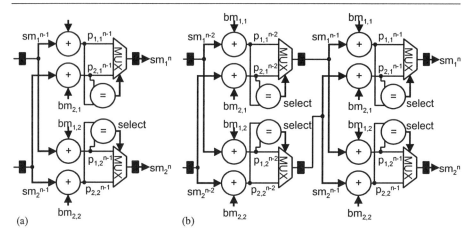

Figure 6. Implementations of the Viterbi algorithm for two states. (a) The add-compare-select implementation. (b) Loop unrolling the add-compare-select algorithm.

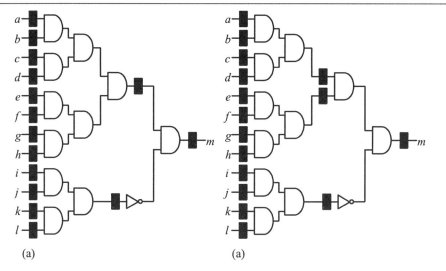

Figure 7. Example of retiming the unbalanced pipeline in (a) to give a smaller clock period for the circuit in (b).

1.7 Retiming

Pipelined logic often has unbalanced pipeline stages. Retiming can balance the pipeline stages, by shifting the positions of the register positions to try and equalize the delay of each stage.

Figure 7(a) shows an unbalanced pipeline. The first pipeline stage has a combinational delay of three AND gates, whereas the second stage has the combinational delay of only one AND gate. The clock period is

(11)
$$\begin{aligned} T_{flip-flops} &= \max_{combinational\ paths} \{t_{comb} + t_{sk} + t_j + t_{CQ} + t_{su}\} \\ &= 3t_{AND} + t_{sk} + t_j + t_{CQ} + t_{su} \\ &= 3 \times 2 + 3 + 1 + 4 + 2 \\ &= 16\ \text{FO4 delays} \end{aligned}$$

Changing the positions of the registers so as to preserve the circuit's functionality, retiming [24], can balance the pipeline stages and reduce the clock period. The clock period of the balanced pipeline in Figure 7(b) is

(12)
$$\begin{aligned} T_{flip-flops} &= \max_{combinational\ paths} \{t_{comb} + t_{sk} + t_j + t_{CQ} + t_{su}\} \\ &= 2t_{AND} + t_{sk} + t_j + t_{CQ} + t_{su} \\ &= 2 \times 2 + 3 + 1 + 4 + 2 \\ &= 14\ \text{FO4 delays} \end{aligned}$$

Retiming the registers may increase or decrease the area, depending on whether there are more or less registers after retiming. Synthesis tools support retiming of edge-triggered flip-flop registers [31]. The reduction of clock period by retiming is not large, if the pipeline stages were designed to be fairly balanced in the RTL. The speed penalty of unbalanced pipeline stages in two ASICs is estimated in Section 6.2.1 of Chapter 3.

1.8 Tools for Microarchitectural Exploration

Synthesis tools are available that will perform retiming on gate net lists [31]. Some RTL synthesis tools can perform basic pipelining and parallelization, particularly when given recognizable structures for which known alternative implementations exist such as adders.

An ASIC designer can examine pipelining and parallelization of more complicated logic by modifying the RTL to change the microarchitecture. Other techniques such as loop unrolling and retiming logic (rather than registers) cannot be done by current EDA software, and require RTL modifications. Design exploration with synthesizable logic is much easier than for custom logic, as RTL modifications are quick to implement in comparison with custom logic.

While retiming does not affect the number of instructions per cycle (IPC), pipelining may reduce the IPC due to pipeline stalls and other hazards, and duplicated logic may not be fully utilized. All of these techniques modify the positions of registers with respect to the combinational logic in the circuit. For this reason verifying functional equivalence requires more difficult sequential comparisons. There are formal verification tools for sequential verification, but their search depth is typically limited to only several clock cycles.

Retiming is performed at the gate level with synthesis tools, whereas pipelining, parallelization, and loop unrolling are usually done at the RTL level. It is possible to do pipelining at the gate level, though there is less software support. Some synthesis tools duplicate logic to reduce delays when driving large capacitances, but do not otherwise directly support parallelization or duplication of modules. Table 1 summarizes the techniques and their trade-offs.

The memory hierarchy is also part of the microarchitecture. The cache access time takes a substantial portion of the clock period, and can limit performance.

2. MEMORY ACCESS TIME AND THE CLOCK PERIOD

Traditionally, caches have been integrated on chip to reduce memory access times. Custom chips have large on-chip caches to minimize the chance of a cache miss, as cache misses incur a large delay penalty of multiple cycles. Frequent cache misses substantially reduces the advantage of higher clock frequencies [15].

ASICs do not run as fast, and thus are not as desperate for high memory bandwidth and don't need as large caches to prevent cache misses. The off-chip memory access time is not substantially faster for custom designs, hence larger caches on-chip are needed to compensate. Large on-chip caches are expensive in terms of real estate and yield. A larger die substantially increases the cost of the chip as the number of die per wafer decreases and the yield also decreases. The design of the memory hierarchy is discussed in detail by Hennessy and Patterson [16].

Technique	Granularity	Method	Cost	Benefit
retiming	gate	EDA tools	may decrease or increase number of registers	balances pipeline
pipelining	functional block	modify RTL	may reduce IPC, more registers	reduces the clock period
parallelization, loop unrolling	functional block	modify RTL	k times the area and power for k duplicates, same energy/calculation	increases throughput

Table 1. Overview of microarchitectural techniques to improve speed

If a processor's clock frequency is a multiple of the clock frequency for off-chip memory, the synchronization time to access the off-chip memory is reduced. A study by Hauck and Cheng shows that a 200MHz CPU with 100MHz SDRAM has better performance than a 250MHz CPU with 133MHz SDRAM if the cache miss rate is more than 1% [15].

Memory access time is a substantial portion of the clock cycle. For example, cache logic for tag comparison and data alignment can take 55% of the clock period at 266MHz in 0.18um technology [15]. However, the cache access time is also a substantial portion of the clock cycle in an ASIC, and it can be very difficult to fit the cache logic and cache access in a single cycle. This can limit the clock period. Lexra devoted an entire pipeline stage for the cache access, thus the cache access time was not a critical path. This allowed larger 32K caches and a higher clock frequency [33].

STMicroelectronics iCORE also devoted two stages to reading memory. The first stage is for tag access and tag comparison, and the second stage is for data access and alignment. Chapter 16 discusses the iCORE memory microarchitecture in detail.

We will now examine the speedup by pipelining in more detail. Chapter 15 gives further examples analyzing the other architectural techniques.

3. SPEEDUP FROM PIPELINING

Consider some combinational logic between flip-flop registers shown in Figure 8(a), with critical path delay of t_{comb}. As discussed in Section 1.3, the delay of the critical path t_{comb} through the combinational logic limits the minimum clock period $T_{flip-flops}$, as given in (6).

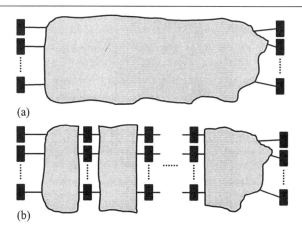

Figure 8. Diagram of logic before and after pipelining, with combinational logic shown in gray. Registers are indicated by black rectangles.

The latency of the pipeline is the time from the arrival of the pipeline inputs to the pipeline, to the exit of the pipeline outputs corresponding to this given set of inputs, after calculations in the pipeline. There is only a single pipeline stage, so the latency of the path $T_{latency}$ is simply the clock period:

(13) $\quad T_{latency} = t_{comb} + t_{register} + t_{clocking}$

Suppose this logic path is pipelined into n stages of combinational logic between registers as shown in Figure 8(b). If the registers are flip-flops, the pipeline stage with the worst delay limits the clock period according to Equation (6) (where $t_{comb,i}$ is the delay of the slowest path in the ith stage of combinational logic):

(14) $\quad T_{flip-flops} = \max_i \{t_{comb,i}\} + t_{register} + t_{clocking}$

The latency is then simply n times the clock period, as the delay through each stage is the clock period.

(15) $\quad T_{latency} = nT$

3.1 Ideal Clock Period after Pipelining

Ideally, the pipeline stages would have equal delay and the maximum delay $t_{comb,i}$ of combinational logic in each pipeline stage i is the same:

(16) $\quad t_{comb,i} = t_{comb,average} = \dfrac{t_{comb}}{n}$

Thus the minimum possible clock period after pipelining with flip-flops is

(17) $\quad T_{min} = \dfrac{t_{comb}}{n} + t_{register} + t_{clocking}$

And with this ideal clock period, the latency would be

(18) $\quad T_{latency} = nT_{min} = t_{comb} + n(t_{register} + t_{clocking})$

3.2 Clock Period with Retiming Balanced Pipeline

By retiming the positions of the registers between pipeline stages, the delay of each stage can be made nearly the same. If retiming is possible, a clock period close to the ideal clock period is possible with flip-flop registers.

Improving Performance through Microarchitecture 47

Observation: If retiming is possible, the combinational delay of each pipeline stage can be balanced. Retiming can balance the pipeline stages to within a gate delay of the average combinational delay:

(19) $\quad t_{comb,i} < \dfrac{t_{comb}}{n} + t_{gate} = t_{comb,average} + t_{gate}$

Proof:

Suppose the jth pipeline stage has combinational delay more than a gate delay t_{gate} slower than the average combinational delay.

(20) $\quad t_{comb,j} \geq \dfrac{t_{comb}}{n} + t_{gate}$

Suppose there is no other pipeline stage with delay less than the average combinational delay. Then

(21) $\quad t_{comb} = \sum\limits_{i=1}^{n} t_{comb,i} \geq (n-1)\dfrac{t_{comb}}{n} + \dfrac{t_{comb}}{n} + t_{gate} > t_{comb}$

This is a contradiction. Hence, there must be some other pipeline stage with delay of less than the average combinational delay. That is, there exists some pipeline stage k, such that

(22) $\quad t_{comb,k} < \dfrac{t_{comb}}{n}$

If $j < k$, then all the registers between pipeline stages j and k can be retimed to be one gate delay earlier. If $j > k$, then all the registers between pipeline stages j and k can be retimed to be one gate delay later. This balances the pipeline stages better:

(23) $\quad \begin{aligned} t'_{comb,j} &= t_{comb,j} + t_{gate}, \\ t'_{comb,k} &= t_{comb,k} + t_{gate} < \dfrac{t_{comb}}{n} + t_{gate} \end{aligned}$

In this manner, all pipeline stages with combinational delay more than a gate delay slower than the average combinational delay can be balanced by retiming to be less than a gate delay more than the average combinational delay:

(24) For all i, $t_{comb,i} < \dfrac{t_{comb}}{n} + t_{gate}$

Thus the level of granularity of retiming is a gate delay, though it is limited by the slowest gate used. It is not always possible to perform retiming, as there may be obstructions to retiming (e.g. there may be a late-arriving external input that limits the clock period), but in most cases it is possible to design the sequential circuitry so that the pipeline can be balanced.

If retiming to balance the pipeline stages is possible, from (17) and (24), the clock period of a balanced pipeline with flip-flop registers is bounded by

$$(25) \quad \frac{t_{comb}}{n} + t_{register} + t_{clocking} < T \leq \frac{t_{comb}}{n} + t_{gate} + t_{register} + t_{clocking}$$

Correspondingly, the latency is bounded by

$$(26) \quad t_{comb} + n(t_{register} + t_{clocking}) \leq T_{latency} \leq t_{comb} + n(t_{gate} + t_{register} + t_{clocking})$$

The delay of a gate t_{gate} is small relative to the other terms. Retiming thus reduces the clock period and latency, giving a fairly well-balanced pipeline.

3.3 Estimating Speedup with Pipelining

Having pipelined the path into n stages, ideally the pipelined path would be fully utilized. However, there are usually dependencies between some calculations that limit the utilization, as discussed in Section 1.4.1. To quantify the impact of deeper pipelining on utilization, we consider the number of instructions per cycle (IPC).

The average calculation time per instruction is

$$(27) \quad T/IPC$$

Suppose the number of instructions per cycle is IPC_{before} before pipelining and IPC_{after} after pipelining. Assuming no additional microarchitectural features to improve the IPC, limited pipeline utilization results in

$$(28) \quad IPC_{after} \leq IPC_{before}$$

For example, the Pentium 4 has 10% to 20% less instructions per cycle than the Pentium III, due to branch misprediction, pipeline stalls, and other hazards [20]. From (6), (25) and (27), the increase in performance by pipelining with flip-flop registers is (assuming minimal gate delay and well-balanced pipelines)

$$(29) \quad \frac{IPC_{after}}{IPC_{before}} \times \frac{T_{before}}{T_{after}} \approx \frac{IPC_{after}}{IPC_{before}} \times \frac{t_{comb} + t_{register} + t_{clocking}}{\left(\frac{t_{comb}}{n} + t_{register} + t_{clocking}\right)}$$

The Pentium 4 was designed to be about 1.6 times faster than the Pentium III microprocessor in the same technology. This was achieved by increasing the pipelining: the branch misprediction pipeline increased from 10 to 20 pipeline stages [17]. Assuming the absolute value of the timing overhead remains about the same, the relative timing overhead increased from about 20% of the clock period in the Pentium III to 30% in the Pentium 4.

The pipelining overhead consists of the timing overheads of the registers and clocking scheme, and any penalty for unbalanced pipeline stages that can't be compensated for by slack passing. The pipelining overhead is typically about 30% of the pipelined clock period for ASICs, and 20% for custom designs (see Section 6 of Chapter 3 for more details). However, super-pipelined custom designs (such as the Pentium 4) may also have a pipelining overhead of about 30%.

We estimate the pipelining overhead in FO4 delays for a variety of custom and ASIC processors, and the speedup by pipelining. The ASIC clock periods are 58 to 67 FO4 delays. This is shown in Table 2 and Table 3. FO4 delays for reported custom designs may be better than typical due to speed binning and using higher supply voltage to achieve the fastest clock frequencies.

Custom	Frequency (GHz)	Technology (um)	Effective Channel Length (um)	Voltage (V)	Pipeline Stages	FO4 delays/stage	30% Pipelinging Overhead (FO4 delays)	Unpipelined Clock Period (FO4 delays)	Pipelining Overhead % of Unpipelined Clock Period	Increase in Clock Frequency by Pipelining
Pentium 4 (Willamette)	2.000	0.18	0.10	1.75	20	10.0	3.0	143	2.1%	×14.3
ASICs										
Tensilica Xtensa (Base)	0.250	0.25	0.18	2.50	5	61.7	18.5	235	7.9%	×3.8
Tensilica Xtensa (Base)	0.320	0.18	0.13	1.80	5	66.8	20.0	254	7.9%	×3.8
Lexra LX4380	0.266	0.18	0.13	1.80	7	57.8	17.4	301	5.8%	×5.2
Lexra LX4380	0.420	0.13	0.08	1.20	7	59.5	17.9	310	5.8%	×5.2
ARM1020E	0.325	0.13	0.08	1.20	6	64.1	19.2	288	6.7%	×4.5

Table 2. Characteristics of ASICs and a super-pipelined custom processor [2][8][25][27][35], assuming 30% pipelining overhead. The ARM1020E frequency is for the worst case process corner; LX4380 and Xtensa frequencies are for typical process conditions. The ARM1020E and LX4380 frequencies are for worst case operating conditions; the Xtensa frequency is for typical operating conditions.

Custom	Frequency (GHz)	Technology (um)	Effective Channel Length (um)	Voltage (V)	Pipeline Stages	FO4 delays/stage	30% Pipelinging Overhead (FO4 delays)	Unpipelined Clock Period (FO4 delays)	Pipelining Overhead % of Unpipelined Clock Period	Increase in Clock Frequency by Pipelining
Alpha 21264	0.600	0.35	0.25	2.20	7	13.3	2.7	77	3.4%	×5.8
Pentium III (Katmai)	0.600	0.25	0.15	2.05	10	22.2	4.4	182	2.4%	×8.2
Athlon	0.600	0.25	0.16	1.60	10	20.8	4.2	171	2.4%	×8.2
IBM Power PC	1.000	0.25	0.15	1.80	4	13.3	2.7	45	5.9%	×3.4
Pentium III (Coppermine)	1.130	0.18	0.10	1.75	10	17.7	3.5	145	2.4%	×8.2
Athlon XP	1.733	0.18	0.10	1.75	10	11.5	2.3	95	2.4%	×8.2

Table 3. Custom design characteristics [5][8][9][10][11][12][26][27][32], assuming 20% timing overhead. Effective channel length of Intel's 0.25um process was estimated from the 18% speed increase from P856 to P856.5 [3][4].

Pipeline stages listed are for the integer pipeline; the Athlon floating point pipeline is 15 stages [1]. The Pentium designs also have about 50% longer floating point pipelines. The Tensilica Xtensa clock frequency is for the Base configuration [35]. The total delay for the logic without pipelining ('unpipelined clock period') is calculated from the estimated timing overhead and FO4 delays per pipeline stage.

Table 2 and Table 3 show the estimated FO4 delays per pipeline stage. The FO4 delays were calculated from the effective gate length, as detailed in Section 1.1. For example, Intel's 0.18um process has an effective gate length of 0.10um, and the FO4 delay is about 50ps from Equation (2) [18]. The FO4 delays for custom processes were calculated using Equation (2). The estimated FO4 delay for these custom processes may be higher than the real FO4 delay in fabricated silicon for these chips, because of speed-binning, unreported process improvements, and better than worse case operating conditions. As a result, the custom FO4 delays/stage may be underestimated.

From (29), if we specify the register and clock overhead as a fraction k of the total clock period if it wasn't pipelined,

$$(30) \quad k = \frac{t_{register} + t_{clocking}}{t_{comb} + t_{register} + t_{clocking}}$$

The fraction k of the timing overhead of the unpipelined delay is shown in the 'Pipelining Overhead % of Unpipelined Clock Period' column of

Table 2 and Table 3. The increase in speed by having n pipeline stages (assuming a well-balanced pipeline for pipelines with flip-flop registers), substituting (30) into (29), is

$$(31) \quad \frac{IPC_{after}}{IPC_{before}} \times \frac{T_{before}}{T_{after}} = \frac{IPC_{after}}{IPC_{before}} \times \frac{1}{\left(\frac{1-k}{n}+k\right)}$$

For maximum performance, the timing overhead and instructions per cycle limit the minimum delay per pipeline stage. A study considered a pipeline with four-wide integer issue and two-wide floating-point issue [19]. The study assumed timing overhead of 2 FO4 delays and the optimal delays for the combinational logic per pipeline stage were: 6 FO4 delays for in-order execution; and 4 to 6 FO4 delays for out-of-order execution. The authors predict that the optimal clock period for performance will be limited to between 6 and 8 FO4 delays [19]. This scenario is similar to the super-pipelined Pentium 4 with timing overhead of about 30%.

Based on the limits of pipelining for maximum performance, and the results in Table 2 and Table 3, the practical range for the fraction of timing overhead k is between 2% (rough theoretical limit and the Pentium 4) and 8% (Xtensa) for high-speed ASIC and custom designs.

From Equation (31), the performance increase by pipelining can be calculated if the timing overhead and reduction in IPC are known.

3.3.1 Performance Improvement with Custom Microarchitecture

It is difficult to estimate the overall performance improvement with microarchitectural changes. Large custom processors are multiple-issue, and can do out-of-order and speculative execution. ASICs can do this too, but tend to have simpler implementations to reduce the design time.

The 520MHz iCORE fabricated in STMicroelectronics' HCMOS8D technology (0.15um L_{eff} [34]) has about 26 FO4 delays per pipeline stage, with 8 pipeline stages (see Chapter 16 for details). The iCORE had clock skew of 80ps, which is about 1 FO4 delay. Slack passing would reduce the clock frequency to about 23 FO4 delays (as detailed in Chapter 3, Section 6.2.1). Using high speed flip-flops or latches might further reduce this to 21 FO4 delays. Thus the iCORE would have about the same delay per pipeline stage as the Pentium III, and the Pentium 4 has 28% better performance than the Pentium III.

The iCORE is the fastest ASIC microprocessor we have come across, so we estimate a factor of 1.3× between high-speed ASIC and custom implementations (e.g. the Pentium 4). It seems unlikely that ASICs will achieve substantially better microarchitecural performance than this. The improvement by pipelining is limited by the larger timing overhead and difficulties in maintaining high IPC for ASICs. The microarchitectural complexity (wiring and layout for more forwarding and branch prediction logic, etc., to avoid a large IPC penalty) to maintain high IPC for super-pipelined designs does not seem feasible or cost effective for ASICs: there is only a moderate additional improvement in performance at a substantial cost in area, power, and design time.

How much slower is a typical ASIC because of microarchitecture? Other ASIC embedded processors have between 5 and 7 pipeline stages. The iCORE has instruction folding, branch prediction, and a number of other sophisticated techniques to maintain a high IPC of 0.7. Conservatively, a typical ASIC with 5 pipeline stages, without this additional logic to reduce the effect of pipeline hazards, may also have an IPC of 0.7. The iCORE has a timing overhead fraction of about 0.05 of the unpipelined delay, assuming 30% timing overhead. Thus going from 5 to 8 pipeline stages maintaining the IPC, there is a factor of

$$(32) \quad \frac{IPC_{after}}{IPC_{before}} \times \frac{\left(\frac{1-k}{n_{before}} + k\right)}{\left(\frac{1-k}{n_{after}} + k\right)} = \frac{0.7}{0.7} \times \frac{\left(\frac{1-0.05}{5} + 0.05\right)}{\left(\frac{1-0.05}{8} + 0.05\right)} = 1.4$$

In other words, a high performance ASIC compared to a typical ASIC may be 1.4× faster. A super-pipelined custom design may be a further factor of 1.3× faster. Overall, custom microarchitecture may contribute a factor of up to 1.8× compared to a typical ASIC.

3.3.2 ASIC and Custom Examples of Pipelining Speedup

For a custom example, consider the 20% reduction in instructions per cycle (IPC) with 20 pipeline stages (rather than 10 for the Pentium III) for the Pentium 4 [20]. There is 2% timing overhead as a fraction of the total unpipelined delay. From Equation (31):

$$(33) \quad \frac{IPC_{after}}{IPC_{before}} \times \frac{\left(\frac{1-k}{n_{before}} + k\right)}{\left(\frac{1-k}{n_{after}} + k\right)} = 0.8 \times \frac{\left(\frac{1-0.02}{10} + 0.02\right)}{\left(\frac{1-0.02}{20} + 0.02\right)} = 1.37$$

This estimates that the Pentium 4 has only about 37% better performance than the Pentium III in the same process technology, despite having twice the number of pipeline stages. The relative frequency target for the Pentium 4 was 1.6× that of the Pentium III [17]. With the 20% reduction in IPC, the actual performance increase was only about 28% (1.6×0.8) – so our estimate is reasonable.

From (18), the latency of the pipeline path of the Pentium 4 increases from 143 FO4 delays without pipelining, to 170 FO4 delays if it had 10 pipeline stages, to 200 FO4 delays with 20 pipeline stages.

An ASIC example is the 520MHz iCORE ASIC processor has eight pipeline stages. See Chapter 16 for the details of this high performance ASIC. Assuming 30% timing overhead, the increase in clock frequency by pipelining was

$$
\begin{aligned}
\frac{f_{after}}{f_{before}} &= \frac{T_{before}}{T_{after}} \\
&= \frac{n(T - t_{timing\ overhead}) + t_{timing\ overhead}}{T} \\
&= \frac{8 \times (1 - 0.3) + 0.3}{1} \\
&= 5.9
\end{aligned}
\tag{34}
$$

The clock frequency increased by a factor of 5.9×. However, the instructions per cycle was only 0.7, which was optimized using branch prediction and forwarding. Thus the increase in performance was only a factor of 4.1×:

$$
\frac{IPC_{after}}{IPC_{before}} \times \frac{T_{before}}{T_{after}} = \frac{0.7}{1} \times 5.9 = 4.1
\tag{35}
$$

In general, this approach can be used to estimate the additional speedup from more pipeline stages.

Figure 9 shows the estimated speedup by pipelining, assuming the instructions per clock cycle are unchanged (which is overly optimistic). With flip-flops the timing overhead for ASICs is between about 0.06 and 0.08, and this larger timing overhead substantially reduces the benefit of having more pipeline stages. Thus to gain further benefit from pipelining, ASICs need to reduce the timing overhead.

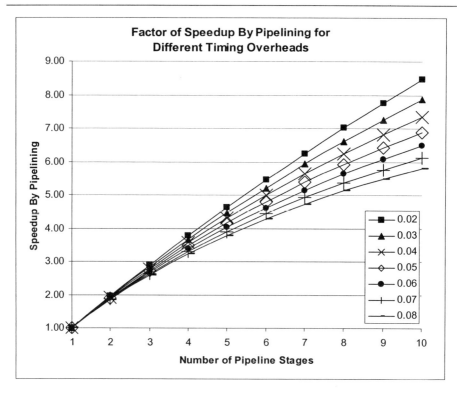

Figure 9. Graph of speedup by pipelining for timing overhead as a fraction of the unpipelined delay. One pipeline stage corresponds to the circuit being unpipelined. For simplicity, is it assumed the instructions per clock cycle are unaffected by pipelining.

4. REFERENCES

[1] AMD, AMD Athlon Processor – Technical Brief, December 1999, http://www.amd.com/products/cpg/athlon/techdocs/pdf/22054.pdf
[2] ARM, ARM1020E and ARM1022E - High-performance, low-power solutions for demanding SoC, 2002, http://www.arm.com/
[3] Bohr, M., et al., "A high performance 0.25um logic technology optimized for 1.8 V operation," *Technical digest of the International Electron Devices Meeting*, 1996, pp. 847-850.
[4] Brand, A., et al., "Intel's 0.25 Micron, 2.0 Volts Logic Process Technology," *Intel Technology Journal*, Q3 1998. http://developer.intel.com/technology/itj/q31998/pdf/p856.pdf.
[5] De Gelas, J. AMD's Roadmap. February 28, 2000. http://www.aceshardware.com/Spades/read.php?article_id=119
[6] Diefendorff, K., "The Race to Point One Eight: Microprocessor Vendors Gear Up for 0.18 Micron in 1999," *Microprocessor Report*, vol. 12, no. 12, September 14, 1998.

[7] Duarte, D., et al., "Evaluating the Impact of Architectural-Level Optimizations on Clock Power," *Proceedings of the 14th Annual IEEE International ASIC/SOC Conference*, 2001, pp. 447-51.
[8] Ghani, T., et al., "100 nm Gate Length High Performance / Low Power CMOS Transistor Structure," *Technical digest of the International Electron Devices Meeting*, 1999, pp. 415-418.
[9] Golden, M., et al., "A Seventh-Generation x86 Microprocessor," *IEEE Journal of Solid-State Circuits*, vol. 34, no. 11, November 1999, pp. 1466-1477.
[10] Gronowski, P., et al., "High-Performance Microprocessor Design," *IEEE Journal of Solid-State Circuits*, vol. 33, no. 5, May 1998, pp. 676-686.
[11] Hare, C. 586/686 Processors Chart. http://users.erols.com/chare/586.htm
[12] Hare, C. 786 Processors Chart. http://users.erols.com/chare/786.htm
[13] Harris, D., and Horowitz, M., "Skew-Tolerant Domino Circuits," *IEEE Journal of Solid-State Circuits*, vol. 32, no. 11, November 1997, pp. 1702-1711.
[14] Harris, D., et al. "The Fanout-of-4 Inverter Delay Metric," unpublished manuscript. http://odin.ac.hmc.edu/~harris/research/FO4.pdf
[15] Hauck, C., and Cheng, C. "VLSI Implementation of a Portable 266MHz 32-Bit RISC Core," *Microprocessor Report*, November 2001.
[16] Hennessy, J., Patterson, D. *Computer Architecture: A Quantitative Approach*. 2nd Ed. Morgan Kaufmann, 1996.
[17] Hinton, G., et al. "A 0.18-um CMOS IA-32 Processor With a 4-GHz Integer Execution Unit," *IEEE Journal of Solid-State Circuits*, vol. 36, no. 11, November 2001, pp. 1617-1627.
[18] Ho, R., Mai, K.W., and Horowitz, M., "The Future of Wires," *Proceedings of the IEEE*, vol. 89, no. 4, April 2001, pp. 490-504.
[19] Hrishikesh, M.S., et al., "The Optimal Logic Depth Per Pipeline Stage is 6 to 8 FO4 Inverter Delays," *Proceedings of the 29th Annual International Symposium on Computer Architecture*, May 2002.
[20] Intel, Inside the NetBurst Micro-Architecture of the Intel Pentium 4 Processor, Revision 1.0, 2000. http://developer.intel.com/pentium4/download/netburst.pdf
[21] Intel, The Intel Pentium 4 Processor Product Overview, 2002, http://developer.intel.com/design/Pentium4/prodbref/
[22] Kawaguchi, H.; Sakurai, T., "A Reduced Clock-Swing Flip-Flop (RCSFF) for 63% Power Reduction," *IEEE Journal of Solid-State Circuits*, vol.33, no.5, May 1998. pp. 807-811.
[23] Kessler, R., McLellan, E., and Webb, D., "The Alpha 21264 Microprocessor Architecture," *Proceedings of the International Conference on Computer Design*, 1998, pp. 90-95.
[24] Leiserson, C.E., and Saxe, J.B. "Retiming Synchronous Circuitry," in *Algorithmica*, vol. 6, no. 1, 1991, pp. 5-35.
[25] Lexra, Lexra LX4380 Product Brief, 2002, http://www.lexra.com/LX4380_PB.pdf
[26] MTEK Computer Consulting, AMD CPU Roster, January 2002. http://www.cpuscorecard.com/cpuprices/head_amd.htm
[27] MTEK Computer Consulting, Intel CPU Roster, January 2002. http://www.cpuscorecard.com/cpuprices/head_intel.htm
[28] Naffziger, S., "A Sub-Nanosecond 0.5um 64b Adder Design," *International Solid-State Circuits Conference Digest of Technical Papers*, 1996, pp. 362-363.
[29] Posluszny, S., et al., "Design Methodology of a 1.0 GHz Microprocessor," *Proceedings of the International Conference on Computer Design*, 1998, pp. 17-23.
[30] Partovi, H., "Clocked storage elements," in *Design of High-Performance Microprocessor Circuits*, IEEE Press, Piscataway NJ, 2000, pp. 207-234.

[31] Shenoy, N., "Retiming Theory and Practice," *Integration, The VLSI Journal*, vol.22, no. 1-2, August 1997, pp. 1-21.

[32] Silberman, J., et al., "A 1.0-GHz Single-Issue 64-Bit PowerPC Integer Processor," *IEEE Journal of Solid-State Circuits*, vol.33, no.11, November 1998. pp. 1600-1608.

[33] Snyder, C.D., "Synthesizable Core Makeover: Is Lexra's Seven-Stage Pipelined Core the Speed King?" *Microprocessor Report*, vol. 139, July 2, 2001.

[34] STMicroelectronics, "STMicroelectronics 0.25µ, 0.18µ & 0.12 CMOS," slides presented at the annual Circuits Multi-Projets users meeting, January 9, 2002.
http://cmp.imag.fr/Forms/Slides2002/061_STM_Process.pdf

[35] Tensilica, *Xtensa Microprocessor – Overview Handbook – A Summary of the Xtensa Microprocessor Databook*. August 2001. http://www.tensilica.com/dl/handbook.pdf

[36] Thompson, S., et al., "An Enhanced 130 nm Generation Logic Technology Featuring 60 nm Transistors Optimized for High Performance and Low Power at 0.7 – 1.4 V," *International Electron Devices Meeting*, 2001.
http://www.intel.com/research/silicon/0.13micronlogic_pres.pdf

Chapter 3

Reducing the Timing Overhead
Clock Skew, Register Overhead, and Latches vs. Flip-Flops

David Chinnery, Kurt Keutzer
Department of Electrical Engineering and Computer Sciences,
University of California at Berkeley

There are two components of delay on a sequential path in a circuit: the combinational logic delay, and the timing overhead for storing data in registers between each set of combinational logic. Pipelining can break up a long combinational path into several smaller groups of combinational logic, separated by registers. However, pipelining is limited by the timing overhead. The more pipeline stages there are, the more cycle time taken by the timing overhead, as there are more registers. This chapter discusses the timing overhead, and some methods of reducing it.

Digital circuits use synchronous clocking schemes to synchronize calculations and transfer of data at a local level. Synchronizing events to a given clock simplifies design, and avoids the need for circuits to signal the completion of an operation. The logic is designed such that each step in a calculation will take at most one clock cycle.

Circuits with high clock frequency can require asynchronous communication between regions of the chip, because clock skew makes it difficult to distribute a global clock to all regions of the chip. Higher speed custom chips require more carefully designed clocking schemes than ASICs. This chapter only discusses synchronous clocking, which is of primary importance to ASICs.

Section 1 provides an introduction to the properties of registers and the clock signal. This is followed by analysis of the minimum clock period with edge-triggered flip-flop registers and with level-sensitive latch registers.

There is an example of where level-sensitive latches reduce the timing overhead, compared to edge-triggered flip-flops for registers, in Section 2. Then Section 3 discusses the optimal positions of the latches for the latch inputs to arrive within the window when the latches are transparent. Section

4 gives a contrasting example where edge-triggered flip-flops are faster than level-sensitive latches.

Section 5 compares the speedup by pipelining with level-sensitive latches versus pipelining with edge-triggered flip-flops. Section 6 concludes with a summary of the timing overhead in typical ASIC and custom designs.

1. CHARACTERISTICS OF SYNCHRONOUS SEQUENTIAL LOGIC

A synchronous register stores its input after the arrival of a rising or falling clock edge. In Chapter 2, we discussed pipelining using only D-type flip-flop registers that only sample the input value at the rising or falling clock edge. For the rest of the clock period, D-type flip-flops are opaque, and the input of the flip-flop cannot affect the output. In contrast, a latch register is transparent for a portion of the clock period, and stores the input on the clock edge that causes the latch to become opaque.

Flip-flops are edge sensitive, and latches are level sensitive [14]. Positive edge-triggered flip-flops store the input at a rising clock edge. Negative edge-triggered flip-flops store the input at a falling clock edge. Active high, transparent high, latches are transparent when the clock is high and store the input on the falling clock edge. Active low, transparent low, latches are transparent when the clock is low and store the input on the rising clock edge. To simplify discussion, we confine our discussion to rising edge flip-flops – the properties of falling edge flip-flops are the same, with respect to the opposite clock edge.

Both flip-flops and latches have a setup time t_{su} before the clock edge arrives at which the register stores the input, where the input must be stable. The input must also remain unchanged during the hold time t_h after the arrival of the clock edge. The setup time limits the latest possible arrival of the input. The hold time limits the earliest possible arrival of the next input.

A flip-flop's output changes after at most t_{CQ}, the clock-to-Q propagation delay, after the arrival of the triggering clock edge. Similarly, if a latch is opaque when its input arrives, its output Q will change t_{CQ} after the clock edge causes the latch to become transparent. If the latch input D arrives while the latch is transparent, the latch behaves as a buffer and the propagation delay is t_{DQ}.

The diagrams on the left-hand side of Figure 1 illustrate t_{CQ}, t_{su}, and t_{DQ} assuming an ideal clock. As shown in Figure 1, the setup time t_{su} is relative to the clock edge that the register stores the input value – the rising clock edge for positive-edge triggered flip-flops and active low latches, and the falling clock edge for active high latches.

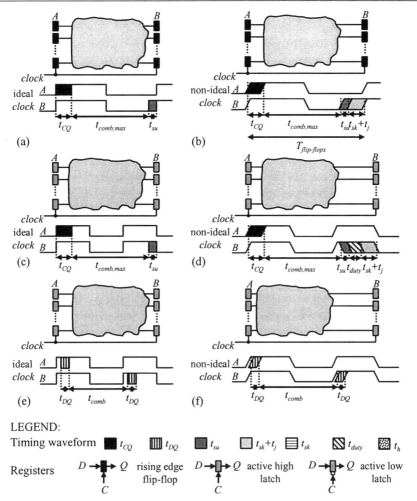

Figure 1. These diagrams display the register propagation delays and setup times. On the left an ideal clock is assumed, and on the right a non-ideal clock is considered. (a) and (b) show positive edge-triggered flip-flops, where the register inputs must arrive t_{su} before the rising clock edge. (c) and (d) present active high latches, and assume the inputs at A arrive before the *rising* clock edge, and the outputs of the combinational logic must arrive t_{su} before the *falling* clock edge at B. (e) and (f) show active high latches, and assume the register inputs arrive while the latches are transparent. In (e) and (f), the setup time, clock skew and jitter do not affect the clock period, providing that the latch inputs arrive while the latch is transparent and $t_{su}+t_{duty}+t_{sk}+t_j$ before the nominal falling clock edge arrival time (multi-cycle jitter is considered later).

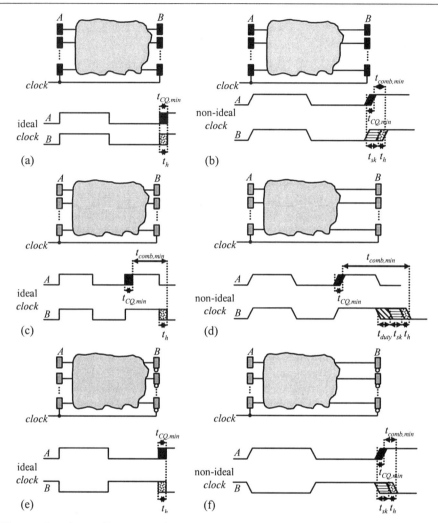

Figure 2. These diagrams show the hold time for registers. On the left an ideal clock is assumed, and on the right a non-ideal clock is considered. (a) and (b) show positive edge-triggered flip-flops, the other diagrams have active high latches. In (a) and (e), $t_{CQ,min} > t_h$ and there is no possibility of a hold time violation. The latches in (c) and (d) have active high latches triggered by the same clock phase, and there is a long period of time where hold time violations may occur. (e) and (f) show how to reduce the chance of a hold time violation by using latches that are active on opposite clock phases (the same can be achieved by using active high latches and two clock phases).

If the latch inputs arrive while the latches are transparent and t_{su} before the earliest possible arrival of the clock edge causing the latches to become opaque, then the clock period does not need to account for the setup time (see Figure 1(e)).

The minimum clock period with D-type flip-flops must account for the setup time, as D-type flip-flops cannot take advantage of an early input arrival. The input must be stable from t_{su} before the arrival of the rising clock edge. The output will change by t_{CQ} after the arrival of the rising clock edge (see Figure 1(a)).

Figure 2 shows the register hold time. The *minimum* clock-to-Q propagation delay $t_{CQ,min}$ must be used to calculate if there is a hold time violation. This is because races on the shortest paths cause hold time violations. In Figure 2(c) and (d), latches that are active on the same clock phase make it very easy to have hold time violations. Active high and active low latches with the same clock, or active high latches with two clock phases, reduce the potential for hold time violations, as shown in Figure 2(e) and (f).

To avoid violating setup and hold times, the arrival time of the clock edge must be considered. The arrival time of the clock edge is affected by clock skew and clock jitter.

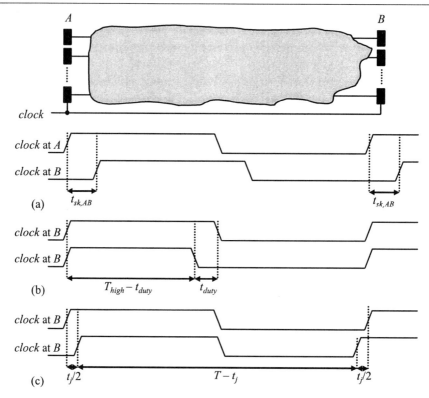

Figure 3. Timing diagram showing (a) clock skew $t_{sk,AB}$ between the arrival of the clock edge at A and at B, (b) duty cycle jitter t_{duty} between rising and falling clock edges at the same point on the chip, and (c) edge jitter t_j between consecutive rising edges at the same point on the chip. Combinational logic is shown in grey.

1.1 Properties of the Clock Signal

Ideally, each register on the chip would receive the same clock edge at the same time and clock edges would arrive at fixed intervals. A rising clock edge would arrive exactly T, the nominal clock period, after the previous clock edge. If the clock is high for a length of time T_{high}, then the falling clock edge would arrive exactly T_{high} after the rising clock edge. The nominal duty cycle is

(1)   Duty Cycle = T_{high}/T

In practice, the exact arrival of the clock edges varies. There is cycle-to-cycle *edge jitter*, t_j, the maximum deviation from the nominal period T between consecutive rising (or falling) clock edges. There is *duty cycle jitter*,

Reducing the Timing Overhead

t_{duty}, the maximum difference from the nominal interval T_{high} between consecutive rising and falling clock edges. There is also *clock skew*, t_{sk}, the maximum difference between the arrival times of the clock edge at different points on the chip. Figure 3 illustrates these deficiencies, and Figure 1 and Figure 2 show their impact on setup and hold time constraints.

Figure 4 shows the range of possible arrival times of clock edges, with respect to a reference rising clock edge arriving at A at time zero. This assumes that clock jitter is additive over several clock periods, as there can be long-term jitter [14]. Clock skew between locations depends on the clock tree and their positions. The load at the clock distribution points varies, which also affects the clock skew.

It is possible to carefully tailor clock skew by changing the buffering in the clock tree. This can help balance pipeline stages. Positive clock skew can give a pipeline stage more time between consecutive rising clock edges, but another pipeline stage must have less time as a result. This method of slack passing by adjusting the clock skew is known as *cycle stealing*. Chapter 8 discusses adjusting the clock skew to increase a design's speed.

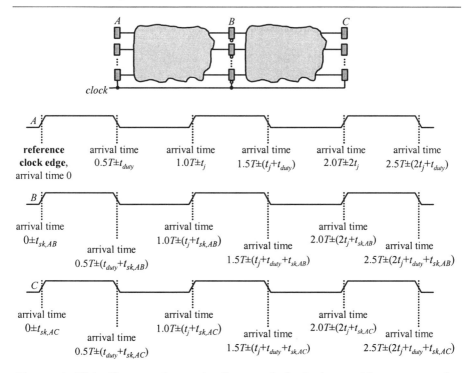

Figure 4. This diagram shows the jitter and clock skew with respect to the reference clock edge that arrives at A. $t_{sk,AB}$ is the clock skew between A and B. $t_{sk,AC}$ is the clock skew between A and C.

For simplicity in this chapter, we assume a maximum clock skew of t_{sk} between locations. If more than one clock is used, there can be some additional skew between the clocks. We assume that all the clock skew is accounted for in t_{sk}.

Jitter and clock skew have random components due to variation in the supply voltage and noise. The clock tree distributes the clock signal across the chip to the registers. Unbalanced delays in the clock tree add to the clock skew. The clock signal is generated by a phase-lock loop with reference to an external oscillator, typically a crystal oscillator. The phase-lock loop (PLL) jitters around some multiple of the reference frequency. A phase detector controls the voltage controlled oscillator that generates the clock signal in the PLL [15]. Noise in the PLL supply voltage contributes to the duty cycle and edge jitter. Process variation and temperature variation during operation also affect jitter and skew [14]. The jitter and clock skew give maximum deviations of the arrival time of the clock edge from its expected arrival time.

If the clock skew and jitter are such that the clock edge arrives late at the register, this gives more time for the pipeline stage to complete, so it is not accounted for when considering the setup time constraint. However, a late arrival of the clock edge at the next stage does increase the window for hold time violations, as shown in Figure 2(b), (d) and (f).

Latches are subject to duty cycle jitter, as their behaviour depends on the arrival times of both clock edges. A circuit with only rising edge flip-flops only needs to consider the arrival time of the rising clock edge, and thus is immune to duty cycle jitter. Latches in particular are more subject to races that violate hold time constraints, because there less logic between registers. Latch-based designs have about half the combinational logic between registers compared with flip-flop based designs [14].

1.2 Avoiding Races with Latches

As shown in Figure 2(b), only a very short path can violate the hold time constraint with flip-flops. The constraint is [14]

(2) $\quad t_{comb,\min} > t_{sk} + t_h - t_{CQ,\min}$

Edge jitter does not affect the hold time constraint, because the hold time constraint is only for a path that propagates from the preceding flip-flops on the same clock edge. Additional caution is required in designs with *multi-cycle paths*, where paths through combinational logic have more than one clock cycle to propagate.

1.2.1 Ordering Latches to Reduce the Window for Hold Time Violations

Comparing Figure 2(d) and Figure 2(f), it is advantageous to use latches that are active on opposite clock phases to avoid races with latch-based designs. Ensuring that consecutive sets of latches are active on opposite clock phases reduces the hold time constraint to that given by Equation (2).

In general, designs may have a mixture of flip-flops and latches. There are also circuit inputs and outputs that are referenced to some clock edge. **To reduce the window in which races can occur, the latches must go opaque on the same clock edge that *inputs* change to the combinational logic preceding the latches.** This gives the following rules for good design:

- **Active low latches**, which go opaque on the rising clock edge, should **follow** inputs that can change on the rising clock edge from:
  - **Active high latches**
  - **Rising edge flip-flops**
  - **Inputs with respect to the rising clock edge**
- **Active high latches**, which go opaque on the falling clock edge, should **follow** inputs that can change on the falling clock edge from:
  - **Active low latches**
  - **Falling edge flip-flops**
  - **Inputs with respect to the falling clock edge**

Figure 5 shows that there is also a large window for possible hold time violations when rising edge flip-flops follow transparent low latches. This can be avoided by ensuring that rising edge flip-flops are preceded by transparent high latches. In general, **to reduce the window in which races can occur, the latches must become transparent on the same clock edge that the *outputs* store the values**. If the latches become transparent on the earlier clock edge, there is a much larger window for hold time violations. Consequently:

- **Active low latches**, which become transparent on the falling clock edge, should **precede** outputs that are with respect to the rising clock edge:
  - **Active high latches**
  - **Falling edge flip-flops**
  - **Outputs with respect to the falling clock edge**
- **Active high latches**, which become transparent on the rising clock edge, should **precede** outputs that are with respect to the rising clock edge:
  - **Active low latches**
  - **Rising edge flip-flops**
  - **Outputs with respect to the rising clock edge**

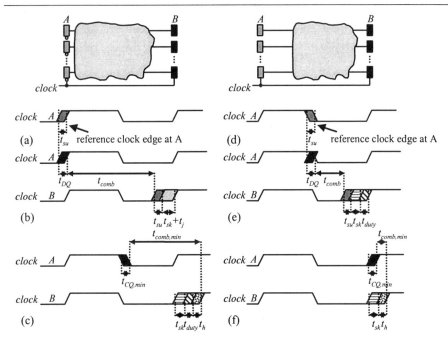

Figure 5. (a), (b) and (c) show that having transparent low latches followed by rising edge flip-flops give a large window for hold time violations. In comparison, (d), (e) and (f) show the small window for hold time violations when transparent high latches are followed by rising edge flip-flops. (a) and (d) show the reference clock edge when the latches at A store their inputs. (b) and (e) illustrate the combinational delay after the inputs at A propagate through the latches (the delay can be more if the latch inputs arrive earlier), and the clock edge when the flip-flops at B store their inputs. (c) and (f) show the window for hold time violations.

Similar rules apply to two phase clocking schemes for latches. The left side of Figure 6 illustrates a two-phase clocking scheme that ensures there are no races violating the hold time constraints at B.

In the remainder of this chapter, we assume configurations with latches that reduce the window for hold time violations.

1.2.2 Non-Overlapping Clocks or Buffering to Further Reduce the Window for Hold Time Violations

Races can be completely avoided by using *non-overlapping clocks*, as shown in Figure 6(a). With 50% duty cycle, two clock signals of the same period will overlap due to clock skew. From Equation (2), to avoid races, the clocks must not overlap by at least

Reducing the Timing Overhead

(3) $T_{non-overlap} > t_{sk} + t_h - t_{CQ,min}$

Equation (3) assumes that there is no additional skew between the two clocks; otherwise this should be added to the t_{sk} term. The additional clock skew between the two non-overlapping clocks can be minimized if the clocks are locally generated from a single global clock [14].

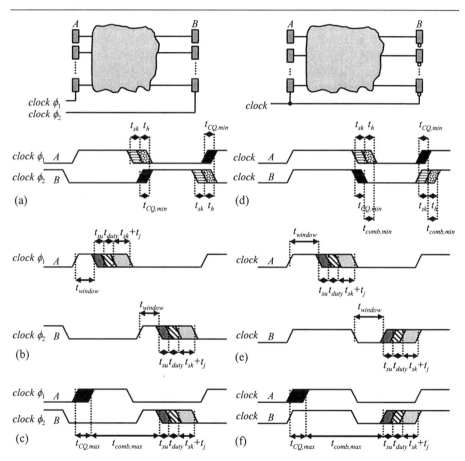

Figure 6. (a) shows the advantage of using non-overlapping clock phases to avoid races, but this reduces the window t_{window} in which the input can arrive while the latch is transparent as shown in (b). In comparison, (d) shows the possibility of races by using the same clock for active high and active low latches, but there is a greater time window, shown in (e), for the input arrival while the latch is transparent. In addition, a smaller duty cycle reduces the maximum possible combinational delay between latches, as can be seen by comparing (c) and (f) carefully.

Using non-overlapping clocks reduces the portion of time T_{high} that each clock phase is high:

(4) $\quad T_{high} = \dfrac{T}{2} - T_{non-overlap}$

The examples in Figure 9 and Figure 10 have clock phases with duty cycles of 50% and 40% respectively (T_{high} is $0.5T$ and $0.4T$ respectively). For example, the ARM7TDMI devoted 15% of the clock period of each clock phase to avoid overlap (see Chapter 17), which is a 42.5% duty cycle, with $T_{high} = 0.425T$.

Unfortunately, using non-overlapping clocks also reduces the window for the input to arrive while the latch is transparent. The latch's active window is smaller because T_{high} is reduced [14]. The input must arrive before the clock edge that makes the latch opaque, so the time window t_{window} is

(5) $\quad t_{window} = T_{high} - (t_{sk} + t_j + t_{duty} + t_{su})$

An alternative solution to using non-overlapping clocks is buffer insertion. CAD tools can analyze the circuit to find short paths that could violate hold times, and insert buffers to increase the path delays to ensure that the hold time constraints are not violated [14]. As buffers take additional area and power, it is preferable to increase the path delay by using minimally sized gates that are slower. Sometimes slower gates cannot be used on the short paths, because these paths also coincide with critical paths (e.g. if an intermediate value on the critical path is stored). Buffer insertion does not reduce the time window when the latches are transparent for the inputs. This can be a substantial benefit compared with using non-overlapping clocks.

Using active high and active low latches with the same clock avoids additional skew and wiring overhead for distributing two non-overlapping clocks. Only the *clock* signal needs to be distributed, rather than *clock* ϕ_1 and *clock* ϕ_2 (the inverse of the clock signal still needs to be generated locally).

Given the timing characteristics of latches, we can now calculate the minimum clock period for both a single clock scheme and two non-overlapping clocks.

1.3 Minimum Clock Period

Chapter 2, Section 1.3 discussed the clock period with D-type flip-flop registers – see Figure 3 therein for a timing diagram showing the minimum clock period calculated from the critical path. The minimum clock period with flip-flops $T_{flip-flops}$ is also shown in Figure 1(b), and it is given by [14]

Reducing the Timing Overhead

(6) $T_{flip-flops} = \max\{t_{CQ} + t_{comb} + t_{su} + t_{sk} + t_j\}$

With D-type flip-flops, the minimum clock period is simply the maximum delay of any pipeline stage, $t_{comb}+t_{CQ}$, plus the time needed to avoid violating the setup time constraint $t_{su}+t_{sk}+t_j$. In comparison, the delay of a pipeline stage does not limit the minimum clock period when using latches, as there is flexibility in when the latch inputs arrive within t_{window}.

1.3.1 Slack Passing and Time Borrowing with Latches

Figure 6(c) and (f) show the maximum combinational delay between two sets of latches. This is the delay from the arrival of the clock edge causing the first set of latches to become transparent, to the arrival of the clock edge causing the second set of latches to become opaque, taking into account the clock-to-Q propagation delay and setup time constraint. The delay between these two edges is $T_{high}+T/2$. Thus the maximum combinational logic delay with latches is

(7) $t_{comb,max,opaque\ input\ latches} = \dfrac{T}{2} + T_{high} - (t_{CQ} + t_{su} + t_{sk} + t_j + t_{duty})$

If the duty cycle is 50%, T_{high} is $T/2$. The maximum combinational logic delay between latches assumes that the inputs of the first set of latches arrive before they become transparent. If some inputs of the first set of latches arrive $t_{arrival}$ after the clock edge that makes the latches transparent, the arrival time and latch D-to-Q propagation delay t_{DQ} must be accounted for. This gives maximum delay for the following logic of

(8) $t_{comb,max,transparent\ input\ latches} = \dfrac{T}{2} + T_{high} - (t_{DQ} + t_{arrival} + t_{su} + t_{sk} + t_j + t_{duty})$

Each latch stage takes about $T/2$ to compute, including the propagation delay through the latch. The flexibility in the time window for a latch's input arrival allows slack passing and time borrowing between pipeline stages. Slack passing and time borrowing allow some stages to take longer than $T/2$, if other stages take less time. If the output of a stage arrives early within this time window, the next stage has more than $T/2$ to complete – *slack passing*.

In comparison, when using flip-flops each pipeline stage has exactly T to compute. If the pipeline stage takes less than T, the slack cannot be used elsewhere. With latches there are twice as many pipeline stages, and pipeline stages have about half the amount of combinational logic. Latch stages are not required to use only $T/2$. Latch stages may take up to $T_{high}+T/2$, if slack is available from other pipeline stages. If the pipeline is unbalanced, slack passing with latches allows a smaller clock period than flip-flops, as slack passing effectively balances the delay.

Slack passing also gives latch-based designs some tolerance to inaccuracy in wire load models and process variation. If one pipeline stage is slower than expected, time can be borrowed from other pipeline stages to reduce the penalty on the clock period. In comparison, the hard clock edge with flip-flops limits the clock period to the delay of the worst pipeline stage.

While a substantial portion of the process variation is systematic, longer paths have less percentage degradation in speed from process variation. One study shows that a circuit with 25 logic levels has about 1% less degradation that a circuit with 16 logic levels [13]. With latches, the clock period is determined by the delay of multi-cycle paths, so the impact of process variation is reduced by using latches.

There is an alternative to using level-sensitive latches for slack passing. By changing the delays in the clock tree, the clock skew can be adjusted to allow time borrowing between pipeline stages. Chapter 8 discusses this in more detail.

If a pipeline stage with latches takes the maximum time to finish computation, the next stage has only $T/2$ to complete. This is illustrated in Figure 7. Timing with latches depends on the delay of preceding and following stages. In general, a *critical loop* through the sequential logic may need to be considered to determine the minimum clock period.

1.3.2 Critical Loops in Sequential Logic

When retiming flip-flops, a path p through n pipeline stages of sequential logic, with delay $d(p)$, limits the minimum clock period T to $d(p)/n$ (see Chapter 2, Section 1.7, for a brief description of retiming). Retiming is often used to balance pipeline stages, if registers can be moved so that the delay $d(p)$ is evenly distributed amongst the n stages. Conceptually, timing with latches is very similar.

If the latches are transparent when their inputs arrive, the latch is treated as a buffer with delay t_{DQ} and the calculation of timing on the sequential path p must be calculated to the next set of registers. Of course, each set of latches imposes setup time constraints that must not be violated. Eventually, the sequential path end at a point where either (a) the setup time is violated, (b) it arrives at an output, or (c) it arrives at an opaque latch or flip-flop. This sequential path can go through the same pipeline stage several times if there is sequential feedback to earlier pipeline stages.

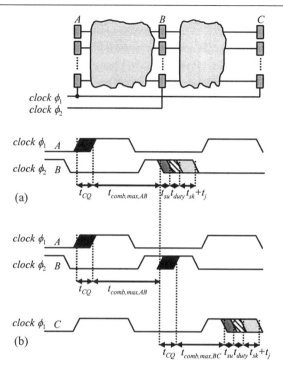

Figure 7. This figure illustrates the impact of the pipeline stage between A and B, borrowing time from the pipeline stage between B and C. (a) shows the maximum combinational delay for the pipeline stage between A and B, assuming that inputs to latch registers at A arrive before the latches become transparent. (b) illustrates how this maximum delay reduces the computation time allowed for the logic between B and C. Duty cycle jitter is included in (a), as duty cycle jitter on clock phase ϕ_2 affects the portion of time that ϕ_2 is high.

If there is a setup time violation, the clock period is too small. Otherwise when the sequential path arrives at an opaque latch, flip-flop, or output, there is a "hard" boundary ending the delay calculation for the sequential path. In general, outputs also have setup time constraints, or *output constraints*, which require that the skew and jitter be considered. It is not straightforward to calculate the delay through all such paths by hand, but calculating the timing with latches is fully supported by current CAD tools [2][18].

The next section gives an example of a sequential critical loop with latches.

1.3.3 Example of Sequential Critical Loop for a Design with Latches

For the examples in this chapter, we use units of FO4 delays, as discussed in Chapter 2, Section 1.1. Consider the circuit in Figure 8 with the following timing characteristics:

- flip-flop and latch setup time t_{su} = 2 FO4 delays
- flip-flop and latch clock-to-Q delay of t_{CQ} = 4 FO4 delays
- latch propagation delay t_{DQ} = 2 FO4 delays
- clock skew of t_{sk} = 3 FO4 delays
- edge jitter of t_j = 1 FO4 delay
- duty cycle jitter of t_{duty} = 1 FO4 delays
- combinational logic critical path delays of
  - $t_{comb,1}$ = 12 FO4 delays between A and B
  - $t_{comb,2}$ = 18 FO4 delays between B and C
  - $t_{comb,3}$ = 13 FO4 delays between C and D, and between C and B

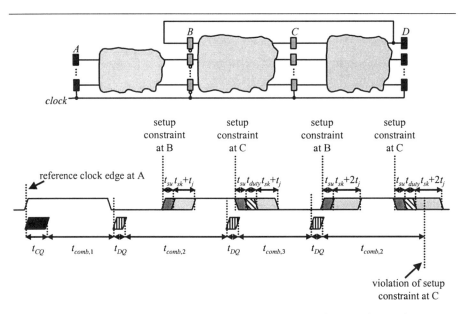

Figure 8. This shows the sequential path $ABCBC$ that violates the setup time constraint at C. Delays and constraints are shown to the same scale. The inputs at A come from rising edge flip-flops and the outputs at D go to rising edge flip-flops. In-between, there are active low latches at B followed by active high latches at C.

The path $ABCD$ has a total delay of 51 FO4 delays from the arrival of the rising clock edge at A:

$$
\begin{aligned}
(9) \quad & t_{CQ} + t_{comb,1} + t_{DQ} + t_{comb,2} + t_{DQ} + t_{comb,3} \\
& = 4 + 12 + 2 + 18 + 2 + 13 \\
& = 51 \text{ FO4 delays}
\end{aligned}
$$

The setup time constraint at D requires that the sequential path $ABCD$ arrive $t_{su}+t_{sk}+2t_j$ before the rising clock edge $2T$ later at D:

$$
\begin{aligned}
(10) \quad & t_{su} + t_{sk} + 2t_j \\
& = 2 + 3 + 2 \times 1 \\
& = 7 \text{ FO4 delays}
\end{aligned}
$$

One might assume that a clock period of 30 FO4 delays would suffice for this circuitry to work correctly, as

(11) $\quad 2T = 2 \times 30 = 60 > 51 + 7 = 58$ FO4 delays

However, there is a loop BCB through the transparent latches that has path delay of $t_{DQ}+t_{comb,2}+t_{DQ}+t_{comb,3}$, which is 35 FO4 delays. The loop BCB should take at most one clock period, 30 FO4 delays, to avoid a setup constraint violation. The sequential path $ABCBC$ violates the setup constraint at C, as shown in Figure 8.

The total delay on path $ABCBC$ is

$$
\begin{aligned}
(12) \quad & t_{CQ} + t_{comb,1} + t_{DQ} + t_{comb,2} + t_{DQ} + t_{comb,3} + t_{DQ} + t_{comb,2} \\
& = 4 + 12 + 2 + 18 + 2 + 13 + 2 + 18 \\
& = 71 \text{ FO4 delays}
\end{aligned}
$$

The corresponding setup constraint at C is

$$
\begin{aligned}
(13) \quad & 2.5T - (t_{su} + t_{sk} + 2t_j + t_{duty}) \\
& = 75 - (2 + 3 + 2 + 2) \\
& = 66 \text{ FO4 delays}
\end{aligned}
$$

Thus there is a setup constraint violation.

In order to calculate the clock period in a latch-based design, all the sequential critical paths must be examined, as shown in this example. The clock period may be bounded by a sequential critical loop, or a sequential critical path that does not have a loop.

1.3.4 Latch Clock Period bounded by a Critical Loop

If each set of latches are active on opposite clock phases, there is $T/2$ between the clock edges when successive sets of latches become opaque. Thus a loop through k pipeline stages, with k sets of latches, has $kT/2$ for computation. The sequential path through the loop violates the clock period if the loop has delay greater than $kT/2$. Therefore static timing analysis only needs to consider a sequential loop through the same logic once [private communication with Earl Killian].

Figure 8 shows a sequential loop with two sets of latches that has T to compute. As we've restricted the design to use latches that are active on opposite clock phases to avoid races, the loop must go through an even number of latches. In general, a critical loop through $2n$ stages has nT for computation, but the cycle-to-cycle jitter must be considered. This places a constraint on the delay through the critical loop:

$$(14) \quad 2nt_{DQ} + \sum_{i=1}^{2n} t_{comb,i} \leq nT - nt_j$$

This equation considers the jitter to be additive across clock cycles. The setup constraint places a lower bound on the clock period of

$$(15) \quad T_{latches} \geq \frac{2nt_{DQ} + nt_j + \sum_{i=1}^{2n} t_{comb,i}}{n}$$

Let $t_{comb,average}$ be the average combinational delay per latch pipeline stage. Then

$$(16) \quad T_{latches} \geq 2t_{DQ} + 2t_{comb,average} + t_j$$

The t_j term can be replaced by the n-cycle-to-cycle jitter averaged across n cycles, if the jitter for n clock cycles is known. The same limit holds for the clock period of a long sequential path.

1.3.5 Latch Clock Period bounded by a Sequential Path

Consider an input, with arrival time t_{input}, to a critical sequential path with latches. The input may be from a register or from a primary input of the circuit. As the sequential path is critical, the setup time constraint at the end of that path will be barely satisfied (i.e. the output of the sequential path doesn't arrive with plenty of time to spare). The sequential path output goes to a register or primary output of the circuit.

We assume that the input arrival times are given with respect to the rising clock edge, and that a single-phase clocking scheme with active high and active low latches is used. Consider the delay from the inputs, through n sets of latches to a hard boundary with some setup time constraint t_{su}. This corresponds to $n+1$ pipeline stages.

As discussed in Section 1.2.1, a register should store its input on the same clock edge as the inputs from the previous pipeline stage can change, to avoid races. The first set of latches must store their inputs on the rising clock edge and are active low, as the inputs to the sequential path are with respect to the rising clock edge. The next latches are active high, then active low, and so forth to the output. Figure 8 shows this for two sets of latches, $n = 2$, with three pipeline stages from the input flip-flops to the output flip-flops.

The output setup time constraint is with respect to the rising clock edge if the preceding latches are active high, or with respect to the falling clock edge otherwise. For example, in Figure 8 the last set of latches are active high latches, so rising edge flip-flops must follow them. In either case, from the input arrival after the rising clock edge to the first active low latches in the sequential path there is T between the clock edge when the input arrives and the rising clock edge when the latch becomes opaque. Thereafter, each pipeline stage has $T/2$ from the previous clock edge to the next clock edge.

This is the case in Figure 8:
- T from the rising clock edge at A to the rising clock edge when the first set of active low latches at B store their inputs
- $T/2$ from B to C where the active high latches store their inputs on the falling clock edge
- $T/2$ from C to D where the rising edge flip-flops store their inputs on the rising clock edge

Thus the total delay allowed for the sequential path is

(17) $T + n\dfrac{T}{2}$

The time constraint on this sequential path is

(18)
$$t_{arrival} + nt_{DQ} + \sum_{i=1}^{n+1} t_{comb,i} \le T + n\dfrac{T}{2} - \left(t_{su} + t_{sk} + \dfrac{(n+1)}{2}t_j + t_{duty}\right), n \text{ odd}$$
$$t_{arrival} + nt_{DQ} + \sum_{i=1}^{n+1} t_{comb,i} \le T + n\dfrac{T}{2} - \left(t_{su} + t_{sk} + \dfrac{(n+2)}{2}t_j\right), n \text{ even}$$

where $t_{comb,i}$ is the delay of combinational logic in latch pipeline stage i. Correspondingly, this constraint places a lower bound on the clock period:

(19)
$$T_{latches} \geq \frac{\left(t_{arrival} + nt_{DQ} + t_{su} + t_{sk} + \frac{(n+1)}{2}t_j + t_{duty} + \sum_{i=1}^{n+1} t_{comb,i}\right)}{\left(1 + \frac{n}{2}\right)}, n \text{ odd}$$

$$T_{latches} \geq \frac{\left(t_{arrival} + nt_{DQ} + t_{su} + t_{sk} + \frac{(n+2)}{2}t_j + \sum_{i=1}^{n+1} t_{comb,i}\right)}{\left(1 + \frac{n}{2}\right)}, n \text{ even}$$

To calculate the minimum clock period with latches, this lower bound must be determined over the critical sequential paths through transparent latches. This is not amenable to hand calculation.

For back-of-the-envelope calculations, we can neglect the arrival time, setup time, skew, and duty cycle jitter, which are only a small portion of a sequential path if there are many pipeline stages, n. This reduces the constraint to

(20)
$$T_{latches} > \frac{2nt_{DQ} + 2(n+1)t_{comb,average} + (n+1)t_j}{n+2}, n \text{ odd}$$

$$T_{latches} > \frac{2nt_{DQ} + 2(n+1)t_{comb,average} + (n+2)t_j}{n+2}, n \text{ even}$$

where $t_{comb,average}$ is the average combinational delay per latch pipeline stage. As $n+2 \approx n$, for n much larger than 2,

(21) $T_{latches,min} = \lim_{n \to \infty} T_{latches} = 2t_{DQ} + 2t_{comb,average} + t_j$

This gives a lower bound on the cycle time with latches, but the clock period may need to larger, depending on t_{comb} for each stage, t_{duty}, t_{sk}, and $t_{arrival}$. This is similar to the simplification for the clock period with latches reported by Partovi [14], but it assumes that edge jitter is additive across clock cycles in the worst case. The t_j term can be replaced by more accurate models of worst case jitter average across $1+n/2$ cycles (from (17)), if they are available.

For example, consider $n = 18$ sets of latches. From (17), this corresponds to 10 clock periods. If the worst case jitter for 10 cycles is 10 FO4 delays, then the value averaged over 10 cycles of 1 FO4 delay is used for t_j in (21), rather than the worst case jitter per cycle which may be 2 FO4 delays.

The next example quantifies the speedup that can be achieved by using latches.

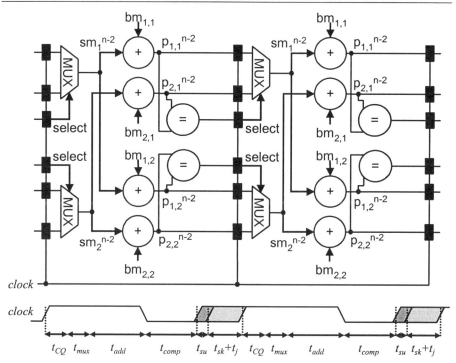

Figure 9. Timing for a two-state add-compare-select with all rising edge flip-flops. FO4 delays are shown to the same scale. At each rising clock edge, the clock skew and edge jitter are relative to the hard boundary at the previous set of flip-flops. The duty cycle is 50%.

2. EXAMPLE WHERE LATCHES ARE FASTER

Consider the unrolled two-state Viterbi add-compare-select calculation, shown in Figure 9. To avoid considering the best position for each latch on a gate-by-gate basis, we have selected nominal delays for functional elements that allow the latches to be well placed when directly between the functional elements. The nominal delays considered in this example are:

- adder delay $t_{add} = 10$ FO4 delays
- comparator delay $t_{comp} = 9$ FO4 delays
- multiplexer delay $t_{mux} = 4$ FO4 delays
- flip-flop and latch setup time $t_{su} = 2$ FO4 delays
- flip-flop and latch hold time $t_h = 2$ FO4 delays
- flip-flop and latch maximum clock-to-Q delay of $t_{CQ} = 4$ FO4 delays

- flip-flop and latch minimum clock-to-Q delay $t_{CQ,min} = 2$ FO4 delays
- latch propagation delay $t_{DQ} = 2$ FO4 delays
- clock skew of $t_{sk} = 4$ FO4 delays
- edge jitter of $t_j = 2$ FO4 delays
- duty cycle jitter of $t_{duty} = 1$ FO4 delay

For the add-compare-select examples in this chapter, we assume the branch metric inputs $bm_{i,j}$ are fixed. In real Viterbi decoders, the branch metric inputs are only updated occasionally, and thus can be assumed to be constant inputs for the purpose of timing analysis.

The clock period with flip-flops is

$$\begin{aligned}T_{flip-flops} &= t_{CQ} + t_{mux} + t_{add} + t_{comp} + t_{su} + t_{sk} + t_j \\ &= 4 + 10 + 9 + 4 + 2 + 4 + 2 \\ &= 35 \text{ FO4 delays}\end{aligned}$$ (22)

Note that each pipeline stage between flip-flops only considers one cycle of edge jitter, as the reference clock edge for edge jitter is the rising clock edge that arrives at the previous set of flip-flops. This is because flip-flops present a "hard boundary" at each clock edge, fixing a reference point for the next stage. In contrast, if a signal propagates through transparent latches, the edge jitter and duty cycle jitter must be considered over several cycles.

Figure 10 shows the circuit with the central flip-flops by latches. The latches are positioned so that the inputs arrive when the latches are transparent, before the setup time constraint. The clock period with latches is

$$\begin{aligned}2T_{latches} &= t_{CQ} + t_{mux} + t_{add} + t_{DQ} + t_{comp} + t_{mux} + t_{DQ} + t_{add} + t_{comp} + t_{su} + t_{sk} + 2t_j \\ &= 4 + 4 + 10 + 2 + 9 + 4 + 2 + 10 + 9 + 2 + 4 + 2 \times 2 \\ &= 64 \text{ FO4 delays}\end{aligned}$$ (23)

$$\therefore T_{latches} = 32 \text{ FO4 delays}$$

Thus replacing the central flip-flops by latches gives a 9% speed increase.

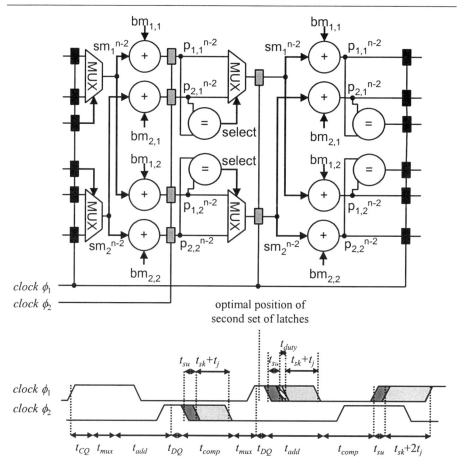

Figure 10. Timing for a two-state add-compare-select with rising edge flip-flop registers at the boundaries and active high latches between. FO4 delays are shown to the same scale. The clock skew, duty cycle jitter and edge jitter are relative to the rising clock edge at the first set of flip-flops. The duty cycle is 40%. The first set of latches is placed optimally, so that the latest inputs arrive in the middle of when the latches are transparent. The second sets of latches are placed a little too early, by 0.5 FO4 delays, and the latest input does not arrive in the middle of when they are transparent.

To avoid races when using latches, non-overlapping clock phases are used. From Equation (4), the clock phases should be high for

(24)
$$\begin{aligned} T_{high} &= \frac{T}{2} - (t_{sk} + t_h - t_{CQ,\min}) \\ &= \frac{32}{2} - (4 + 2 - 2) \\ &= 12 \text{ FO4 delays} \end{aligned}$$

Figure 10 shows the optimal position for the first set of latches: halfway between the arrival of the clock edge that makes the latch transparent and the setup time constraint before the latch becomes opaque. This gives the best immunity to variation, such as process variation and inaccuracy in the wire load models, to ensure the latch inputs will arrive after the latch is transparent without violating the setup time constraint. Chapter 13 discusses a variety of approaches for reducing the impact of design uncertainty, considering clock skew as an example.

D-type flip-flops consist of two back-to-back latches, a master-slave latch pair. Thus a latch cell is smaller than a flip-flop cell. In this example, six transparent high latches have replaced six flip-flops, so there is a slight reduction in area.

In general, consider replacing n sets of flip-flops by latches. Latches are needed on both clock phases to avoid races, so there will be $2n$ sets of latches. The central set of flip-flops in Figure 9 was replaced by two sets of latches in Figure 10. If the average number of cells k in each set of latches or flip-flops is about the same, the total cell areas are about the same, but there will be nk additional wires, as illustrated in Figure 11. Thus on average, latch-based designs may be slightly larger than designs using flip-flops.

In Figure 11, n sets of flip-flop registers break up the combinational logic into $n+1$ pipeline stages from inputs to outputs. Correspondingly, $2n$ sets of latches break up the logic into $2n+1$ pipeline stages. With flip-flops each stage has clock period $T_{flip\text{-}flops}$ to complete computation, so $(n+1)T_{flip\text{-}flops}$ is the total delay from inputs to outputs.

With latches, the total delay from inputs to outputs is $(n+1)T_{latches}$ (compare latch clock phase $clock\phi_1$ and $clock$ for the flip-flops). Between the latches, each stage gets on average $T_{latches}/2$ to compute. This is an average because latches allow slack passing and time borrowing. For the first stage between the inputs and first set of latches, there is about $3T_{latches}/4$ for computation. For the last stage, between the last set of latches and the outputs, there is also about $3T_{latches}/4$. This corresponds to $(n+1)T_{latches}$,

(25) $(n+1)T_{latches} = \frac{3}{4}T_{latches} + \frac{(2n-1)}{2}T_{latches} + \frac{3}{4}T_{latches}$

It is important to note that the optimal positions for latches are not equally spaced from inputs to outputs.

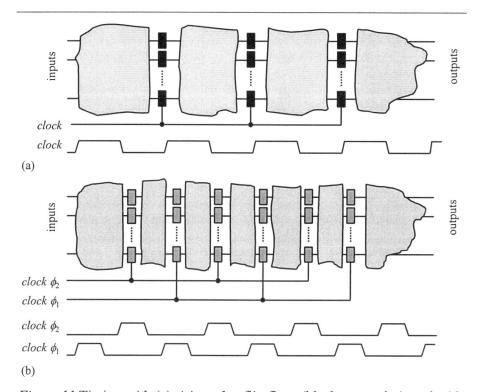

Figure 11. Timing with (a) rising edge flip-flops (black rectangles), and with (b) active high latches (rectangles shaded in grey). Inputs and outputs are with respect to the rising clock edge. The design in (a) has three sets of flip-flops, and with latches the design has six sets of latches in (b). Combinational logic is shown in light grey. Latch positions are optimal with the slowest inputs arriving in the middle of when the latch is active, assuming zero setup time, clock skew, and jitter.

3. OPTIMAL LATCH POSITIONS WITH TWO CLOCK PHASES

We can derive the optimal positions for latches, to ensure that the latest latch input arrival times are when the latches are transparent. We assume the inputs and outputs of the circuitry are relative to a hard rising clock edge boundary (e.g. at edge-triggered flip-flops).

After a set of rising clock edge flip-flops or inputs with respect to the rising clock edge, the first set of latches must be activated by a clock edge that is $T/2$ out of phase with the rising clock edge. Thereafter, each set of latches are activated by a clock edge $T/2$ later. The last set of latches become transparent on a clock edge that is in phase with the rising clock edge to the rising clock edge flip-flops or outputs. This is shown in Figure 11, where the latches are placed optimally so that latest time each input will arrive is in the middle of when each latch is transparent.

Optimal positioning needs to consider the impact of setup time, and clock skew and jitter on the time window t_{window} (see Figure 6). The length of time that the clock phase is high, t_{high}, must also be considered. The latest input should arrive halfway between when the latch becomes transparent and $t_{su}+t_j+t_{sk}+t_{duty}$ before T_{high} later, when the latch becomes opaque.

An example of optimal positions for latches is in Figure 10. The first set of latches become transparent $T/2$ after the rising clock edge at the inputs, so the optimal position p_1 of the first set of latches is at

$$(26) \quad p_1 = \frac{T}{2} + \frac{T_{high} - (t_{su} + t_{sk} + t_j)}{2}$$

The clock edge of the phase triggering the kth set of latches to become transparent arrives $(k-1)T/2$ later. As shown in Figure 8, the edge jitter and duty cycle jitter must be included on successive clock edges, and we assume the edge jitter is additive in the worst case. So the kth set of latches are optimally positioned at

$$(27) \quad \begin{aligned} p_k &= (k-1)\frac{T}{2} + p_1 - \frac{(k-1)}{4}t_j, k \text{ odd} \\ p_k &= (k-1)\frac{T}{2} + p_1 - \left(\frac{(k-2)}{4}t_j + \frac{t_{duty}}{2}\right), k \text{ even} \end{aligned}$$

To simplify things, we use $t_{duty} = t_j/2$, which gives

$$(28) \quad p_k = (k-1)\frac{(T - t_j/2)}{2} + p_1$$

Therefore generally,

$$(29) \quad p_k = k\frac{(T - t_j/2)}{2} + \frac{T_{high} - (t_{su} + t_{sk} + t_j/2)}{2}$$

This derivation assumes that

$$(30) \quad T_{high} \geq t_{su} + t_{sk} + \frac{t_j}{2} + k\frac{t_j}{2}$$

Otherwise, the clock skew and multi-cycle jitter on the sequential path through the k sets of transparent latches is too large for the critical path input at the kth set of latches to arrive while the latch is transparent. The input must arrive *before* the *nominal* time of the clock edge at which the kth latch becomes transparent to ensure the setup time constraint is met. Thus the clock jitter over multiple cycles limits the length of a sequential path through transparent latches. After a few cycles, the propagating signal must still be guaranteed to be synchronized with respect to the clock edge. When the sequential path is too fast with respect to the *actual* clock edge arrival times, it will arrive at a hard boundary provided by an opaque latch or a flip-flop, which synchronizes it. By choosing a sufficiently large clock period, the path is guaranteed not to be too slow, ensuring the setup time is not violated.

Consider Figure 10, where the duty cycle is 40% and the clock period T is 30 FO4 delays, corresponding to T_{high} of $0.4T$ = 12 FO4 delays (see Equation (1)). The optimal position of the latches is

$$\begin{aligned} p_k &= k\frac{(T-t_j/2)}{2} + \frac{T_{high} - (t_{su} + t_{sk} + t_j/2)}{2} \\ &= k \times \frac{(32-1)}{2} + \frac{12 - (2+4+1)}{2} \\ &= 15.5k + 2.5 \end{aligned} \qquad (31)$$

Thus the optimal positions for the latches are at positions of 18.0 and 33.5 FO4 delays relative to the clock edge arrival at the first set of rising edge flip-flops. This is shown in Figure 10.

Note that the optimal latch positions, to ensure the inputs arrive while the latches are transparent, are not power or area optimal. As there is a window when the latches are transparent, there is some flexibility to retime latches in the logic to reduce the number of latches. Reducing the number of latches reduces the power consumption, wiring and area. Wiring and area reductions can lead to a small performance improvement as average wire loads decrease. Careful analysis is required to trade-off the performance, area and power.

4. EXAMPLE WHERE LATCHES ARE SLOWER

Consider the unrolled two-state Viterbi add-compare-select calculation, shown in Figure 12. This includes the feedback sequential loops. The nominal delays considered in this example are:
- adder delay t_{add} = 8 FO4 delays
- comparator delay t_{comp} = 6 FO4 delays
- multiplexer delay t_{mux} = 2 FO4 delays
- flip-flop and latch setup time t_{su} = 1 FO4 delay
- flip-flop and latch maximum clock-to-Q delay t_{CQ} = 3 FO4 delays

- latch propagation delay $t_{DQ} = 3$ FO4 delays
- clock skew of $t_{sk} = 1$ FO4 delays
- edge jitter of $t_j = 1$ FO4 delays
- duty cycle jitter of $t_{duty} = 1$ FO4 delay

The clock period with flip-flops is

(32)
$$\begin{aligned}T_{flip-flops} &= t_{CQ} + t_{mux} + t_{add} + t_{comp} + t_{su} + t_{sk} + t_j \\ &= 3 + 2 + 6 + 8 + 1 + 1 + 1 \\ &= 22 \text{ FO4 delays}\end{aligned}$$

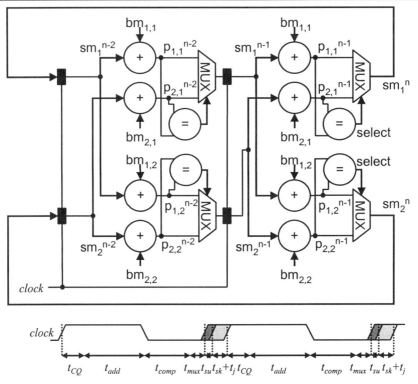

Figure 12. Timing for a two-state add-compare-select with rising edge flip-flop registers and recursive feedback. FO4 delays are shown to the same scale. The duty cycle is 50%.

Now consider replacing the flip-flops with latches as shown in Figure 13.

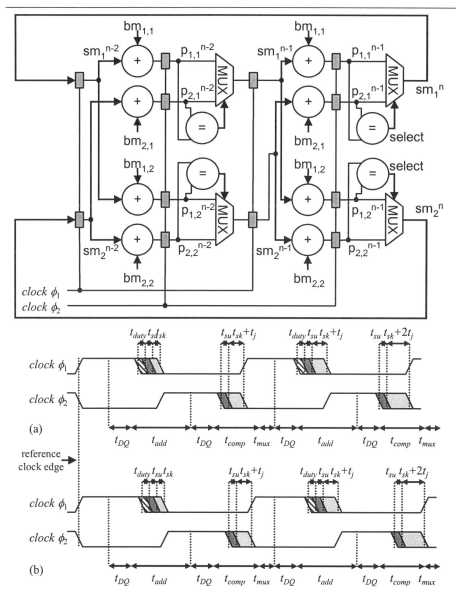

Figure 13. Timing for a two-state add-compare-select with active high latch registers and recursive feedback. FO4 delays are shown to the same scale. The duty cycle is 50%. (a) Shows a clock period of 22 FO4 delays, and (b) has a clock period of 23 FO4 delays. In (a), the arrival time after two clock periods is closer to violating the setup time constraint, and over a few cycles it will be violated. Thus the clock period is too small.

From (21), the lower bound on the clock period with latches is

$$
\begin{aligned}
T_{latches,min} &= 2t_{DQ} + 2t_{comb,average} + t_j \\
&= 2\times 3 + 2\times \frac{(8+6+2)}{2} + 1 \\
&= 23 \text{ FO4 delays}
\end{aligned}
\tag{33}
$$

A duty cycle of 50% is used for both the flip-flop and latch versions of this example. Instead of using non-overlapping clock phases, buffering can be used to fix hold time constraints, as discussed in Section 1.2.1.

The correct clock period is shown in Figure 13(b). Over multiple cycles, the latch inputs arrive closer to when the latch becomes transparent to account for worst case clock jitter.

In comparison, Figure 13(a) shows a clock period of 22 FO4 delays. After two cycles, the latch inputs arrive closer to when the latch becomes opaque, and after several more cycles there will be a setup time violation.

For example, suppose the input at the first set of latches arrives before they become transparent. The combinational delay of each pipeline stage is the same, 8 FO4 delays, which simplifies analysis. Then with respect to the reference clock edge, the delay of the sequential path through k stages is

$$
\begin{aligned}
t_k &= t_{CQ} + (k-1)t_{DQ} + kt_{comb} \\
&= 3 + 3(k-1) + 8k \\
&= 11k + 2 \text{ FO4 delays}
\end{aligned}
\tag{34}
$$

After k stages, the setup time constraint at the kth set of latches is

$$
\begin{aligned}
t_k &\leq \frac{T}{2} + k\frac{T}{2} - (t_{su} + t_{sk} + t_{duty} + \frac{(k-1)}{2}t_j), k \text{ odd} \\
t_k &\leq \frac{T}{2} + k\frac{T}{2} - (t_{su} + t_{sk} + \frac{k}{2}t_j), k \text{ even}
\end{aligned}
\tag{35}
$$

which for a clock period of 22 FO4 delays is

$$
\begin{aligned}
t_k &\leq 11 + 11k - (1+1+1+\frac{(k-1)}{2}) = 8.5 + 10.5k, k \text{ odd} \\
t_k &\leq 11 + 11k - (1+1+\frac{k}{2}) = 9 + 10.5k, k \text{ even}
\end{aligned}
\tag{36}
$$

Thus there is a setup constraint violation for $k \geq 19$. At $k = 19$,

$$
11\times 19 + 2 = 211 > 8.5 + 10.5\times 19 = 208
\tag{37}
$$

Therefore, the correct clock period for the latch-based two-state add-compare-select shown in Figure 13 is more than 22 FO4 delays. This is

Reducing the Timing Overhead

slower than a flip-flop based design. See Chapter 15, Section 4, for more examples of appropriate situations in which to use latches or flip-flops.

Assuming a flip-flop based pipeline is balanced, we can determine when flip-flops or latches are better for a critical loop of n cycles. Comparing Equations (6) and (16),

(38) $T_{latches} \geq T_{flip-flops}$, if
$$2t_{DQ} + t_{j,over\ n\ cycles} \geq t_{CQ} + t_{su, flip-flop} + t_{sk} + t_j$$

where $t_{j,over\ n\ cycles}$ is the n-cycle-to-cycle jitter averaged over n cycles. In this example,

(39) $T_{latches} \geq T_{flip-flops}$, as
$$2 \times 3 + 1 = 7 \geq 3 + 1 + 1 + 1 = 6$$

In general, in circuitry with tight sequential feedback loops, such as in Figure 13, it may not be appropriate to use latches. The main advantage of latches that they reduce the impact of clock skew and setup time constraints, and allowing slack passing and time borrowing. Sequential loops of two clock cycles do not allow significant slack passing and time borrowing, unless the two pipeline stages are poorly balanced. There is obviously no slack passing in a critical sequential loop with single-cycle feedback. Latches in a tight sequential feedback loop can still reduce the effects of clock skew, but there are also high speed flip-flops with clock skew tolerance (see Chapter 15, Section 4.3).

In the example above, the clock-to-Q delay of the flip-flops and D-to-Q propagation delays of the latches were the same. Unfortunately, standard cell libraries often lack high speed latches, or latches with sufficient drive strength. As a result, flip-flop and latch delays can be similar – despite flip-flops being composed of a pair of master-slave latches. See Section 6.2.2 for some more discussion of registers in standard cell libraries.

The next section analyzes when latches can reduce the clock period of a pipeline.

5. PIPELINE DELAY WITH LATCHES VS. PIPELINE DELAY WITH FLIP-FLOPS

If the inputs arrive sufficiently early within the latch's transparent window, the setup time and clock skew have less effect on the clock period of a pipeline as compared to a pipeline with flip-flops.

To compare the clock period of pipelines with flip-flops or latches, we consider a pipeline with $k+1$ stages separated by k flip-flops, and a pipeline with $2k+1$ stages separated by $2k$ latches. The inputs to the pipeline come from a rising edge flip-flop and have a latest arrival time of t_{CQ} with respect to the rising clock edge. The outputs of the pipeline have worst case setup constraints of t_{su} with respect to the rising clock edge, and go to a rising edge flip-flop. With either the flip-flops or latches, the pipeline has $k+1$ clock periods to complete computation.

From (6), for flip-flops the clock period is

(40)
$$T_{flip-flops} \geq t_{CQ} + t_{comb,i} + t_{su} + t_{sk} + t_j$$
$$\therefore T_{flip-flops} = t_{CQ} + t_{comb,max} + t_{su} + t_{sk} + t_j$$

where $t_{comb,i}$ is the delay of the ith stage of combinational logic, and $t_{comb,max}$ is the maximum of these combinational delays. If the flip-flop pipeline is balanced perfectly, the combinational delay of each stage is the same (the average combinational delay t_{comb}) and

(41) $T_{flip-flops} = t_{su} + t_{sk} + t_{CQ} + t_{comb} + t_j$

Providing that the inputs arrive while the latches are transparent, from (19), the clock period with latches is

(42)
$$T_{latches} = \frac{t_{CQ} + 2kt_{DQ} + t_{su} + t_{sk} + (k+1)t_{j,over\ k+1\ cycles} + (k+1)t_{comb}}{(k+1)}$$
$$= \frac{t_{CQ} - 2t_{DQ} + t_{su} + t_{sk}}{(k+1)} + 2t_{DQ} + t_{j,over\ k+1\ cycles} + t_{comb}$$

where $t_{j,over\ k+1\ cycles}$ is the average edge to edge jitter over $k+1$ cycles. The latch D-to-Q propagation time is about half of the clock-to-Q propagation time of a flip-flop, as a D-type flip-flop is a master-slave latch pair. So we approximate t_{CQ} by $2t_{DQ}$, giving

(43) $T_{latches} = \dfrac{t_{su} + t_{sk}}{(k+1)} + t_{CQ} + t_{comb} + t_{j,over\ k+1\ cycles}$

Comparing (41) and (43), the major advantage of latches is only considering the setup time and clock skew once for the entire pipeline, rather than for each pipeline stage. This reduces the impact of clock skew and setup time to a factor of $1/(k+1)$, where $k+1$ is the number of clock periods for the pipeline to complete computation. The edge-to-edge jitter over $k+1$ cycles is less than the edge-to-edge jitter over one cycle, so the effect of jitter is also reduced.

As discussed in Section 4, latches are not always useful, particularly when there are sequential loops of only one or two pipeline stages. In such cases, the impact of clock skew and setup time are not reduced substantially.

The other advantage of latches over flip-flops is balancing the delay of pipeline stages by slack passing and time borrowing. Flip-flops are limited by the maximum delay of any pipeline stage, as given in (40), whereas slack passing and time borrowing with latches can allow a pipeline stage to take up to the amount of time given by Equation (8).

While retiming flip-flops can balance pipeline stages, in some cases this is not possible. For example, accessing cache memory is a substantial portion of the clock period, and limits the clock period of the pipeline stage. There also needs to be additional logic for tag comparison and data alignment [4]. With flip-flops, the only method for increasing the speed may be to give the cache access an additional pipeline stage to complete [4], if it is the critical path limiting the clock period.

Using latches can also reduce the latency through the pipeline, as the clock period can be reduced, and the latency is $nT_{latches}$ (from Equation (15) in Chapter 2).

Consider pipelining a combinational path with total delay of t_{comb}. With $2(n-1)$ sets of latches between inputs and outputs, there are $2n-1$ pipeline stages with $nT_{latches}$ for the pipeline to complete computation. The clock period is

$$(44) \quad T_{latches} = \frac{t_{comb} + t_{su} + t_{sk}}{n} + 2t_{DQ} + t_{j, over\ n\ cycles}$$

From Equation (29) in Chapter 2, and Equation (44), the speedup by pipelining is

$$(45) \quad \frac{IPC_{after}}{IPC_{before}} \times \frac{t_{comb} + t_{su} + t_{sk} + t_{CQ} + t_j}{\left(\frac{t_{comb} + t_{su} + t_{sk}}{n} + 2t_{DQ} + t_{j, over\ n\ cycles}\right)}$$

This assumes the inputs to the pipeline arrive with maximum delay t_{CQ} with respect to the rising clock edge from rising clock edge inputs. Even if the pipeline is perfectly balanced, with flip-flops the speedup by pipelining is (from Chapter 2, Equation (29))

$$(46) \quad \frac{IPC_{after}}{IPC_{before}} \times \frac{t_{comb} + t_{su} + t_{sk} + t_{CQ} + t_j}{\left(\frac{t_{comb}}{n} + t_{su} + t_{sk} + t_{CQ} + t_j\right)}$$

Comparing Equations (45) and (46), latches reduce the total register and clock overhead per pipeline stage. Thus latches increase the performance improvement by pipelining. This may make more pipeline stages worthwhile.

6. CUSTOM VERSUS ASIC TIMING OVERHEAD

Custom chips typically have manually laid out clock trees. The clock trees may be designed with phase detectors and programmable buffers to reduce skew. Filters are used to reduce the supply voltage noise and shielding is used to reduce inter-signal interference, which in turn reduce the clock jitter. Custom designs also typically use higher speed flip-flops on critical paths. This substantially reduces the timing overhead per clock cycle.

In comparison, ASICs generally use D-type flip-flops from a standard cell library with automatic clock tree generation. Let's examine the timing overhead for custom and ASIC chips to see how they compare.

6.1 Custom Chips

Custom microprocessors have used latches, high speed pulsed flip-flops, and latches with a pulsed clock to reduce the timing overhead. These techniques are often restricted to critical paths, because there is a greater window for hold time violations, or they have higher power consumption.

Even in latch-based custom designs, flip-flops are still used where it is important to guarantee that the inputs to the next logic stage only change at a given clock edge. For example, inputs to RAMs are usually registered, but this is also typically a critical path [private communication with Earl Killian].

There are a variety of high speed registers that have been used in custom designs:
- Latches (two latches per cycle)
- Latches incorporating combinational logic (two latches per cycle, reduced register overhead)
- Latches with pulsed clock input (one latch per cycle)
- Pulsed flip-flops (one flip-flop per cycle)
- Pulsed flip-flops incorporating combinational logic (one flip-flop per cycle, reduced register overhead)

A number of techniques are typically used in custom designs for reducing the clock skew. In addition, clock skew to registers can be selectively adjusted to allow slower pipeline stages more time to compute. These techniques are listed in increasing ability to reduce skew:
- Balanced clock trees, which balance delays of the clock tree after clock tree synthesis
- Balanced clock trees with paired inverters at each leaf of the tree. One inverter drives the clock signal to the registers and is resized for different loads to maintain the same signal delay, and hence reduce skew relative to other signals, to the registers. The other inverter does not drive anything, and is used to balance the delay of the

inverter that is being resized to drive the registers, so that the higher portions of the clock tree see the same load at each leaf [21].
- Balanced clock trees with phase detectors to set programmable delays in registers on the clock tree to deskew the signal across the chip. This compensates for process variation affecting the clock skew [19].

The advantages and disadvantages of different types of registers, and clocking schemes used in custom processors are discussed below.

6.1.1 The Alpha Microprocessors

The Alpha 21164 uses dynamic level-sensitive pass-transistor latches [1], where the charge after the clocked transmission gate stores the input value. Simple combinational logic was combined with the latch input stage to reduce the latch overhead to the delay of a transmission gate. Also, combining combinational logic with registers reduces the additional wiring to the registers and reduces the area. This is shown in Figure 14. The stored charge is prone to noise, making this latch style inappropriate for many deep submicron applications. These fast latches are subject to races, which are avoided by minimizing the clock skew and requiring a minimum number of gate delays between latches [1].

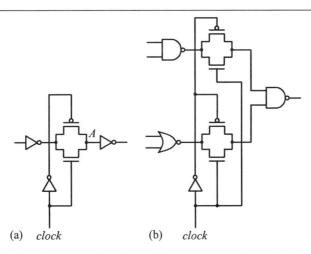

Figure 14.(a) The dynamic level-sensitive pass-transistor latches used in the Alpha 21164. The charge at node A stores the state of the previous input when the pass transistors are off. (b) Logic incorporated with the pass-transistor latch to reduce the effective latch delay to the delay of a pass transistor. [1]

There were additional concerns about races in the Alpha 21264, because of clock-gating, which introduces additional delays in the gated clock. Gated clocks are used to reduce the power consumption by turning off the clock to modules that are not in use. The Alpha 21264 uses high speed edge-triggered dynamic flip-flops to reduce the potential for races that violate hold time constraints.

In both the Alpha 21164 and Alpha 21264, the registers have an overhead of about 15% per clock cycle, or 2.9 and 2.0 FO4 delays respectively. The 600MHz Alpha 21264 has 75ps global clock skew, which is about 0.6 FO4 delays. The Alpha 21164 distributes one clock over the chip, whereas the Alpha 21264 distributes a global reference clock for the buffered and gated local clocks [3].

A 1.2GHz Alpha microprocessor was implemented in 0.18um bulk CMOS with copper interconnect, and standard and low threshold transistors [8]. It is a large chip with a higher clock frequency, and uses four clocks over the chip [21]. One of the clocks is a reference clock generated from a phase-locked loop. The other three clocks are generated by delay-locked loops (DLLs) from this reference clock. To further reduce the impact of skew on the memory and network subsystem, pairs of inverters were fine-tuned to the capacitive load of the clock network they were driving [21]. The worst global clock skew is about 90ps (1.4 FO4 delays), and the inverter pairs reduce the local skew to about 45ps (0.69 FO4 delays). To reduce the effects of supply voltage noise on the jitter, a voltage regulator was used to attenuate the noise by 15dB. The cycle-to-cycle edge jitter of the PLL is about 0.13 FO4 delays [20], and the maximum phase error (duty cycle jitter) is about 30ps or 0.46 FO4 delays [21].

6.1.2 The Athlon Microprocessor

The Athlon uses a pulsed flip-flop with small setup time and small clock-to-Q delay, but long hold time [17]. The first stage of the flip-flop has a dynamic pull-down network, and combinational logic can be included in the first stage to reduce the register overhead [14]. The dynamic pull-down network uses the same principle used in high-speed domino logic, which is discussed in Chapter 4. CAD tools can avoid violations of the long hold time, but this introduces additional delay elements, and it is sub-optimal when the reduced latency is unnecessary and normal flip-flops can be used [14]. Thus the high-speed pulsed flip-flop was used only on critical paths, where it reduced the delay by up to 12% [17]. A 12% reduction in delay corresponds to about 2.8 FO4 delays (the Athlon had about 20.8 FO4 delays per pipeline stage from Table 3 of Chapter 2).

6.1.3 The Pentium 4 Microprocessor

As discussed in Chapter 2, Section 3.3, the timing overhead is about 30% of the clock period in the Pentium 4, which is 3.0 FO4 delays. The Pentium 4 uses pulsed clocks derived from the clock edges of a normal clock for the domain. The duty cycle of the normal clock is adjusted from a 50% duty cycle, so that the rising clock edge is one inverter delay later, to compensate for the additional inversion to generate a pulse to V_{DD} (the supply voltage) from the falling clock edge [12].

6.1.3.1 Clock Distribution in the Pentium 4

The 100MHz system reference clock of the Pentium 4 is a differential low-swing clock, which goes to sense amplifier receivers to restore the signal to full ground and supply voltage levels [11]. *Low swing* signals reduce power consumption, as capacitances are not charged and discharged from V_{DD} to ground, and they cause less electromagnetic interference noise when they switch. As power consumption for the clock grid may be around 45% of the total power consumption [9], using low-swing clocking can substantially reduce power consumption. The sense amplifiers are designed to have high tolerance to process, voltage and temperature variation. The sense amplifiers are followed by a high-gain stage to drive the output clock, and this configuration reduces the impact of voltage supply noise [11].

A phase-lock loops generates the 2GHz core clock frequency from the system reference clock of 100MHz. The 2GHz core frequency is distributed across the chip to 47 domain buffers, with three sets of binary trees with 16 leaf nodes each [11]. Each domain buffer has a 5 bit programmable register to remove skew from the clock signal in that domain, compensating for clock skew caused by process variation. The skew in each domain is reduced by using 46 phase detectors to compare with a reference domain [12]. Jitter in the buffer clock signal, caused by supply voltage noise, is reduced using a low-pass RC filter. The clock wires are shielded to reduce jitter caused by signals from cross-coupling capacitance [11].

In the worst case after delay matching with the phase detectors, the cycle-to-cycle jitter t_j is 35ps, the long term jitter is 90ps, and the skew is 16ps. These numbers correspond to 0.70, 1.8, and 0.32 FO4 delays respectively. The clock skew and jitter together take about 1.0 FO4 delay per clock cycle.

6.1.3.2 Pulsed Latches in the Pentium 4

The pulsed clocks go to latches, which effectively store their inputs at a hard clock edge like a master-slave flip-flop because of the short pulse duration. Individually, latches take less area and are lower power than flip-flops, thus replacing flip-flops by latches with a pulsed clock reduces area and reduces power consumption [12]. Using latches in this manner also effectively halves the t_{CQ} delay, as a master-slave flip-flop comprises two latches.

If the timing overhead is about 3.0 FO4 delays for the Pentium 4, and the clock skew and clock jitter are about 1.0 FO4 delays together, the latches with pulsed clocks have a register overhead of about 2.0 FO4 delays per clock cycle.

6.1.4 Pulsed Flip-Flops are Faster than D-Type Flip-Flops

Pulse-triggered flip-flops such as the hybrid latch flip-flop (HLFF) [6] and semi-dynamic flip-flop (SDFF), which is similar to the flip-flops used in the Athlon, are substantially faster than normal master-slave latch flip-flops which are used in ASICs. Table 1 shows the comparison of SDFF and HLFF flip-flops with a transmission gate based master-slave latch flip-flop in 0.25um technology given by Klass et al. [10]. A summary of their results is presented in Table 1. The SDFF has a clock-to-Q delay of about 2.1 FO4 delays, zero setup time, and hold time of 1.4 FO4 delays. The register overhead for the SDFF can be further reduced to 1.3 to 0.8 FO4 delays by combining combinational logic with its input stage. In comparison, the master-slave flip-flop built with transmission gates (SFF) has clock-to-Q delay of 3.3 FO4 delays.

	SFF	HLFF	SDFF	SDFF with 2-input AND	SDFF with 2-input OR	SDFF with AB+CD
Clock-to-Q Delay t_{CQ} (ps)	300	194	188	208	196	228
Setup Time t_{su} (ps)			0	20	8	40
Hold Time t_h (ps)			130			
Delay of separate gate and SDFF flip-flop (ps)				280	286	348
Effective SDFF Register Latency (ps)				116	98	68

Table 1. Comparison of master-slave latch flip-flop (SFF) with high speed pulse-triggered flip-flops such as the hybrid latch flip-flop (HLFF) and semi-dynamic flip-flop (SDFF) in 0.25um technology [10]. Integrating combinational logic with the SDFF reduces the overall delay, and thus reduces the register overhead. Setup times for the SDFF combined with combinational logic are simply calculated from the latency.

6.2 ASICs

Many standard cell ASICs use only rising edge flip-flops for sequential logic, though register banks may use latches to achieved higher density and lower power consumption. A major reason for not using latches has been lack of tool support, though latches are supported by EDA tools now. Chapter 7 describes some of the issues that have limited use of latches in ASIC designs, and an approach to converting flip-flop based designs to use latches.

Timing characteristics of the Tensilica Base Xtensa microprocessor configuration are discussed in detail in Chapter 7. From the Tensilica numbers, typical timing parameters for a high speed low-threshold voltage ASIC standard cell library are:
- 4 FO4 delays clock skew and edge jitter
- 4 FO4 delays clock-to-Q delay for flip-flops
- 4 FO4 delays D-to-Q propagation delay for latches
- 2 FO4 delays flip-flop setup time
- 0 FO4 delays latch setup time
- 1 FO4 delay hold time for latches
- 0 FO4 delays hold time for D-type flip-flops

The flip-flop and latch delays are for a large fanout load. Flip-flop propagation delays are typically 3 to 4 FO4 delays for standard cell libraries.

Lexra reports worst-case duty cycle jitter of ±10% of T_{high} [5], which is about ±3 FO4 delays. Standard cell ASICs usually have automatically generated clock trees, with poor jitter and skew compared to custom.

6.2.1 Imbalanced ASIC Pipelines and Slack Passing

The STMicroelectronics iCORE, discussed in Chapter 16, is an ASIC design with well balanced pipeline stages. Figure 4 in Chapter 16 shows the worst case delay for each pipeline stage. The design uses flip-flops, so there will be some penalty for the small imbalance between stages. Suppose slack passing was possible in this design, whether by using latches or by cycle stealing. Comparing Figure 1 and Figure 4 in Chapter 16, we determine that the critical sequential loop is IF1, IF2, ID1, ID2, OF1 back to IF1 through the branch target repair loop. This loop has an average delay of about 90% of the slowest pipeline stage (ID1), which has the worst stage delay and limits the clock period. Thus slack passing would give at most a 10% reduction in the clock period.

Converting the Tensilica Xtensa flip-flops to latches improved the speed by up to 20% (see Chapter 7). Between 5% and 10% of this speed increase was from reducing the effect of setup time and clock skew on the clock period. The remainder is slack passing balancing pipeline stages. The slack passing in the latch-based design gave up to a 10% improvement in clock speed.

ASICs with poorly balanced pipeline stages would benefit more from slack passing, if retiming cannot better balance the pipeline stages. From the iCORE and Xtensa examples, we estimate about a 10% improvement in clock period by slack passing for ASICs.

6.2.2 Deficiencies of Latches in Standard Cell Libraries

Both flip-flops and latches are available in standard cell libraries, though often there is a greater range of flip-flops. Scan flip-flops for testing are available in any standard cell library, but scan latches [16] are available in only a few libraries currently [7]. Scan latches are required for verification of latch-based designs. There are often more drive strengths for flip-flops, and there are sometimes a wider range of flip-flops integrating simple combinational logic functions.

Flip-flops are composed of a master-slave latch pair, thus latches should have a smaller delay than flip-flops. However, standard cell latches often have additional buffering, which makes them slower. For example, latches may have input buffers to isolate the input of the latch from actively driven feedback nodes within the latch [6]. This guard-banding improves the robustness of the library. The cell driving the latch input would have to provide large drive strength to drive both a long wire and fight with the internal driver of the active node to change the node's value. However, faster variants, without the buffering, can improve performance on critical paths. For further discussion of problems with buffered combinational cells, see Section 3.4 of Chapter 16.

High speed flip-flops that are often used in custom processors are not typically available in standard cell libraries. High speed latches, such as the dynamic level-sensitive pass-transistor latch, are not included in standard cell libraries for ASICs, because of the difficulty of ensuring that noise does not affect the dynamically stored charge. Custom designs have also used latches and flip-flops that incorporate combinational logic to reduce the register delay (see Section 6.1.1 for an example). Some standard cell libraries are now including latches and flip-flops that have combinational logic.

6.3 Comparison of ASIC and Custom Timing Overhead

Table 2 compares ASIC and custom timing overhead per clock cycle. Custom designs achieve about 3 FO4 delays per clock cycle. In comparison ASICs have a timing overhead of about 10 FO4 delays per clock cycle. These values assume that the pipelines are well-balanced. t_{DQ} for poor latches is for libraries with insufficient latch drive strengths, and latches with too much guard-banding.

To reduce the timing overhead, fast custom designs have used latches, or pulse-triggered flip-flops incorporating logic with the flip-flop, or latches with a pulsed clock. Pulse-triggered flip-flops have about zero setup time, but have longer hold times, like latches. The longer hold times of latches and pulse-triggered flip-flops require careful timing analysis with CAD tools, and buffer insertion where necessary, to avoid short paths violating the hold time. ASICs can use pulse-triggered flip-flops if they are characterized for the standard cell flow (e.g. if a standard cell library includes these high speed flip-flops) – this was done in the SP4140 (see Chapter 15). High speed pulsed flip-flops are not generally available in standard cell libraries. D-type flip-flops can't include combinational logic with the first stage of the flip-flop to reduce the register overhead, whereas pulsed flip-flops can [14].

	ASICs		Custom		
FO4 delays	Poor Latches	Good Latches	Alphas	Pentium 4	SDFF
Clock-to-Q Delay t_{CQ}	4.0	4.0	2.0	2.0	2.1
D-to-Q Latch Propagation Delay t_{DQ}	4.0	2.0	1.3		
Flip-Flop Setup Time t_{su}	2.0	2.0	0.0	0.0	0.0
Latch Setup Time t_{su}	0.0	0.0	0.0		
Flip-Flop Hold Time t_h	0.0	0.0			1.4
Latch Hold Time t_h	1.0	1.0			
Edge Jitter t_j			0.13	0.70	
Clock Skew t_{sk}			0.70	0.32	
Clock Skew and Edge Jitter $t_{sk} + t_j$	4.0	4.0	0.83	1.0	
Duty Cycle Jitter t_{duty}	3.0	3.0	0.46		
Timing Overhead per Cycle with Flip-Flops	10.0	10.0	2.8	3.0	
Timing Overhead per Cycle with Latches	9.0	5.0	2.6		

Table 2. Comparison of ASIC and custom timing overheads, assuming balanced pipeline stages. Alpha and Pentium 4 setup times were estimated from known setup times for latches and pulse-triggered flip-flops. Other values used are discussed in 6.1 and 6.2. The clock-to-Q delay for the Pentium 4 is the estimated delay of the latches with a pulsed clock. Multi-cycle jitter of 1.0 FO4 delays is assumed for ASICs. Blanks are left where information is not readily available.

If the clock skew and setup time are small, latches with pulsed clocks, or pulsed flip-flops incorporating logic into the input stage, have the smallest timing overhead. If the skew is very small, using level-sensitive latches (with a normal clock) may not be as good, because generally $2t_{DQ,latches}$ will be greater than the t_{CQ} of a single pulsed latch or flip-flop. Current clock tree synthesis tools are not able to reduce the clock skew sufficiently, but designs with small clock skew from manual clock tree layout should carefully compare using pulsed flip-flops as well as latches.

If the clock skew and setup time are larger, latches can substantially reduce the timing overhead, by as much as 50% for the numbers in Table 2. Latches significantly reduce the impact of the clock skew and setup time over multi-cycle paths.

Heo et al. carefully compare choices of latches and flip-flops to increase performance and reduce a chip's power consumption. They conclude that different choices for registers should be considered for the variety of activation patterns and slacks within the circuit [6].

A typical ASIC might have 60 FO4 delays per pipeline stage (see Table 2 in Chapter 2 for delays per pipeline stage of embedded ASIC processors). A difference of 7 FO4 delays, gained by custom quality timing overhead, reduces the clock period by a factor of about 1.1× for a slow ASIC.

The timing overhead of a typical ASIC with flip-flops is 10 FO4 delays (see Table 2), and about an additional 10% for unbalanced pipeline stages. The total timing overhead is about 30% of the clock period for an ASIC with clock period of 40 to 60 FO4 delays (35% for a clock period of 40 FO4 delays, 27% for a clock period of 60 FO4 delays). In contrast, the custom timing overhead is only 2.8 FO4 delays for the Alpha 21264, 20% of the clock period of 13.3 FO4 delays.

A very fast ASIC such as the Texas Instruments SP4140 disk drive read channel has about 20 FO4 delays per stage. The SP4140 achieved a clock frequency of 550MHz in a 0.18um effective channel length process using custom techniques: high speed pulsed flip-flops, and manual clock tree design (see Chapter 15 for more details). The clock skew of 60ps was about 0.7 FO4 delays, and the pulsed flip-flops would have delay of around 2 FO4 delays. In total about 3 FO4 delays of timing overhead. If the SP4140 was limited to typical ASIC D-type flip-flops and clock tree synthesis, the additional 7 FO4 delays of timing overhead would increase the clock period by a factor of 1.35×, reducing the clock frequency to 410MHz.

Custom designs may be a further 1.1× faster by using slack passing, compared to ASICs that can't do slack passing to balance pipeline stages. If the SP4140 had unbalanced pipeline stages without slack passing, the unabalanced pipeline stages might increase the clock period by 10% (of 20

FO4 delays), which is 2 FO4 delays. Combining this with the impact of reduced timing overhead, gives an overall factor of 1.45×.

Chapter 7 examines an automated approach to changing flip-flop based gate netlists to use latches, in a standard cell ASIC flow, achieving a 10% to 20% speed improvement. It also details the problems that have impeded use of latches in ASIC flows and solutions to these problems.

The Texas Instruments SP4140 disk drive read channel uses modified sense-amplifier flip-flops based on a pulse-triggered design. It also uses latches on the critical paths where there is not tight sequential recursive feedback. This is discussed in detail in Sections 4 and 5 of Chapter 15.

Comparing the absolute differences in clock skews, there is about a 10% increase in speed of designs using flip-flops with custom quality clock tree distribution to reduce clock skew and jitter. Clock tree synthesis tools are improving. Chapter 8 discusses new approaches in detail.

The combinational delay of each pipeline stage can also be reduced by a variety of different techniques. The next Chapter explores the differences between the combinational delay in standard cell ASIC and custom methodologies.

7. REFERENCES

[1] Benschneider, B.J., et al., "A 300-MHz 64-b Quad-Issue CMOS RISC Microprocessor," *IEEE Journal of Solid-State Circuits*, vol. 30, no. 11, November 1995, pp. 1203-1214.
[2] Cadence, "Theory of Operation: Transparent Latches," *Pearl User Guide*, 1998, pp. 6.2–6.13.
[3] Gronowski, P., et al., "High-Performance Microprocessor Design," *IEEE Journal of Solid-State Circuits*, vol. 33, no. 5, May 1998, pp. 676-686.
[4] Hauck, C., and Cheng, C. "VLSI Implementation of a Portable 266MHz 32-Bit RISC Core," *Microprocessor Report*, November 2001.
[5] Hays, W.P., Katzman, S., and Hauck, C., "7 Stages: Lexra's New High-Performance ASIC Processor Pipeline," June 2001.
http://www.lexra.com/whitepapers/7stage_Pipeline_Web.pdf
[6] Heo, S., Krashinsky, R., and Asanovic, K., "Activity-Sensitive Flip-Flop and Latch Selection for Reduced Energy," *Proceedings of the Conference on Advanced Research in VLSI*, 2001, pp. 59-74.
[7] IBM, ASIC SA-27 Standard Cell/Gate Array. December 2001. http://www-3.ibm.com/chips/products/asics/products/sa-27.html
[8] Jain, A., et al., "A 1.2 GHz Alpha Microprocessor with 44.8 GB/s Chip Pin Bandwidth," *Digest of Technical Papers of the IEEE International Solid-State Circuits Conference*, 2001, pp. 240-241.
[9] Kawaguchi, H., and Sakurai, T., "A Reduced Clock-Swing Flip-Flop (RCSFF) for 63% Power Reduction," *IEEE Journal of Solid-State Circuits*, vol.33, no.5, May 1998. pp.807-811.
[10] Klass, F., et al., "A New Family of Semidynamic and Dynamic Flip-flops with Embedded Logic for High-Performance Processors," *IEEE Journal of Solid-State Circuits*, vol. 34, no. 5, May 1999, pp. 712-716.

[11] Kurd, N.A, et al., "A Multigigahertz Clocking Scheme for the Pentium® 4 Microprocessor," *IEEE Journal of Solid-State Circuits*, vol. 36, no. 11, November 2001, pp. 1647-1653.

[12] Kurd, N.A., et al., "Multi-GHz clocking scheme for Intel® Pentium® 4 Microprocessor," *Digest of Technical Papers of the IEEE International Solid-State Circuits Conference*, 2001, pp. 404-405.

[13] Orshansky, M., et al., "Impact of Systematic Spatial Intra-Chip Gate Length Variability on Performance of High-Speed Digital Circuits," *Proceedings of the International Conference on Computer Aided Design*, 2000, pp. 62-67.

[14] Partovi, H., "Clocked storage elements," in *Design of High-Performance Microprocessor Circuits*. IEEE Press, Piscataway NJ, 2000, pp. 207-234.

[15] Rabaey, J.M., *Digital Integrated Circuits*. Prentice-Hall, 1996.

[16] Raina, R., et al., "Efficient Testing of Clock Regenerator Circuits in Scan Designs," *Proceedings of the 34th Design Automation Conference*, 1997, pp. 95-100.

[17] Scherer, A., et al., "An Out-of-Order Three-Way Superscalar Multimedia Floating-Point Unit," *Digest of Technical Papers of the IEEE International Solid-State Circuits Conference*, 1999, pp. 94-95.

[18] Synopsys, "Optimization and Timing Analysis: 11. Timing Analysis in Latches," in *Design Compiler Reference Manual* v.2000.11, pp. 11.1-11.34.

[19] Tam, S., et al., "Clock generation and distribution for the first IA-64 microprocessor," *IEEE Journal of Solid-State Circuits*, vol.35, no. 11, November 2000, pp. 1545-1452.

[20] von Kaenel, V.R., "A High-Speed, Low-Power Clock Generator for a Microprocessor Application," *IEEE Journal of Solid-State Circuits*, vol.33, no. 11, November 1998, pp. 1634-1639.

[21] Xanthopoulos, T., et al., "The Design and Analysis of the Clock Distribution Network for a 1.2 GHz Alpha Microprocessor," *Digest of Technical Papers of the IEEE International Solid-State Circuits Conference*, 2001, pp. 402-403.

Chapter 4

High-Speed Logic, Circuits, Libraries and Layout

Andrew Chang, William J. Dally
Computer Systems Laboratory,
Stanford University

David Chinnery, Kurt Keutzer, Radu Zlatanovici
Department of Electrical Engineering and Computer Science,
University of California at Berkeley

1. INTRODUCTION

The quality of a design is greatly influenced by the characteristics of its constituent pieces. In both ASIC and custom chips, the organization of the logic, the choice of circuit styles, and the composition and layout of the cell libraries all affect the ultimate performance, density, energy efficiency and implementation effort. In Chapter 2, the impact of pipelining was discussed. In this chapter, we turn our attention to the impact of circuit styles, differences between datapath and random logic cell libraries, and variations between "standard" libraries. We identify key differences between the logic, circuits, libraries and layout of custom chips and ASICs. Then the impact of these differences is quantified.

This chapter is organized into eight sections. Section 2 defines a set of technology independent metrics useful in comparing ASIC and custom designs across processes. These metrics are applied to an example set of technologies and "flagship" designs. The third section analyzes the performance differences between custom chips and ASICs due to logic style, circuit design, libraries, and layout. The fourth section explores the causes of the area and density differences between custom and ASIC chips. The fifth section discusses the impact of circuit styles on energy efficiency. Section 6 discusses future trends and the impact of technology scaling. The final section summarizes the key points.

2. TECHNOLOGY INDEPENDENT METRICS

The quality of a design can be evaluated in many different ways. We focus on four criteria: performance, area efficiency, energy, and design effort. With these metrics, we can quantify the differences between the quality of custom and ASIC cells, and the design effort required to create them.

2.1 Performance – Fanout-of-4 Inverter Delay

As discussed previously in Chapter 2, Section 1.1, we adopt the fanout-of-4 inverter (FO4) delay, proposed by Horowitz et al. [29], as the technology independent performance metric. It is useful to have a common metric to enable the comparison of different designs. It can be challenging to sort and grade the performance of successive generations of fabrication processes from the same foundry and across different foundries. The capability of a process for digital systems is a function of four basic characteristics: the effective channel length (L_{eff}), the oxide thickness (t_{ox}), the nominal operating voltage, and the metal pitch rules. The first three parameters affect gate delay and energy, while the last affects achievable density. Foundries continuously tune their processes and optimize both the transistors (front-end-of-line – FEOL) and the metal pitch rules (back-end-of-line – BEOL) separately and concurrently. FEOL improvements usually lead BEOL updates. Therefore, within the lifetime of a major process generation, it is possible to have up to three successive improvements in transistors and two different sets of metal pitches. For example, in the 0.25um process generation, a foundry might start with 0.25um L_{eff} transistors and older 0.35um generation metal rules, progress to 0.25um metal rules and then provide two successive transistor shrinks to 0.18um L_{eff} and 0.13um L_{eff}. Designs benefit from significant improvements in both speed and density by migrating to each updated process point. A detailed discussion of the effects of process variations and operation conditions is provided in Chapter 5.

Designers can make quick initial estimates of achievable performance and design quality using the fanout-of-4 inverter delay metric. A rule of thumb can be used to approximate the FO4 delay. An FO4 delay is approximately half the process L_{eff} when picoseconds are substituted for nanometers. If, L_{eff} equals 0.5um (500nm) the corresponding FO4 delay is 250ps. Similarly, the FO4 delay for an L_{eff} of 0.13um is approximately 65ps. This estimate assumes operation at a process, voltage and temperature (PVT) point of typical n-channel MOSFET transistors, typical p-channel MOSFET transistors, 90% of the supply voltage V_{DD}, and 100°C. This PVT point is also referred to as TTLH (typical n-channel MOSFET, typical p-channel MOSFET, low voltage, high temperature).

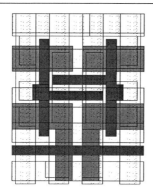

Figure 1. SRAM bit-cell 0.5um and 0.18um.

2.2 Area Efficiency – χ^2/T Cost

Efficient VLSI layout results from balancing the transistor density and the routing resource usage. Micron based distance and area metrics do not scale with technology. Historically, λ (one half the minimum device length) has been used as a normalized size metric [42]. However, beginning with 0.5um technology, the drawn (L_{drawn}) and effective (L_{eff}) transistor lengths have continued to diverge with each process generation. Also, the density and the performance of designs are increasingly limited by interconnect. Thus, the intermediate metal pitch more clearly reflects the constraints in modern design. We propose using the intermediate metal (M2) pitch (χ) as the technology independent measure of size and distance.

While the absolute size of gates shrinks with each successive improvement in technology, the χ^2 area for gates remains relatively constant over a wide range of process generations. For example, as shown in Figure 1, a six transistor SRAM bit cell in 0.5um CMOS, with χ equal to 1.8um, is approximately 58um². The same SRAM bit requires only 5.7um² in a 0.18um process, with χ equal to 0.56um. Note, in both cases, the ram bit requires $18\chi^2$.

Design	χ^2/transistor	1,000 gates/mm²
SRAM (array only)	3.0	285
SRAM (full)	3.3 - 6.0	143 - 260
full custom	6.0 - 8.0	107 - 143
ASIC (with regions)	11.0 - 16.0	50 - 70
ASIC (standard)	16.0 - 25.0	34 - 50

Table 1. Transistor cost and equivalent gates/mm² for a typical 0.18um process.

Using this size metric, we introduce *transistor cost* (with units of χ^2/T) as a technology independent VLSI density metric. Table 1 presents typical values of transistor cost and the resulting *realizable* gates/mm^2 in a 0.18um technology for a range of design styles. In this context, a "gate" is a static CMOS 2-input NAND. Also, as this is a *cost* metric, a *denser* design has a *lower* transistor cost.

2.3 Energy – E_{bit}, E_{nand}, E_{inv}

The energy required to switch a gate can be modeled as CV_{DD}^2, where C is the gate's capacitance. The dynamic power dissipation $P_{dynamic}$ is then calculated as this energy multiplied by the operating frequency f and an activity factor α:

(1) $P_{dynamic} = CV_{DD}^2 \alpha f$

We select three different gates as energy references: E_{bit}, E_{nand}, E_{inv}. The E_{bit} metric is the minimum energy necessary to write a 1 or a 0 into the SRAM bit-cell. Thus, it is a gauge of the minimum energy required to preserve state and serves as a reference point for flip-flops, latches, registers and memories. The E_{nand} metric is the minimum energy required to modify/transform state and thus is a reference point for combinational logic. The E_{inv} metric is defined as the energy required for switching a minimally sized inverter driving an FO4 load and is proportional to the energy needed to transfer a bit of state. It is useful in benchmarking the costs of moving data across blocks.

2.4 Measuring Design Effort – t_{unique}/wk

In 1998, an industry study examining predominantly ASIC flows, [13] reported a factor-of-14 difference in design productivity across 21 chip projects from 14 companies. However, the results of this study relied on estimating a statistical normalization factor to compensate for the high variation in complexity between the individual chips and projects. Through our own experiences in three successive semi-custom chip projects [17][18][37], we have only seen a 50% productivity difference within the same methodology between excellent designers and average designers. Based on our experiences in these projects, we have derived the *unique transistors per week* (t_{unique}/wk) metric. Average designers can produce 22 t_{unique}/wk, while excellent designers can produce 33 t_{unique}/week. This metric was derived historically from the design of large custom blocks. In contrast to the full chip results in [13], the t_{unique}/wk metric is most useful for module, block, and cell creation. Our definition of a "unique transistor" includes circuit design, schematic entry, timing and functional simulation, layout, backend verification, and integration overhead. Assuming the basic gate is a 2-input NAND composed of four unique transistors, our estimate for excellent and average designers corresponds to a productivity of eight gates and five gates per week respectively. Our results are comparable to the 2.7 gates/day reported in 1987 [49]. This indicates that very little progress has been made in improving the productivity of custom VLSI designers in the intervening 15 years.

Applying our metric to both a basic 2-input NAND gate and an 8-KB SRAM block yields the following: a superior designer can create and install a new 2-input NAND cell in under five hours, while the average designer takes over seven hours. On the larger example, consisting of 333 unique transistors organized as 22 distinct reusable cells, the superior designer takes 10 weeks from concept to verified layout while the average designer takes 17 weeks. In both cases the basic six transistor SRAM bit-cell was already provided as a building block. Note that a supplemental set of metrics needs to be applied to properly estimate global chip assembly effort and integration complexity.

2.5 Survey of Semiconductor Processes from 1995-2002

The continuous innovations in process technology fuel the growth of innovations in information technology. From 1995 through 2002, drawn transistor lengths shrunk from 0.5um to 0.13um and effective device lengths shrunk from 0.5um to 0.06um. Correspondingly, processor clock frequencies increased from 180MHz (HP PA-8000) to 2.2GHz (Intel Pentium 4). While most vendors present their process generations in nice quantum steps, within each major generation there can be multiple sub-generations. Table 2 summarizes the key features for a set of representative processes [7][10][23] [25][35][57][58][59][60][61][64].

The table shows six major process generations. The second column lists the advertised maximum number of gates per chip provided by an ASIC vendor. The common definition of gate is the four transistor 2-input static CMOS NAND gate. The next two columns list the FO4 delays in the process. The next two columns show the size of the average SRAM bit-cell for the process. The final five columns provide the absolute values for five technology independent metrics – FO4 delay, χ^2 SRAM bit area, E_{bit} E_{inv} E_{nand}. The predicted and simulated FO4 delays show strong correlation.

Technology Generation L_{drawn} (um)	Maximum Number of Gates (10^6)	Predicted FO4 Delays (ps)	Simulated FO4 Delays (ps)	SRAM bit cell area (um^2)	SRAM bit cell (χ^2)	E_{bit} (fJ)	E_{inv} (fJ)	E_{nand} (fJ)	L_{eff} Range (um)	Maximum Frequency of 20 FO4 Delays (MHz)
0.50	1.0	244	250	58.0	18	57.6	24.5	44.1	0.25-0.50	200-400
0.35	3.5	164	175	24.0	16	40.3	17.8	30.2	0.22-0.35	285-455
0.25	10.0	137	125	12.0	17	16.5	7.0	12.7	0.14-0.25	400-715
0.18	20.0	90	90	5.6	18	6.2	2.6	4.7	0.10-0.18	555-1000
0.15	32.0	75		4.6	17	4.2	1.6	3.0	0.11-0.15	665-910
0.13	40.0	58	65	2.1	18	2.3	1.0	1.8	0.06-0.13	770-1665

Table 2. Reference performance across processes.

Table 2 also presents the achievable system clock frequency assuming a logic depth of 20 fanout-of-4 delays per cycle. It is a useful reference for the following comparison between flagship products in each process generation.

2.6 Applying Metrics to Flagship Designs

The previous subsection provided a snapshot of the potential for each process generation. Table 3 compares high performance custom chips [4][15][22][27][44] and ASICs [12][14][40][48][50][51][63] fabricated in similar processes.

Several trends are discernable in Table 3. First, the data confirms the benefit of custom techniques in the design of high-end microprocessors. Also, it confirms that custom designs can meet or even exceed the reference clock rate in Table 2 and the maximum available gate capacity offered by ASIC vendors in Table 2. Recent microprocessors have been designed with fewer than 15 FO4 delays per cycle. The combination of lower χ^2/T, (leading to higher density) and large on-chip SRAMs (used for L1 through L3 caches) enables custom chips to exploit the large number of available transistors effectively. Also, the maximum size of custom chips is less constrained compared to ASICs. Custom designs are not required to fit into generic die images and sizes. Second, many "flagship" ASICs are not microprocessors. Instead, they tend to be graphics and communication chips whose primary purpose is to move high bandwidth data. Third, high quality ASIC designs vary in their ability to approach the gate counts advertised by foundries. Since many of these chips are not microprocessors, they incorporate fewer large SRAM arrays. The large number of transistors in the Sony PS2 chip is a notable exception. However, the Sony chip is an extreme design point in 0.18um, as most of its 280-million transistors are contained in embedded DRAM arrays.

With the exception of the Sony PS2 chip, the ASICs make far less use of the maximum chip area. The ASICs presented in Table 3 have much larger areas than the embedded ASIC processors presented in Table 1 of Chapter 1.

2.7 Summary of Metrics

In this section we have introduced four main technology independent metrics. The fanout-of-4 inverter (FO4) delay is a metric for performance, where an FO4 delay is estimated by converting half the effective channel length from nanometers to picoseconds. The transistor cost measures area efficiency, with a lower bound of 3 derived from SRAM bit-cells. E_{bit}, E_{nand}, E_{inv} are a set of reference points proportional to the energy required to store, transform and move one bit of data. t_{unique}/wk is a measure of designer productivity, with superior designers capable of creating 33 t_{unique}/wk in our experience.

Technology (um)	L_{eff} (um)	Custom Microprocessor	Clock Frequency (MHz)	Number of Transistors (10^6)	Area (mm^2)	L_{eff} (um)	ASIC	Clock Frequency (MHz)	Number of Transistors (10^6)	Area (mm^2)
0.50	0.28	HP PA8000	180	3.8	338	0.35	IBM PPC AS A10	77	4.7	213
0.35	0.25	Intel Pentium II	300	7.5	209	0.35	Infineon	57	2.4	97
0.25	0.14	HP PA8500	500	120.0	469	0.24	nVidia GeForce256	120	23.0	
						0.18	Lucent	200	63.0	221
0.18	0.15	Sun Ultra-III	600	23.0	242	0.16	Broadcom BCM5632	125	70.0	240
	0.10	Alpha 21264	1200	152.0	397	0.16	Sony PS2	150	287.0	462
	0.10	Intel McKinley	1000	221.0	464	0.13	nVidia GeForce2	200	25.0	270
						0.13	Broadcom BCM3351	200	25.0	
0.15	0.10	IBM Z900	1100	34.0	177	0.11	nVidia GeForce4	350	63.0	
						0.11	Fujitsu FRV-8way	533	10.4	61
0.13	0.06	Intel Pentium 4	2200	54.0	145	0.06	Broadcom BCM5424	125	16.0	

Table 3. Flagship custom and ASIC designs. The Sony PS2 had embedded DRAM, and thus a very large number of transistors on the chip.

3. PERFORMANCE PENALTIES IN ASIC DESIGNS FROM LOGIC STYLE, LOGIC DESIGN, CELL DESIGN, AND LAYOUT

In this section, we focus primarily on the impact of the circuit styles used to build custom chips and ASICs, and quantify the resulting variation in combinational logic delay (t_{comb} and $t_{total\ comb,average}$) between the two styles. Our analysis is applied in turn to four different implementation styles, two custom, one semi-custom, and the ASIC approach. Figure 2 provides an overview of example circuit styles in each category. The full-custom styles are single-rail and differential domino logic, and differential static logic. The semi-custom style employs structured single-ended static logic. The ASIC style employs single-ended static CMOS logic.

Domino circuits are clocked logic elements. All domino circuits are initialized (pre-charged) to a known logic state by a clocking signal (*precharge*) before the function *f* is evaluated. They require monotonic inputs. After initialization, they generate monotonic outputs. The single-ended domino logic family is incomplete, since it can not perform logical inversions. Differential domino logic families provide both the true and complement outputs but require additional transistors to implement the complementary function. Differential static circuits, as exemplified by the DCVSL, do not need to be pre-charged. Also, their cross-coupled outputs improve noise immunity by providing a built-in latch. However, overcoming the hysteresis of this latch incurs a delay penalty and can reduce performance. Differential domino and differential static logic also offer the advantage of implicit inversion. Finally, while both these styles employ extra transistors to implement the complementary logic function, no parasitic p-channel MOSFETs are required.

Standard static CMOS is a robust logic style with good noise immunity. Also, both the synthesis and functional testing of static CMOS logic is well understood. Static CMOS gates employ p-channel MOSFET pull-up and n-channel MOSFET pull-down trees to evaluate the function *f*. The performance of wide NOR functions in standard CMOS logic is slow, as long p-channel MOSFET pull-up chains are required. The performance difference between standard static CMOS and custom circuit styles derives from the restriction of logic depth imposed by the need for PMOS pull-up trees and the substantial additional parasitic loading imposed by the PMOS trees.

High-Speed Logic, Circuits, Libraries and Layout

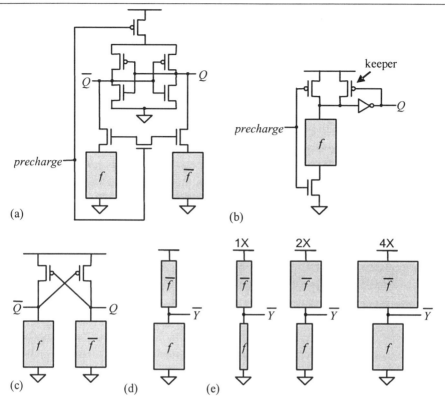

Figure 2. An overview of four logic styles implementing the function *f*. The output *Y* or *Q* evaluates to true (V_{DD}) if *f* evaluates to true (i.e. the network implementing *f* conducts). (a) shows differential domino logic. (b) shows domino logic, which is more compact, but needs a logical inversion. Both domino logic styles need a keeper or restoring logic to ensure that they are not subject to noise on the floating precharged node. The circuit in (c) is differential cascode voltage switch logic (DCVSL). (d) and (e) show static CMOS logic, where (e) shows a library with three drive strengths (1X, 2X, and 4X), and (d) shows a single cell that has been custom sized for a specific design, with a 1X pull-up network and 4X pull-down network.

One of the most obvious differences between custom and ASIC designs is the speed, as indicated by cycle time and corresponding clock frequency. To frame the analysis of the causes of this difference, it is useful to start with basic equations for the cycle time derived in Chapter 3 for both flip-flop based designs ($T_{flip-flops}$) and latch based designs ($T_{latches}$).

(2) $\quad T_{flip-flops} = \max\{t_{CQ} + t_{su} + t_{comb} + t_{sk} + t_j\}$

(3) $\quad T_{latches} \geq 2t_{DQ} + t_{total\ comb,average} + t_j$

(4) $\quad t_{total\ comb,average} = 2t_{comb,average}$

where $t_{comb,average}$ is the average logic delay per phase with active high and active low latches partitioning combinational logic.

The cycle time is composed of three parts: the *register overhead*, either $t_{CQ} + t_{su}$ for flip-flops or $2t_{DQ}$ for latches; the *timing margin*, $t_{sk} + t_j$ for flip-flop-based designs or just t_j for latch-based designs; and the *combinational logic delay* within each cycle, either a fixed t_{comb} for flip-flop-based designs or a $t_{total\ comb,average}$ for latch-based designs (as time borrowing is possible between cycles). In the remainder of this section we explore the difference in combinational logic delay between implementation styles. We also expand on the analysis of register overhead and timing margin from Chapter 3. In Chapter 3, the difference in timing overhead between custom chips and ASICs was found to be about 6 FO4 delays for flip-flop-based ASICs (assuming balanced pipeline stages), and 2-4 FO4 delays for latch-based ASICs. Also in Chapter 3, the difference in typical timing margins between the two styles was found to be 3 FO4 delays. In Section 3, we discuss how these three components of cycle time grow in each implementation style.

3.1 Comparing Custom and ASIC Implementations of the Same Logic

A custom differential domino design is the reference point for our analysis of logic delay. The high performance microprocessors in the last five technology generations have employed custom techniques, single-phase clock, and latched based timing to achieve cycle times equal to or less than 20 FO4 delays. We now further divide the 20 FO4 delays amongst the three categories: register overhead, timing margin, and logic delay.

Historically, high performance designs have allocated between 8%-15% of their cycle time to the latching elements (3 FO4 delays) [3][27]. Previous work [53] confirms this for several of the commonly employed latches. Clock skew and jitter account for 5-10% of the cycle time, which correspond to 1 FO4 delay. This leaves 16 FO4 delays for the logic within each stage. In summary, for custom domino logic, register overhead equals 3 FO4 delays, timing margin equals 1 FO4 delay and logic equals to 16 FO4 delays. This results in a cycle time of 20 FO4 delays.

Using this starting point we now examine other implementations. While accounting for the impact on the timing overhead, this chapter focuses on the performance difference between custom and ASIC combinational logic. A primary limitation is the choice of logic style, which we examine first. Additionally, the quality of the cells composing the logic and the quality of the layout affect the speed.

3.1.1 Different Logic Styles from Dynamic to Static Logic

Custom differential static circuit techniques [6] such as DCVSL, DSL, and CNTL are high speed alternatives to domino logic. Like domino logic, custom differential static techniques allow the compression of the number of required logic stages by facilitating the deep and dense logic structures. Also, they minimize the effect of the parasitic p-channel MOSFETs, and remove the overhead of inverters, as both true and complement versions of the signal are always available. Domino logic is, however, generally 30% to 40% faster than differential static logic styles such as DCVSL due to the elimination of the fight to overpower the cross coupled p-channel MOSFETs. To illustrate this, five circuit-implementations [30] of the 1-bit full adder function are compared in Table 4. The single ended static CMOS version is over a factor of two slower than the domino version. The fastest static design, a hybrid DCVSL circuit (HDCVSL), is still 30% slower. This data correlates well with the results reported in [1][2][66].

Accounting for the 30% to 40% slow down for implementing the same logic function in DVCSL expands the original 16 FO4 delays to 21 FO4 delays. The latches available in this technique are the same as in the previous so there is no growth in register overhead. The timing margin is also the same as the first case and equals 1 FO4 delay. In summary, our model for custom differential static logic yields register overhead equal to 3 FO4 delays, timing margin equal to 1 FO4 delay, and logic delay equal to 21 FO4 delays. This results in a cycle time of 25 FO4 delays (an increase of 25% due exclusively to the "fight" in the DCVSL latch).

Circuit Style	Delay (FO4s)
Domino Logic	2.2
Hybrid DCVSL	3.0
DCVSL	4.0
CPL	4.6
Static	4.8

Table 4. Relative performance of a 1-bit full adder for five circuit styles. CPL is complementary pass-transistor logic.

Semi-custom single-ended static techniques include detailed floorplanning, detailed placement regions, matched drivers and loads, manual clock tuning, and the creation of design specific cells as needed. Despite the semi-custom approach to placement, routing and circuits, all three timing parameters degrade when the transition is made from custom differential static to single-ended static. Semi-custom static approaches do not achieve the same timing margins as custom designs, so this parameter increases by 1 FO4 delay. In addition, structured static approaches use more conservative latches requiring 2 or more FO4 delays *each*. Finally, singled-ended static requires the complementary pull up p-channel MOSFET tree for logic, and thus, does not allow the efficient compression of logic stages through the use of only deep and wide n-channel MOSFET pull down trees. Also, it does not provide the "free" inversion of logic. Taken together these can result in 25% degradation in performance of the logic, resulting in an additional 5 FO4 delays. In summary, our timing model for semi-custom single-ended static logic yields register overhead equal to 2×2.0 FO4 delays, timing margin equal to 2 FO4 delays and logic delay equal to 26 FO4 delays. This gives a cycle time of 32 FO4 delays.

3.1.2 The Additional Effects of Logic Design, Cell Design and Layout

A large gap still remains between what is achievable with structured static single-ended logic and what occurs in practice with most ASIC designs. The cycle times of most ASIC designs are between 40 to 65 FO4 delays (e.g. see Chapter 2, Table 2). This discrepancy of up to about 33 FO4 delays from semi-custom static is the result of several factors.

ASIC designs use flip-flops instead of latches. Flip-flops are commonly implemented as back-to-back latches and are an additional 2-3 FO4 delays slower than the pair of latches used in custom and semi-custom designs. Flip-flops do not allow slack passing, costing up to another 4 FO4 delays as the clock period is limited by the delay of the worst pipeline stage – suffering a penalty for an unbalanced pipeline. The time-borrowing and slack passing allowed by latches gives typically a 10% (4 FO4 delays) further improvement in delay (see Chapter 3, Section 6.2.1).

The larger device sizes, higher fan-out, parasitic loading due to added scan devices, and the inability to fold-in logic all contribute to the slower registers. Also, ASIC designs have much coarser control over clock routing, which is reflected by an additional 2 FO4 delay penalty for timing margin.

The remaining 25 FO4 delays comes from differences in the design and layout of the combinational logic. The logic design may not be implemented optimally for speed. For example, faster carry-select logic might not be used to speed up addition, because of the additional area needed [62]. The structure of the design may not be exploited in the placement and routing (e.g. poor layout of bit-sliced logic), which increases the area and parasitic wire capacitance. Mismatched cell drive strengths and loads increases the delay and power consumption. Inaccurate wire load models can cause incorrect sizing and too many or too few buffers before a load. Additional levels of logic may be required to implement complex functions due to the lack of custom cells.

3.1.3 Summary of the Comparison

Table 5 summarizes the approximate clock periods for different logic and design styles, breaking this down into the components of the clock period.

Dynamic logic functions used in the IBM 1.0 GHz design are 50% to 100% faster than static CMOS combinational logic with the same functionality [47][Kevin Nowka, personal communication]. Thus for a complete design, custom domino combinational logic is about a factor of 1.5× faster than semi-custom static combinational logic – the factor for differences in logic style. Taking into account the timing overhead, this is a factor of up to 1.4× the clock period.

Cell design and layout issues account for the remaining 2× difference between the delay of combinational logic in semi-custom static and mediocre static ASIC implementations.

Logic Style, Logic Design, Cell Design, and Layout	Total Critical Path Delay	Register Overhead	Timing Margin	Slack Passing	Combinational Logic Delay
Custom Domino	20	3	1		16
Custom Static (DCVSL)	25	3	1		21
Semi-Custom Static	32	4	2		26
ASIC Static (Good)	40	6	4	4	26
ASIC Static (Poor)	65	6	4	4	51

Table 5. Summary of the timing margin, register overhead, and combinational logic delay for the implementation styles discussed above.

	Technologies (um)											
	0.80		0.50		0.35		0.25		0.18		0.13	
	Fast	Slow	Fast	Slow	Fast	Slow	Fast	Slow	Fast	Slow	Fast	Slow
1 FO4 (ps)	250	400	120	250	110	175	70	125	50	90	30	65
	Clock Rates (MHz)											
Custom Domino	200	125	417	200	455	286	714	400	1000	556	1667	769
Custom Static	154	96	321	154	350	220	**549**	308	769	427	1282	592
Semi-Custom Static	118	74	245	118	267	168	420	235	**588**	327	980	452
ASIC Static (Good)	100	63	208	100	227	143	357	200	500	278	833	385
ASIC Static (Poor)	67	42	139	67	152	95	238	133	333	185	**556**	256

Table 6. Corresponding expected performance ranges for different implementation styles that were listed in Table 5.

Table 6 summarizes the resulting calculated performance of the chips designed with each of the four styles. The data demonstrates two key points. The performance variations within a process generation and across implementation styles are both substantial. Aggressive circuit techniques can compensate for process differences and, similarly, aggressive process choices can compensate for implementation choices

The differences in performance between custom domino (20 FO4 delays) and static CMOS ASIC (40-65 FO4 delays) implementations are summarized in Section 3.7. The timing overhead (register overhead, timing margin, and capability for slack passing) accounts for about 10 FO4 delays. Circuit style accounts for 10 FO4 delays and exploitation of design structure accounts for the remaining 25 FO4 delays, which results in the factor of 2 range in the combinational logic delay.

A process generation step can significantly increase the effectiveness of a particular design style by reducing cycle time (typically by about 30%). This corresponds to a 10-15 FO4 delay reduction for static single-ended implementations. For example, an average ASIC design fabricated in a third generation 0.13um technology (0.06um L_{eff}) could perform comparably to a full-custom domino design with a 0.18um L_{eff} (i.e. a first generation 0.18um process, a second generation 0.25um process or a third generation 0.35um (0.22um L_{eff}) technology. However, the poorer ASIC density, due to the factor of 3× worse χ^2/T transistor cost, results in a larger 0.13um die size than a 0.25um custom design. This is despite the factor of two density advantage of the 0.18um metal pitch rules.

3.2 Comparison of Dynamic and Static 64-Bit Adders

Table 4 compared full adder cells, and showed that domino logic was more than factor of 2× faster than static logic for a single bit full adder cell. However, a larger circuit should be considered to have a more accurate comparison between dynamic and static logic.

We will compare dynamic domino logic and static logic style implementations of a high speed adder. We present the high-speed carry-lookahead style used in many adders, and the Ling adder variant of the carry-lookahead. The static and dynamic circuits were both Ling adders.

3.2.1 Logic Design of a 64 Bit Adder

Summing two numbers A and B of 64 bits, the ith output bit Sum_i is:

(5) $Sum_i = A_i \oplus B_i \oplus C_{i-1}$

where C_{i-1} is the carry from the $(i-1)$ bit. The ith carry is given by

(6) $C_i = A_i B_i + A_i C_{i-1} + B_i C_{i-1}$

The critical path through the adder is the carry chain [62]. Carry-lookahead adders speed up the calculation by compressing the logic. The carry is calculated from local generate g_i and propagate p_i signals [20]:

(7) $g_i = A_i B_i$

(8) $p_i = A_i + B_i$

g_i determines if a carry is generated at the ith bit from the ith bits of A and B. The propagate term p_i determines if a carry from the previous bit will propagate. This gives grouped generate and propagate signals [20]:

(9) $G_i = g_i + p_i G_{i-1}$

(10) $P_i = p_i P_{i-1}$

The sum is given by

(11) $Sum_i = p_i \oplus G_{i-1}$

For four bits, Equations (9) and (10) can be expanded to

(12) $G_{3:0} = g_3 + p_3(g_2 + p_2(g_1 + p_1 g_0))$

(13) $P_{3:0} = p_3 p_2 p_1 p_0$

With these four stages of lookahead, the carry for the fourth bit (subscript 3) is [45]

(14) $C_3 = G_{3:0} + P_{3:0}C_{in}$

where C_{in} is the input carry to this set of four bits of the adder, and bit 3 is the most significant bit. It is common for adders to be implemented with four bit (radix 4) carry-lookahead blocks, by grouping generate and propagate signals into larger blocks [45]:

(15) $G_{15:0} = G_{15:12} + P_{15:12}(G_{11:8} + P_{11:8}(G_{7:4} + P_{7:4}G_{3:0}))$

(16) $P_{15:0} = P_{15:12}P_{11:8}P_{7:4}P_{3:0}$

(17) $C_{15} = G_{15:0} + P_{15:0}C_{in}$

(18) $G_{63:0} = G_{63:48} + P_{63:48}(G_{47:32} + P_{47:32}(G_{31:16} + P_{31:16}G_{15:0}))$

(19) $P_{63:0} = P_{63:48}P_{47:32}P_{31:16}P_{15:0}$

(20) $C_{63} = G_{63:0} + P_{63:0}C_{in}$

The other $G_{j+15:j}$ and $P_{j+15:j}$ signals are generated in the same manner as in Equations (15) and (16). This approach results in a delay on the order of $\log_4 k$ for a k-bit adder. A 64 bit adder with four bit carry lookahead requires $\log_4 64 = 3$ levels of generates ($G_{j+3:j}$, $G_{j+15:j}$, and $G_{63:0}$).

A more compact version of the carry-lookahead adder is the Ling adder [41]. The same local generate and propagate signals are used, but instead the sum of two "global" generates, H_i, is calculated:

(21) $H_i = G_i + G_{i-1} = g_i + p_{i-1}H_{i-1}$

The logic to compute H_i is more compact, e.g. H_3 is given by

(22) $H_{3:0} = g_3 + g_2 + p_2(g_1 + p_1g_0)$

However, calculating the sum is more complicated [20]:

(23) $Sum_i = p_i \oplus H_i + g_i p_{i-1} H_{i-1}$

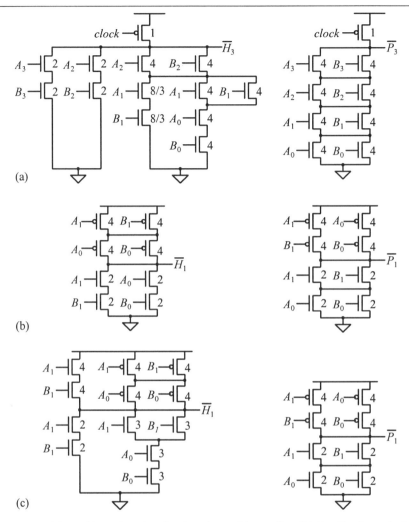

Figure 3. (a) Radix-4 domino logic Ling adder cells. (b) Radix-2 static logic Ling adder cells. (c) Radix-2 static logic carry-lookahead adder cells, which are slower than (b). Relative widths of transistors are shown. In these diagrams, the static CMOS logic is sized so that pull up and pull down drive strengths are the same, but the fastest static implementation skews the drive strengths.

The H and P signals can be grouped in a similar manner to the carry lookahead adder in Equations (15) to (20). The circuitry to implement H_3 and P_3 in a domino logic style is shown in Figure 3(a). This implements the functions:

(24) $H_{3:0} = a_3 b_3 + a_2 b_2 + (a_2 + b_2)(a_1 b_1 + (a_1 + b_1) a_0 b_0)$

(25) $P_{3:0} = (a_3 + b_3)(a_2 + b_2)(a_1 + b_1)(a_0 + b_0)$

It is not possible to implement the same circuitry efficiently in static logic, because of the large pull-up network. Instead, the static logic design uses a radix-2 encoding. Figure 3(b) shows the radix-2 static logic for:

(26) $H_{1:0} = a_1 b_1 + a_0 b_0$

(27) $P_{1:0} = (a_1 + b_1)(a_0 + b_0)$

Figure 3(c) shows the static implementation of the traditional carry-lookahead generate. Compared to H_1 in Equation (26), G_1 is

(28) $G_{1:0} = a_1 b_1 + (a_1 + b_1) a_0 b_0$

3.2.2 Comparison of Adder Logic Designs and Logic Styles

The input capacitance for A and B inputs are the same. For the static logic Ling adder in Figure 3(b), the $A_{1:0}$ inputs drive a load of 24 units. In the static logic carry-lookahead adder in Figure 3(c), the $A_{1:0}$ inputs drive a load of 32 units. The Ling adder implementation in Figure 3(b) loads the inputs less, and the cell calculating H is smaller.

For the domino logic Ling adder in Figure 3(a), the $A_{3:0}$ inputs drive a load of 34.7 units of transistor width. The domino is radix-4 and calculates $P_{3:0}$ and $H_{3:0}$, whereas the static logic only calculates $H_{1:0}$ and $P_{1:0}$. To calculate $H_{3:0}$ and $P_{3:0}$, requires calculation of $H_{3:2}$ and $P_{3:2}$ with cells identical to those in Figure 3(b), and then $H_{3:0}$ and $P_{3:0}$ need to be calculated. Thus the inputs $A_{3:0}$ to the static logic Ling adder drive a total load of 48 units of transistor width, and there needs to be an additional level of logic to calculate $H_{3:0}$ and $P_{3:0}$ from $H_{1:0}$, $P_{1:0}$, $H_{3:2}$ and $P_{3:2}$.

The typical P:N ratio used in static logic is 2:1. The load presented by the logic in Figure 3(b) can be reduced by reducing the widths of transistors in the gate and skewing the drive strength of the cell. Skewing the drive strengths improved the speed by 25%, by allowing the input capacitance of the cells to be reduced. The cells with skewed drive strength are also smaller and have lower power consumption. The drive strengths ratios used in the fastest radix-2 Ling adder implementation were:
- 1:1 for the first logic level (half the PMOS widths), generating $\overline{H}$ and $\overline{P}$.
- 2:0.5 for the second logic level (half the NMOS widths) that calculates the $H_{j+3,j}$ and $P_{j+3,j}$ terms.
- Then alternating 1:1 and 2:0.5 for the remaining logic levels.

These skewed ratios reduce the load of 24 units to 16 units for 1:1, and to 20 units for the 2:0.5 level. The logic is skewed to optimize a pulse of the carry of 1 transition through the logic. There were a total of $\log_2 64 = 6$ levels of generate and propagate calculation.

The domino and static logic adders were implemented in 0.18um with Kogge-Stone carry trees [39]. The Kogge-Stone tree generates the carry signals from blocks generating the grouped H and P signals. The sum is calculated after the carry tree with a sum select block. The 64 bit radix-4 Ling adder implemented with domino logic has a delay of about 9.5 FO4 delays. The domino gates include keepers to avoid leakage draining the stored charge and inverters at the outputs. The timing analysis accounts for wire capacitance and resistance.

In comparison, the radix-2 Ling adder with static logic circuitry has a delay of 11.5 FO4 delays, when the pull-up and pull-down drive strengths are skewed optimally. These calculations assume the domino logic and static logic adders were driving the same load. The domino logic implementation is about 20% faster. Note that the static implementation uses *pulsed static logic*, where the inputs must return to zero before the next calculation. Pulsed static logic is not used in traditional ASIC designs.

There are several significant points in this comparison. Firstly, the choice of logic style changes which logic design is best: radix-4 for domino and radix-2 for static logic. This is because domino logic can implement more complex logic in a compact cell, with less loading of the inputs, which reduces the number of logic levels. Secondly, cells with skewed drive strength can substantially (25% in this case) speed up a design, but cells with skewed drive strength are not typically available in a standard cell library. Thirdly, the logic design affects the speed, area, and power. For example, a carry look-ahead adder would have larger cells than the Ling adder, and these cells are slower, with the larger input capacitance increasing the power consumption.

3.3 A Comparison of Skew Tolerant Domino Logic with Synthesized Static Logic

A comparison between custom domino logic and synthesized static logic implementations of a 32-bit execution unit in 0.18um showed that the custom domino logic was at least 2.1× faster [34]. The execution unit included a register file, adder, shifter, Boolean operations, and multiplier. Both the domino and static logic were placed and routed using Physical Compiler and Silicon Ensemble.

The skew tolerant domino logic used a four phase clocking scheme with semi-custom clock trees. This allows time borrowing to balance the pipeline stages. The skew tolerant domino logic style is less affected by clock skew, so the additional skew from only using automatically generated clock trees was not a substantial penalty [Razak Hossain, personal communication]. The static logic used flip-flops, so no slack passing was possible in the static implementation. The four phase clocking scheme with domino logic substantially reduced the impact of clock skew and setup time on the timing overhead. Overall, the clocking scheme for the domino logic contributed a factor of 1.2× of the improvement for the custom logic [34].

The remaining factor of 1.8× between the implementations is the difference between using domino logic or static logic, and custom-designed logic or synthesized logic. We have estimated that domino logic can be as much as 1.5× faster than static combinational logic, or about a factor of 1.4× considering the combinational logic and timing overhead. Thus custom sizing, manual choices of the cells, and custom logic design contributed about a factor of 1.3× (1.8/1.4) to the performance difference. The majority of this factor of 1.3× was due to the logic design of the modules in the execution unit. The logic was hand-designed for the domino implementation, whereas the static logic was synthesized from RTL [Razak Hossain, personal communication]. There is no difference due to layout, as both implementations were place-and-routed in the same manner.

3.4 Logic Design

Poor logic design can severely penalize a design. A ripple carry adder would be much slower than a carry-lookahead adder. We are trying to quantify how much faster good logic design for a module would be compared to a typical synthesized ASIC design.

In Sections 3.2 and 3.3 above we have discussed the impact of logic design, along with the choice of logic style and cells used. The execution unit discussed in Section 3.3 gained a factor of about 1.3× by custom crafting of the logic design. In Chapter 12, Section 4.4, the wrong path in a 24 bit adder was optimized in its initial version. When this was fixed, the adder speed improved by 25%.

It has been shown that synthesizing arithmetic circuits with carry-save adder logic design optimizations can increase the speed by up to 30% [56]. This optimization was automated.

These comparisons suggest that good logic design may improve the performance by 20% to 30%. Thus we attribute a factor of up to 1.3× between a typical ASIC and a high performance design. Automated approaches have been used to improve logic design, so a good ASIC should be able to avoid this speed penalty (1.0× for custom vs. an ASIC using logic design optimization techniques).

3.5 The Performance Impact of Cell Libraries

While custom designs and ASIC designs are both built from libraries of components, key differences exist in the composition of these libraries that ultimately affect design quality. Custom libraries possess eight main advantages. The cells are designed for specific and limited applications. The cells encompass greater functionality and can consist of hundreds or even thousands of transistors. Custom cells can employ a wide range of logic styles including domino, DCVSL, and CPL. Few restrictions are placed on the layout templates for each cell. The cell heights are taller, generally between 12χ and 18χ. allowing more access to the ports within a cell and allowing more routing to pass through the cell. The drive strengths of cells are designed to match their intended loads. Individual transistor sizes can be tuned to maximize performance. Also, the P:N ratios for gates can be individually tuned to accelerate preferential edge transitions. As already discussed each of these features directly impacts performance, area, energy and effort.

ASIC designers have attempted to close the gap with custom chips by adding an increasing number of "custom" features to their cell libraries and their implementation flows. New libraries are increasingly similar to custom libraries as the number of different drive strengths, the range of additional functionality, the use of device tapering, and P:N ratio tuning [46] become increasingly prevalent. However, ASIC libraries still have not adopted aggressive circuit techniques due to the limited support in synthesis tools and possible restrictions in the available functional test coverage.

3.5.1 Performance of Static CMOS Standard Cell Libraries

One element of the performance degradation of ASIC designs, relative to custom, is due to the limits of the ability of ASIC libraries to approximate custom designs in their transistor-level design and in their transistor sizing. ASIC cells typically include design guard banding, such as buffering flip-flops, which introduce overhead. More fundamentally, the discrete transistor sizes of a library only approximate the continuous transistor sizing of a custom design. With a rich library of sizes, the performance impact of discrete sizes may be 2% to 7% or less [26][28].

Many ASICs standard cell libraries do not have enough drive strengths for each gate, and sometimes only one polarity of a gate is available (e.g. 4-input NAND, but no 4-input AND cell). While buffers will help drive loads, a gate with more drive strength is more compact than using a lower drive strength gate and a buffer to drive a load. More compact cells are faster, because they reduce wire lengths. A cell library with only two drive strengths may be 25% slower than an ASIC library with a rich selection of drive strengths and buffer sizes, as well as dual polarities for functions (gates with and without negated output) [52]. A richer library also reduces circuit area [38].

Transistors can be down-sized to reduce power consumption on paths that are not critical. On critical paths, the transistors of gates should be sized optimally to meet speed requirements. By reducing the size of transistors in gates that are not on the critical path, the load on other gates is reduced. Thus the speed is increased as well, because some gates on the critical path have to drive less load capacitance. Sizing the transistors optimally can make a speed difference of 20% or more [21].

If late arriving signals are closer to the gate output, the speed increases as less capacitance needs to be (dis)charged. In liquid cells, transistors can be rearranged within the gate to place late arriving signals closer to the gate output. Resynthesis using liquid cells can move transistors to maximize the adjacent drains and sources for diffusion sharing [32], reducing the size of cells. Iterative transistor resizing and resynthesis can improve speeds by 20% [24]. In the future, tools for wire sizing along with transistor sizing may be available. While buffers can be inserted and gate drive strengths increased to drive larger wires, wire sizing can give a 5% speed improvement [11].

The performance of ASIC libraries for the same process varies. In our experience, designs synthesized with different high speed libraries for the same process may vary in speed by up to 10%. These libraries were process-specific, and not optimized for the designs.

Another library impact on performance is using different cells to implement logic with the same library. For example, scan flip-flops are needed for verifying fabricated chips. Scan flip-flops are slower than D-type flip-flops, because of the additional logic needed to set the value of the flip-flop when testing.

Sometimes some library cells need to be excluded from synthesis (by setting a "set_dont_use" attribute), because they are slower. Synthesis tools may choose to use slower cells, because they are lower power or smaller area, but it turns out that they slow down the critical path.

Synthesis tools will choose the largest drive strength of a cell to drive a large fanout load. In the layout, the wire loads may be more than predicted by the wire load models. The load is larger, but it is not possible to further increase the drive strength of the cell. Buffer insertion may be difficult post-layout. It may be better to synthesize without making the largest drive strength of the cell available, and buffers will be inserted to help drive the capacitance. Then the largest drive strength cell can be used in layout if it turns out that the load is larger. This is advantageous, as the cell with larger drive strength will have a similar footprint to the cell it replaces, and the layout will not be perturbed.

3.5.2 Design-Specific Libraries

Many designs have logic functions that can be better implemented with a cell that is not available in the standard cell library. For example, the latch-based Xtensa results would gain further improvement compared to flip-flop net lists, if a drive strength X8 latch was added to the library (see discussion in Section 4.1 of Chapter 7). The flip-flop net lists could use drive strength X8 flip-flops for registers with large fanout, but there was no equivalent X8 latch available. As discussed in Section 3.2.2, the static implementation of the 64-bit adder was 25% faster by using cells with skewed drive strengths.

The STMicroelectronics iCORE gained a 20% speed improvement by using a library created specifically for this design. This was in comparison to a large generic standard cell library, which had 600 cells. The design-specific library took about 25 man-months to create – see Section 3.4 of Chapter 16 for more details of their optimizations.

After layout, it is also possible to improve the speed of a design by changing the cells while maintaining the same footprint. A prototype flow implemented by Cadabra was able to increase the speed of a bus controller by 13.5%, and the power was reduced by 18%. This optimization created new cells based on the real loads seen in situ in the layout, with accurate knowledge of the parasitics. Cells with skewed drive strengths were created, optimizing the falling or rising transition time as needed. In total, 300 new cells were generated, adding more drive strengths to the original library. Chapter 9 discusses this in more detail.

Zenasis Technologies has developed a tool that generates complex static CMOS cells to speed up a design. They report a 10% to 20% speed improvement for moderate sized blocks of industrial circuits. Chapter 10 examines their flow and results.

3.5.3 Summary of the Impact of Cell Design and Wire Sizing

Optimizing wire sizes is not available in current ASIC synthesis tools. Wire sizing contributes a factor of 1.05×. Creating design specific libraries or including cells optimized for the design can give up to ×1.25 speed improvement. Creating new cells and characterizing these does take time, and this is not part of the typical automated ASIC flow.

Together, wire sizing and design specific cells or libraries contribute a factor of 1.30× between custom static logic and logic implemented using gates from a generic standard cell library. In addition, the speed of high performance libraries for the same process varies by 10%, so overall there can be a factor of 1.45× between custom and generic standard cell implementations.

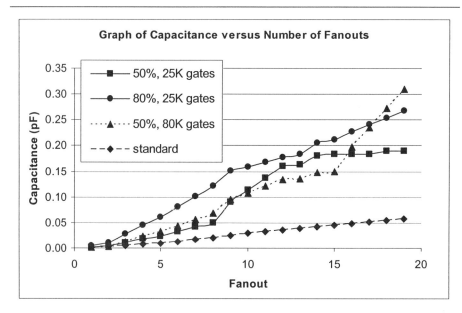

Figure 4. This graph shows wire load models. The standard wire load model is a default wire load model that comes with the standard cell library. The 50% wire load models show the capacitance for which 50% of all wires have less load – the 50th "percentile" wire load model for the respective designs of 25K and 80K gates. Also shown is the 80th percentile wire load model for the 25K gate design.

3.6 Better Performance with Better Layout

A typical ASIC flow does synthesis and place-and-route separately. Synthesis needs to take into account the expected fanout loads of gates, so that the gates can be sized correctly and buffers inserted, where necessary, to drive large loads. The simplest approach in an ASIC flow is to use wire load models to estimate the load for a given fanout. Figure 4 shows some wire load models.

3.6.1 Wire Load Models

The typical load capacitance expected is specified by the wire load model. Standard cell libraries come with a "standard" wire load model, but it is recommended that a wire load model is constructed for the specific design. Once the design has been place-and-routed, capacitances can be extracted from the layout, and wire load models can be constructed. Then the design can be synthesized with the design-specific wire load models, which should improve the results after place-and-route. Wire load models for a design are specified by a percentile – that percentage of gates with a certain fanout have load less than the capacitance given by the wire load model. A larger percentile is more conservative, with a larger capacitance for a given fanout.

Figure 4 shows the wire load models for two designs. The first design is a small embedded processor of 25,000 gates. The 80,000 gate design is a larger processor with a similar architecture. The standard wire load model is also shown. The standard wire load model seriously underestimates the typical load seen in both designs. While the two designs have similar wire load models, the wire load models are different.

Synthesizing with a very conservative wire load model (e.g. 90%) results in oversized gates and too many buffers. The size of the design increases and the performance is worse. The additional capacitance of the larger gates increases the power consumption and slows down the gates attempting to drive the fanout. The wire lengths increase because the design is larger, which also increases the loading. It is also bad to underestimate the wire loads, as then the gates are too small, and it is difficult to fix this in the layout step of the ASIC flow.

Typically, the correct wire load model to use is a wire load model characterized from the layout of the design, with a percentile around 50% to 60%. The best percentile to use depends on the design, and some experimentation is required.

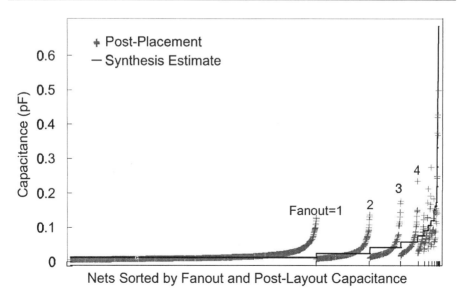

Figure 5. Error in wire load estimation of a 2,000 gate design in 0.5um. (Figure courtesy of Hema Kapadia and Mark Horowitz [36].)

Using a characterized wire load model gives a worse clock period after synthesis, because of the larger fanout capacitances, but the layout results improve because the gates are better sized to drive the loads. Use of the characterized wire load model reduced the clock period by 12% to 14% for designs we examined, in an ASIC flow with Design Compiler for synthesis and Silicon Ensemble for place-and-route.

Using the same wire load model for long nets with large fanout (e.g. global inter-block wires) is not a good approach. Figure 5 shows the inaccuracy of a wire load model for a 2,000 gate design in 0.5um technology [36]. Unfortunately, the typical ASIC flow methodology uses one wire load model and does not model these nets with larger fanout correctly. Possible solutions are partitioning the design hierarchically with separate wire load models for intra-block and inter-block wires, or performing resynthesis and sizing optimization while doing the floorplan and global routing. Partitioning the design into small blocks can increase the area somewhat, as optimization of gates that connect between blocks may not be allowed. Thus it is not advantageous to partition the design into very small blocks of gates. [36].

In the deeper submicron, the load presented by inter-module wires is increasing relative to the load of gates. The wires are taller to reduce their resistance, but this increases cross-talk capacitance. Inductance IR drops and transmission line effects also need to be considered [33]. In earlier generations, wires were only a small portion of the load, and inaccuracies of wire load models could be tolerated, but this is no longer the case.

Technology Generation (um)	Number of Transistors (10^6)	Partition Size (Thousands of Gates)					Speed Increase
		400	200	100	50	25	
		Clock Frequency (MHz)					
0.25	10	295	315	328	342	357	21%
0.18	20	424	449	471	498	518	22%
0.13	40	565	590	620	650	674	19%

Table 7. ASIC clock frequencies for different partition sizes. Clock frequencies were estimated using BACPAC [55]. Parameters typical of ASICs specified were 40% of the transistors were memory, Rent's rule exponent of 0.65 and logic depth of 30. All other parameters were the default for the technology generation. The speed increase is how much faster the design with 25K gate partitions is compared to a design with 400K gate partitioning.

3.6.2 Floorplanning and Global Routing

Better floorplanning can further improve the speed after final placement. Partitioning breaks up the design into smaller blocks, and more accurate wire load models can be developed for each block.

Sylvester et al. proposed a model for the impact of partitioning on interconnect delay and the clock period [54]. Using this model, we compared partitioning designs into partitions of between 25,000 gates and 400,000 gates. Larger ASICs were modeled in more advanced technology generations, to represent increasing integration of system-on-chip designs. The corresponding chip sizes were around $1cm^2$. From Table 7, a design partitioned into blocks of 25K gates is about 20% faster than a chip with 400K gate partitions. Partitioning the design into smaller blocks reduces the number of inter-block wires and the use of limited global routing resources [54]. From this model, we estimate that partitioning a design carefully may give up to a 25% improvement in speed.

After synthesis, Synopsys' Physical Compiler tool performs resizing and some resynthesis as the design is floorplanned and globally routed. After synthesis and before detailed routing and cell placement, using Physical Compiler can improve the speed by 15% compared to the same design with the best choice of characterized wire load model. It can also compensate for a poor choice of wire load models.

Floorplanning also needs to take into account noise. Long wires running side by side have large cross-coupling capacitance. Low swing signaling will reduce the amount of noise generated by a signal transition on the wire, and shielding wires can reduce the impact of noise. Chapter 6 discusses the First Encounter tool, which does floorplanning and takes into account crosstalk. "Twisted pair" wiring can be used to substantially reduce inductive coupling noise [65].

Even with the latest partitioning, floorplanning, and layout tools, ASIC designs still have problems maintaining the RTL-specified hierarchy and timing convergence. IP vendors that sell configurable synthesizable designs cannot manually floorplan each configuration. The synthesizable 266MHz Lexra LX4380 could not use traditional place-and-route tools, because the tools sub-optimally shifted logic between the boundary of modules. Lexra eventually developed an in-house methodology that calculates the cycle time dependency between blocks to constrain the allowable configurations, then builds a floorplan based on this. Lexra requires several iterations for the floorplan to converge [31]. Similar methodologies are used by custom designs to determine the optimal layout. It is hoped that future EDA floorplanning tools will better automate layout of carefully partitioned designs.

Cause	Penalty (FO4 delays)
Timing Overhead (10 FO4 delays)	
High performance registers vs. D-type flip-flops	3
Timing margin	3
Slack passing	4
Logic Style (10 FO4 delays, e.g. domino logic)	10
Tapering, sizing, P:N ratios	
Logic compression (more complex cells)	
Structure (25 FO4 delays)	
Partitioning design into regions	7
Managing wires, accuracy of wire load models	
Global wires	6
Local wires	2
Library Cells	
Good vs. mediocre standard cell library	4
Design specific functions	6

Table 8. Summary of ASIC static CMOS maximum performance penalties compared to custom domino logic.

Overall, we attribute a factor of up to 1.4× between the traditional ASIC flow using wire load models, and a custom flow. This can contribute 15 FO4 delays to the delay of combinational logic in a typical ASIC. Small ASIC designs can avoid this penalty by using more accurate wire load models, and tools that perform partitioning, floorplanning, and resizing well. Larger ASICs may need some manual intervention to improve results.

3.7 Summary of the Performance Differences between ASICs and Custom Chips

Typical high performance microprocessors have a cycle time at or below 20 FO4 delays. Typical ASIC designs have cycle times between 40 and 65 FO4 delays for the same microarchitecture. Table 8 summarizes the causes for this gap, and the most that they typically contribute to the delay. About 20 FO4 delays can be due to the differences in the quality of register elements and clock distribution (10 FO4 delays) and the choice of logic style for the circuit (10 FO4 delays). Exploiting the inherent design structure, with the use of regions and good wire management, accounts for up to 15 FO4 delays. Another 10 FO4 delays can be attributed to the use of a good standard cell library and design-specific cells. In comparison, process variations within a generation and across generations can have a significant impact. Changing to the next process generation reduces the delay by about 10-15 FO4 delays. Starting with an ASIC design, exploiting design structure is roughly equivalent to gaining a process generation. Adding custom circuit techniques and high performance register elements is equivalent to gaining another generation.

4. COMPARISON OF ASIC AND CUSTOM CELL AREAS

Area overhead occurs when datapath functions are mapped to standard cells. This growth in area has been quantified [9] for typical sized cells (between $100\chi^2$ and $400\chi^2$) and is equal to 10%-25% due to the alignment of all cell interface geometries to the global grid; 0.5%-1.0% per extra protection diode contact and between 20%-40% due to both the differences in P:N ratios used; and the differences in the absolute number of p-channel MOSFETs within traditional standard cells as compared with datapath cells. The basis of this data is a comparative study of a 91-cell static CMOS standard cell library and 82-cell static CMOS datapath library. As the circuit style in both cases was static CMOS, these results of a 1.3× to 1.5× area penalty and corresponding increase in χ^2/T cost form a *lower bound* on the actual penalty. Datapath cells employing only NMOS pass gates, or domino logic styles will have even greater density gains over the static CMOS logic in the baseline standard cell set due to the further elimination of parasitic p-channel MOSFETs, and compression of logic depth. Datapath cells implemented in either differential domino or DCVSL styles also benefit from reducing the number of inverters required.

4.1 The Costs of Grid Alignment and Reduced Encompassed Functionality

Standard cells designed for a traditional ASIC flow rely on automated place and route (P&R) tools to position cells and make the final signal connections. To operate efficiently, automated P&R tools typically require input and output ports for each cell to be aligned to the routing grid. In addition, place and route tools require that all edge geometries be "on-grid". Both these requirements exist to reduce the complexity of the placement and routing problem by abstracting each cell's geometry and restricting the number of valid locations seen by the tool. This compatibility requirement results in an inherent handicap in layout density. Figure 6 shows the reduction in area that could be achieved by abutting two cells.

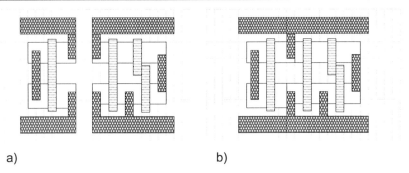

Figure 6. By abutting the two cells in (a), their combined area (b) is reduced.

The left most cell in Figure 6 is constructed to be compatible with the place and route tools and thus is aligned to the grid and must maintain a valid design-rule clearance (typically half the minimum spacing rule for the relevant geometry) along the edge of interest. The rightmost cell demonstrates the potential density improvement when cell geometry can exist up to the edge of the cell.

The depicted situation represents only abutment in the x-direction, the actual potential for lost usable area extends also in the y-direction. Ultimately, the maximum cost is one routing track (χ) in each direction. Small cells are more heavily penalized. For example, the penalty for a cell with a fixed area of 64 χ^2 ranges from 23% to 29%, as cell height is varied from 8χ to 16χ. In contrast, given a fixed cell area of 144χ^2, the penalty varies between 16% and 17% as the height is changed from 8χ to 16χ. The percentage of lost usable area is equal to the increase in χ^2/T cost. Each cell boundary adds inefficiency. Finally, the requirement to be on-grid contributes to area growth especially when several simple standard cells are grouped to implement a more complex datapath function. In Figure 7, three different register elements [9][37] are shown. In each case the same circuit netlist was implemented first in a custom methodology and then with grid-alignment constraints for compatibility with P&R. The cell height in all cases was 14χ. The difference in area and transistor cost varied from 11% to 25%.

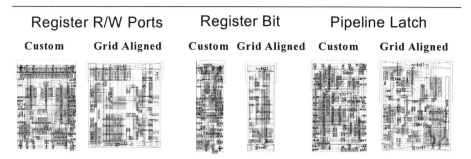

Figure 7. Examples of Grid Aligned and Custom Cell Layouts

4.2 The Impact of Extra PMOS Devices and Differing P:N Sizes and Ratios

In total, P:N ratio differences, transistor sizing differences increase both cell area and χ^2/T transistor cost by 25%-50%. The circuits employed in most traditional static CMOS standard cells are designed with a P:N ratio of between 1.5:1 and 2:1, for either optimum speed or balanced rise and fall times respectively. The structure of the typical standard cell layout template reflects this as the N-well occupies slightly more than half the cell area. This distribution of N-transistor and P-transistor areas is a handicap when traditional standard cells are used to implement custom functions. Specifically, there are two main penalties:

1. **Mapping N-dominated datapaths into the balanced standard cell circuit style results in area inefficiency.** The partition of cell area in balanced CMOS standard cells restricts the size and number of NMOS transistors that can be implemented. The smaller useful area for NMOS transistors further increases both the χ^2/T transistor cost of each cell and the total number of cells needed to emulate a custom functions. The included P-transistors further increase effective transistor cost and decrease layout efficiency

2. **Mapping N-dominated datapaths into the balanced standard cell circuit style results in a degradation in timing.** The extra PMOS transistors present additional parasitic loading. Datapath cells can vary P:N ratios to "prefer" one transition over another. Datapath cells can also take advantage of tapering device sizes to further tune performance.

Library	# Cells	Average # Transistors/Cell	Wp:Wn
Static CMOS Standard Cell	68	8.5	1.50:1
without inverters and buffers	54	9.8	1.80:1
Static CMOS Datapath	66	9.0	1.04:1
without inverters and buffers	52	11.0	0.93:1

Table 9. P:N Ratios for standard cells and datapath cells.

Technology	Foundry A (fJ)	Foundry B (fJ)	Foundry C (fJ)	Custom (fJ)
0.25um	66	60	25	12.7

Table 10. Switching energies for a min-sized 2-input NAND in different foundries.

Static datapath cells attempt to either eliminate PMOS parasitic devices or minimize their size. As a result, it is common to have both the per-cell P:N ratio and the total p-channel MOSFET width (Wp) to total n-channel MOSFET width (Wn) ratio equal to less than 1:1. Table 9 shows the Wp:Wn ratios from the survey of a static CMOS standard cell library and a static CMOS datapath library. The average number of transistors per cell in the datapath library is higher reflecting the aggregation of more complex functions and the opportunity for logic depth compression.

In addition, the average size of transistors in standard cells is greater than that of custom cells implementing the same functionality. Since the specific application environment for custom cells is determined a priori, they are generally smaller and exactly matched to the actual in situ load. No foreknowledge, however, is available for standard cells. The transistor-sizes for the minimum allowable standard cell gates are generally larger than the design rule minimums set by the underlying technology. A survey of the energy per min-sized 2-input NAND gates for a set of contemporary 0.25um processes and related standard cell libraries shows that a factor of 2 to 5 difference is typical. This factor of 2 to 5 beyond the actual technology minimum can degrade performance by increasing input loading and increasing energy dissipation, as shown in Table 10.

4.3 Cell Bit Height – Partitioning Global and Local Resources

In an effort to increase area efficiency, semiconductor foundries provide "high density" cell libraries. These libraries are designed to increase the reportable gates/mm^2 metric. This metric is based on the number of 2-input NAND gates that can be fabricated and not on the number that can be usefully connected. The cell width of a minimum sized 2-input NAND is technology independent and is fixed at either 3.5χ or 4χ based on the configuration of power distribution to the cell. Thus, to maximize the gates/mm^2 metric, the cell height for the library must be minimized, resulting in 7χ and 8χ cell libraries. While the result is an admirable gates/mm^2 value (114Kgates/mm^2 in 0.18um), this metric proves to have little real value as designs typically achieve less than 40% of this maximum. The short cell height greatly reduces the usable inter-cell routing resources and thus has an adverse impact on the block and global assembly. Typically, datapath cells have a cell height of 12χ to 16χ enabling more inter-cell routing per required port connection and allowing more functionality to be collected within the cell. Both of these features reduce the absolute number of required inter-cell connections.

4.4 Extraneous Diode and Substrate Contacts

Standard cells must operate in arbitrary placements and can be driven by unpredictable driving circuits over undetermined distance and through unspecified metal layers. Thus, they must incur the additional overhead of both internal diode contacts on inputs and full well and substrate ties to ensure functionality. In comparison, datapath cells are optimized for a specific placement and a specific set of input drivers, output loads and detailed routing paths. Also, unlike standard cells, the power distribution and required well and substrate ties are usually included in stitch cells that amortize the overhead over 8 to 16 cells.

For standard cells, the requirement of diode contacts on silicon gates contributes a small additional inefficiency to the layout density. Diode contacts are required on each silicon gate that is connected to metal routing beyond a certain length threshold and *which is not already* driven by a transistor source or drain. Traditional standard cells encompass less functionality than datapath cells. More diode contacts are required as the additional input/output ports are exposed at the cell interfaces. If instead, cells encompassed greater functionality, these ports would be subsumed in intra-cell connections. Also, in the custom case, the designer knows exactly which outputs are already driven from the source or drains of transistors, and can be used directly to connect to the gate of another, implicitly forming the required diode. Traditionally, automatic place and route tools, due to the greater number of interfaces required by simpler logic cells and design complexity, have not been able to do this. Avant!'s place and route tools do support diode insertion [5], and standard cell libraries that don't include diodes in each cell can take advantage of this, reducing the cell area. However, not including diodes in each cell causes the library to be incompatible with other placement tools that do not have diode insertion.

The absolute penalty due to extra diode contacts depends on the specific design rules of the selected process. However, it is minimally $1\chi^2$ of routing resource and $1.2\chi^2$ of additional device area per contact. The actual impact can be greater as the additional metal resource requirements can also increase routing congestion. Well and substrate ties have similar impact as the generally require $2\chi^2$ of routing resource per and $2.5\chi^2$ of device for each.

4.5 Summary of Standard Cell Overheads

We have examined a range of area penalties within standard cells. Applying these penalties to an example $200\chi^2$ standard cell results in the following costs: 40% for p-channel MOSFETs and P:N ratios and sizes, 15% for grid alignment, 10% for poor intermediate routing utilization from too narrow cell height, and 1-2% for extraneous diode contacts and well/substrate ties. The cumulative result is a 55% less dense cell (i.e. 1.8× size growth) and a corresponding 80% growth in χ^2/T cost. Further costs can accrue as it may take several $200\chi^2$ standard cells to replicate the functionality of the original custom circuit.

5. ENERGY TRADEOFFS BETWEEN ASIC CELLS AND CUSTOM CELLS

"High performance design is low power design" – Mark Horowitz

This statement captures the essence of the benefits of *efficient* custom circuit styles in minimizing energy dissipation. The energy-delay-product ($E\tau$) of a circuit is a useful metric for efficiency. We use four equations from [16] to define delay (τ), energy (E), energy delay product ($E\tau$) and dynamic power ($P_{dynamic}$). The delay is a function of the load (C_{load}), the power supply voltage (V_{DD}) and the device current (I_{DSS}). The device current (I_{DSS}) is a function of the transistor transconductance (β), the supply voltage and the threshold voltage (V_{TH}). The exponent, α, is equal to 2 for low field transistors and approaches 1.25 in modern n-channel MOSFETs. The transistor transconductance is proportional to the transistor width divided by the transistor length, W/L. The dynamic power dissipation ($P_{dynamic}$) is equal to the energy multiplied by the toggling frequency (f_{toggle}).

$$(29) \quad \tau \propto \frac{C_{load} V_{DD}}{I_{DSS}} = \frac{C_{load} V_{DD}}{\beta (V_{DD} - V_{TH})^\alpha}$$

$$(30) \quad E \propto C_{load} V_{DD}^2$$

$$\text{(31)} \quad E\tau \propto \frac{C_{load}^2 V_{DD}^2}{\beta}\left(1 - 2\frac{V_{TH}}{V_{DD}} + \frac{V_{TH}^2}{V_{DD}^2}\right)$$

$$\text{(32)} \quad P_{dynamic} = C_{load} V_{DD}^2 f_{toggle}$$

Assuming α equal to 2, when the value of V_{DD} is large compared to V_{TH}, the $-2V_{TH}/V_{DD}$ term in (6) dominates and the energy-delay-product decreases linearly with decreases in V_{DD}. The reduction in operating voltage yields a quadratic reduction in energy and power with only a linear decrease in performance. However, as V_{DD} becomes comparable to V_{TH}, the increase in delay cancels the decrease in energy and negates the benefits of reducing V_{DD}. Energy dissipation is still decreasing beneficially, but the energy-delay-product is not. When V_{DD} is equal to V_{TH}, the $E\tau$ approaches infinity as the delay increases rapidly. Linear reductions in the capacitive load cause linear reductions in delay and quadratic reductions in the energy-delay-product. When α is less than 2, the improvement of energy-delay-product with reduced V_{DD} is sub-linear and $E\tau$ becomes larger more slowly as V_{DD} approaches τ.

A lower energy-delay-product enables operation at higher speeds with the same energy, or the same speed with lower energy. Dynamic circuit techniques can produce designs with $E\tau$ more than a factor of two smaller than those of standard static CMOS circuits [1][2]. Dobberpuhl et al. [19][43] further demonstrated that many of the same high-performance dynamic circuit techniques used in the first Alpha 21064 microprocessor could be used to also to achieve energy efficiency in the SA-110 Strong-Arm microprocessor.

As described in Section 3, ASIC designs are between a factor of 2 to 3 slower than custom designs. Much of the difference is due to oversized transistors and inefficient routing, which increases C_{load}. Burns et al. [8] reported a 50% reduction in power, with no loss in performance, after automatically customizing standard cells to match drive strengths with actual loads.

5.1 Circuit Efficiency Leads to Lower Energy Dissipation – An Example

Reducing parasitic capacitance enables a design to run faster and reduces the energy required for each switching event. The ability to run faster enables the reduction in operating voltage to gain the V_{DD}^2 energy and power savings. We illustrate this principle by examining the static and domino-style circuit implementations of a simple logic function: Y = ABCDEF + G in a 0.18um technology at a PVT of TTLH (100°C, 1.65V). The schematics for both circuit implementations are shown in Figure 8. The static version of this function is implemented in 24 transistors organized as four gates: 2-input AND, 4-input AND, AOI21 and 4X INV. The resulting circuit requires 3.85 FO4 delays and 95nW of power. Importantly, none of the transistors can be minimum size to achieve this speed. In contrast, of the 22 transistors in the footless domino circuit (derived from [6]), 8 are minimum sized. Also, while the static CMOS circuit requires 12 p-channel MOSFETs, the domino circuit uses only 9. More importantly, the total Wp for the static gates is 19.8um, while the domino gate has total Wp of 5.3um, representing less than one-third the parasitic loading. At maximum performance, the domino circuit requires 2.3 FO4 delays and dissipates 222nW (1.7× faster for 2.3 times the power). Reducing the operating voltage of the domino circuit to 1.23V allows the gate to operate with the same 95nW of power as the static design while providing an 8% increase in speed. Alternately, reducing V_{DD} further to 1.19V results in an 11% reduction in power, while operating at the same speed. The drawback of the domino circuit is that it is more subject to noise.

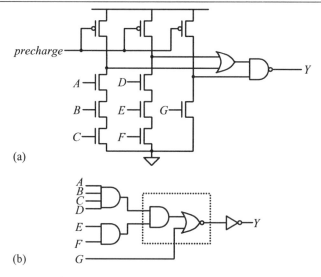

Figure 8. Two Implementations of $Y = (ABCDEF) + G$. (a) shows the footless domino implementation without a keeper. The static CMOS implementation is in (b), with a 4-input AND and 2-input AND, followed by an AOI21 then an inverter.

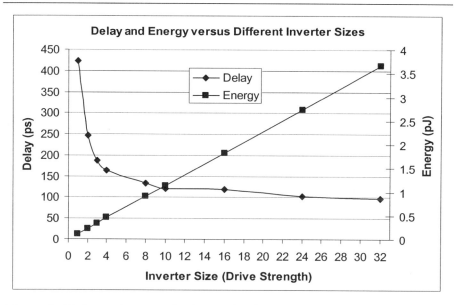

Figure 9. Delay and energy for increasing inverter sizes with a fixed load.

5.2 Energy Efficiency versus Performance

Each circuit has an energy efficient operating point that is a function of its structure, timing, supply voltage and capacitive loading. Custom circuit techniques that minimize parasitic loading increase the maximum *efficient-circuit-speed*. When this efficient-circuit-speed is insufficient and additional performance is required, circuit designers can choose to trade energy efficiency for speed.

Designers choose performance over efficiency when they attempt to reduce delay by simply increasing the drive strength of gates. Figure 9 shows the delay versus energy curve for an inverter in a 0.35um process. While a factor of four increase in drive creates a factor of 2.5 decrease in delay, it requires a factor of four increase in energy dissipation. At the extreme, a factor of 32 increase in drive, with a corresponding factor of 32 increase in energy dissipation only reduces delay by a factor of 4.

5.3 Summary of ASIC and Custom Circuit Energy Differences

The most effective method to reduce both energy and power dissipation is lowering the operating voltage. High-performance custom circuit techniques such as domino and DVCS can have energy-delay-products that are half the value of standard static CMOS logic [2] despite higher relative power requirements. By eliminating inefficient PMOS devices, these custom techniques reduce delay by a greater factor than the corresponding increase in power requirements. Operating these custom circuits at reduced voltages can take advantage of this difference and result in a factor of two improvement for both energy and power over well-optimized static CMOS logic. Due to the mismatch of drivers and loads and the additional routing, the power efficiency of static logic commonly found in ASIC chips can be worse by a further factor of two [8]. In total, ASIC logic may consume 4× the power and has a quarter of the energy efficiency of custom logic.

6. FUTURE TRENDS

The continuation of technology scaling will pose additional challenges in circuit design and library creation. There are four primary trends. First, the combination of aggressive operating voltage scaling and limited threshold voltage scaling will cause increased leakage current and leakage power. The resulting decrease in the V_{DD}/V_{TH} ratio limits transistor stack heights within circuits and reduces the effectiveness of pass-gate logic. Second, interconnect delay will increasingly handicap long wires. Third, the increasing complexity of designs further restricts the ability of designers to customize. Finally, the layout requirements to facilitate Optical-Proximity-Correction (OPC) and Phase-Shift-Masks (PSM) will complicate the design of library components.

Standard static CMOS logic scales well with technology. In contrast, the reduction in the V_{DD}/V_{TH} ratio reduces the benefits of some custom circuit techniques. The performance of pass gate logic styles deteriorates as the difference between V_{DD} and V_{TH} shrinks. Limits in the number of transistors in a stack reduce the allowable depth of the n-channel MOSFET logic trees in domino and DCVSL circuit styles, potentially reducing the amount of logic compression. While some of the advantages of custom circuit techniques will be diminished, they will nevertheless maintain both a performance and area advantage over standard static CMOS. In addition, new logic families such as Current Mode Logic (CML) [1] have been proposed that are energy efficient and possess good performance and result in a low energy-delay product. In the near future, ASIC libraries will still be implemented in static CMOS due to its robustness, scaling properties, and the existing infrastructure for synthesis and functional testing. With higher leakage in future process generations, domino logic will require "keepers" (or wider keepers) to avoid leakage draining the stored charge. Custom designers will increasingly migrate to DCVSL and CML styles as reductions in transistor stack height increasingly penalize domino styles for requiring an additional pre-charge p-channel MOSFET and keeper.

The combination of the importance of interconnect delay, the increasing difficulty to manually customize designs, and the added complexity of OPC and PSM will require ASIC design flows to increasingly adopt custom techniques, and custom design flows to increasingly adopt design automation. High-quality libraries will continue to grow in importance as designers struggle to deal with increasing chip complexity. The differences between ASIC libraries and custom libraries will continue to shrink. Compared with libraries for prior technology generations, current libraries have a larger total number of gates, a wider range of drive strengths, increased functionality, and increased specialization of cells. In addition, some libraries [46] have added automatic device tapering, cell creation, sizing, tunable P:N ratios and the ability to selectively insert low V_{TH} devices to further increase performance. Continuing advances in CAD tools will shrink the difference between ASIC design flows and custom design flows. As a result and in parallel, the quality gap between ASIC and custom design will shrink as well.

6.1 The Future of Domino Logic

Figure 2(b) shows the basic structure of a domino logic gate with a keeper. The *keeper* maintains charge at the precharge node if the output is 1, which stops leakage current draining the stored charge. In future process generations, the leakage current will increase as the supply voltage and threshold voltage are scaled down to reduce the power consumption. The threshold voltage needs to be reduced as the supply voltage is scaled down, in order to minimize the increase in the delay τ, which can be seen from Equation (29).

The leakage current is exponentially dependent on the threshold voltage, so leakage will become a significant issue. Greater leakage current requires that the keepers be larger to combat the drain of the stored charge. In turn, the pull down network must be larger, providing more current, when the gate actually needs to be pulled down to zero. Thus the transistor widths for the keeper and pull down network must increase in size, penalizing domino logic.

7. SUMMARY

In this chapter, we have provided an overview of the differences in performance, density, and energy efficiency between the cells and building blocks used in custom chips and those used in ASICs. We introduced a set of technology independent metrics to help designers compare design quality between chips and across fabrication processes. Calibrating design performance in fanout-of-4 inverter delay enables a quick estimate of relative performance. Examining the χ^2/T transistor provides immediate insight into relative layout efficiency. Comparing the per-bit energy requirement for a circuit to E_{bit}, E_{nand}, and E_{inv} is a measure of relative energy efficiency. Focusing on the number of unique transistors in a design and the t_{unique}/wk productivity provides a more accurate model of required effort.

We have provided a set of benchmarks to enable designers to gauge the relative quality of their designs. The reference custom domino design achieves a cycle time of 20 FO4 delays and has transistor costs between 6-8χ^2/T. In contrast, a high quality ASIC achieves a cycle time of 40 FO4 delays and has transistor costs between 11-16χ^2/T. Finally, an average ASIC design has a cycle time of 60 FO4 delays and a transistor cost of between 16-25χ^2/T. The energy efficiency of the custom designs can be twice that of high quality ASIC designs and over four times that of average ASICs.

The four key differences between the performance of custom designs and ASICs are, with approximate maximal contributions of FO4 delays: the quality of the register elements (3 FO4 delays); the attention and customization of the clock distribution network (3 FO4 delays); circuit logic style (9 FO4 delays); floorplanning and accurate wire load models (15 FO4 delays); and logic design and cell design (10 FO4 delays).

There is typically a factor of 3 difference in area between custom designs and ASICs. The four main causes of the poor relative density in ASIC chips are the need for p-channel MOSFETs and sub-optimal P:N ratios (40%); the requirement for grid alignment (15%); and the additional congestion caused by the short cell heights used in standard libraries (10%). Finally, the remaining factor of 1.6 is attributable to the logic compression gained from employing dedicated custom complex functions instead of collections of simple cells.

It is harder to exactly quantify the differences in energy efficiency and effort. However, the sources of both are clear. The energy efficiency contrast between the two styles is caused by the additional p-channel MOSFETs required in the static CMOS logic used throughout ASIC cell libraries. The energy-delay-product can be 2× smaller for custom circuits compared to optimized standard static CMOS designs. The additional effort required to create a custom chip is the result the significantly larger number of required unique transistors.

As technology scales and design complexity grows, designers need more methods to sift through the large number options in technology, logic, circuit, and library elements to achieve their goals. It will be increasingly important to provide designers with the tools to identify key design structures and create specialized library components. Ultimately, the quality of each design depends on the creativity of the designers and their ability to convert their unique mix of building blocks into an optimal combination for their specific mission.

Another major factor of difference between ASICs and custom designs are the processes that they are produced in, and how conservatively the cell libraries are characterized for these processes. The next chapter examines this.

8. REFERENCES

[1] Allam, M. et al., "Effect of Technology Scaling on Digital CMOS Logic Styles," *IEEE Custom Integrated Circuits Conference*, Orlando, FL, May 2000, pp. 401-408.

[2] Allam, M. et al., "Dynamic Mode Logic (DyCML): A New Low-Power High – Performance Logic Style," *IEEE Journal of Solid-State Circuits*, vol. 36, no 3. March 2001, pp 550-556.

[3] Allen, D.H. et al., "Custom circuit design as a driver of microprocessor performance," *IBM Journal of Research and Development*, vol. 44, no. 6, November 2000, pp. 799-822.

[4] Anderson, C, et al., "Physical Design of A Fourth-Generation POWER GHz Microprocessor," *IEEE International Solid-State Circuits Conference*, 2001.

[5] Avant!, Hercules-II Hierarchical Physical Verification and Mask Generation, 2002. http://www.avanticorp.com/Avant!/SolutionsProducts/Products/Item/1,1500,7,00.html

[6] Berstein, K. et al., *High Speed CMOS Design Styles*, Kluwer Academic Publishers. 1999.

[7] Bohr, M., et al., "A high performance 0.25um logic technology optimized for 1.8 V operation," *Technical Digest of the International Electron Devices Meeting*, 1996, pp. 847-850.

[8] Burns, J., "Cell Libraries – Build vs. Buy; Static vs. Dynamic," *Proceedings of the 36th Design Automation Conference*, New Orleans, LA, June 1999. Panel Discussion.

[9] Chang, A., "VLSI Datapath Choices: Cell-Based Versus Full-Custom," Masters Thesis, Massachusetts Institute of Technology, February 1998.

[10] Chau, R., "30nm and 20nm Physical Gate Length CMOS Transistors," *2001 Silicon Nanoelectronics Workshop*. June 2001. Talk Slides.

[11] Chen, C., Chu, C., and Wong, D., "Fast and Exact Simultaneous Gate and Wire Sizing by Lagrangian Relaxation," *IEEE Transactions on Computer-Aided Design of Integrated Circuits and Systems*, vol. 18, no. 7, July 1999, pp. 1014-1025.

[12] Clark, L. et al., "A Scalable Performance 32b Microprocessor," *IEEE International Solid-State Circuits Conference*, 2001, pp. 186-187, 451.

[13] Collett, R., "Design Productivity: How to Measure It, How to Improve It," Panel Session, at the *35th Design Automation Conference*, San Francisco, CA, June 1998.

[14] Cooley, J. The Surprise Physical Synthesis Tape-Out Census. December 2000.

[15] Curran, B. "A 1.1GHz First 64b Generation Z900 Microprocessor," *IEEE International Solid-State Circuits Conference*, 2001, pp. 194-195, 453-455.

[16] Dally, W. and Poulton, J., *Digital Systems Engineering*, Cambridge Univ. Press, 1998.

[17] Dally, W. et al., "The Message-Driven Processor: A Multicomputer Processing Node with Efficient Mechanisms," *IEEE Micro*, April 1992.

[18] Dally, W. et al., "Architecture and Implementation of the Reliable Router," *Hot-Interconnects II*, Palo Alto, CA, August 1994.

[19] Dobberpuhl, D. et al., "A 200 MHz 64-b dual-issue CMOS microprocessor," *IEEE Journal of Solid-State Circuits*, vol. 27, no 11, November 1992, pp. 1555-1567.

[20] Doran, R.W., "Variants of an Improved Carry Look-Ahead Adder," *IEEE Transactions on Computers*, vol. 37, no. 9, September 1988, pp. 1110-1113.

[21] Fishburn, J., and Dunlop, "A. TILOS: A Posynomial Programming Approach to Transistor Sizing," *Proceedings of the International Conference on Computer-Aided Design*, 1985. pp. 326-328.
[22] Gaddis, N. and Lotz, J., "A 64-b Quad-Issue CMOS RISC Microprocessor," *IEEE Journal of Solid-State Circuits*, vol. 31, no. 11, November 1996, pp. 1697-1702.
[23] Gargini, P., "Intel Process Technology Trends," *Intel Developers Forum*, 2001. Talk Slides.
[24] Gavrilov, S., et al., "Library-Less Synthesis for Static CMOS Combinational Logic Circuits," *Proceedings of the International Conference on Computer-Aided Design*, 1997. pp. 658-663.
[25] Ghani, T. et al., "100nm Gate Length High Performance Low Power CMOS Transistor Structure," *International Electron Devices Meeting*, 1999 Talk Slides.
[26] Grodstein, J., et al., "A Delay Model for Logic Synthesis of Continuously-Sized Networks," *Proceedings of the International Conference on Computer-Aided Design*, 1995, pp. 458-462.
[27] Gronowski, P. et al., "High-Performance Microprocessor Design," *IEEE Journal of Solid-State Circuits*, vol. 33, no. 5, May 1998, pp. 676-686.
[28] Haddad, R., van Ginneken, L., and Shenoy, N., "Discrete Drive Selection for Continuous Sizing," *Proceedings of the International Conference on Computer Design*, 1997, pp. 110-115.
[29] Harris, D. and Horowitz, M., "Skew Tolerant Domino Circuits," *IEEE Journal of Solid-State Circuits*, vol. 32, no. 11, November 1997, pp. 1702-1711.
[30] Hartman, D., "Floating Point Multiply/Add Unit for the M-Machine Node Processor," Masters Thesis, Massachusetts Institute of Technology. May 1996, pp. 47-54.
[31] Hauck, C., and Cheng, C., "VLSI Implementation of a Portable 266MHz 32-Bit RISC Core," in the *Microprocessor Report*. October 22, 2001.
[32] Hill, D., "Sc2: A Hybrid Automatic Layout System," *Proceedings of the International Conference on Computer-Aided Design*, 1985, pp. 172-174.
[33] Ho, R. et al., "The Future of Wires," *Proceedings of the IEEE*, April 2001, pp. 490-504.
[34] Hossain, R., et al., "A Comparison Between Static and Domino Logic," Submitted in March 2002 to the *IEEE Journal of Solid State Circuits*.
[35] IBM Corporation, *SA-27E ASIC Databook*, February 2000.
[36] Kapadia, H., and Horowitz, M., "Using Partitioning to Help Convergence in the Standard-Cell Design Automation Methodology," *Proceedings of the 37th Design Automation Conference*, 1999, pp. 592-597.
[37] Keckler, S. et al., "The MIT Multi-ALU Processor," *Hot Chips IX*, August 1997.
[38] Keutzer, K., Kolwicz, K., and Lega, M., "Impact of Library Size on the Quality of Automated Synthesis," *Proceedings of the International Conference on Computer-Aided Design*, 1987, pp. 120-123.
[39] Kogge, P.M., and Stone, H.S., "A Parallel Algorithm for the Efficient Solution of a General Class of Recurrence Equations," *IEEE Transactions on Computers*, vol. C22, no. 8. August 1973, pp. 786-793.
[40] Lau, M. et al., "A Packet-Memory-Integrated 44Gb/s Switching Processor with a 10Gb Port and 12Gb ports," *IEEE International Solid-State Circuits Conference*, 2002.
[41] Ling, H., "High-speed Binary Adder," *IBM Journal of Research and Development*, vol. 25, no. 2-3, May-June 1981, pp. 156-166.
[42] Mead, C. and Conway, L., *Introduction to VLSI Systems*, Addison Wesley, 1980.
[43] Montanaro et al., "A 160MHz, 32-b 0.5-V CMOS RISC Microprocessor," *IEEE Journal of Solid-State Circuits*, vol. 31, no 11, November 1996, pp 1703-1714.

[44] Naffziger, S. and Hammond G., "The Implementation of the Next-Generation 64b Itanium Microprocessor," *IEEE International Solid-State Circuits Conference*, 2002.
[45] Nikolić, B. EE241 – Spring 2001 Advanced Digital Integrated Circuits – Lecture 18: Adders. 2001. http://bwrc.eecs.berkeley.edu/Classes/icdesign/ee241_s01/Lectures/lecture18-adders-grayscale.pdf
[46] Northrop, G. and Lu, P., "A Semi-custom Design Flow in High-performance Microprocessor Design," *Proceedings of the 38th Design Automation Conference*, Las Vegas, NV, June 2001, pp. 426-431.
[47] Nowka, K., and Galambos, T., "Circuit Design Techniques for a Gigahertz Integer Microprocessor," *Proceedings of the International Conference on Computer Design*, 1998, 11-16.
[48] Okano, H., "An 8-Way VLIW Embedded Multimedia Processor Built in 7-Layer Metal 0.11um CMOS Technology," *IEEE International Solid-State Circuits Conference*, 2002.
[49] Paraskevopoulos, D.E and Fey, C., "Studies in LSI Technology Economics III: Design Schedules for Applications Specific Integrated Circuits," *IEEE Journal of Solid-State Circuits*, vol. 22. April 1987, pp. 223-229.
[50] Rohrer, N. et al., "A 480 MHz RISC Microprocessor in 0.12um Leff CMOS Technology with Copper Interconnects," *IEEE International Solid-State Circuits Conference*, 1998.
[51] Samueli, H., "Designing in the New Millennium – It's Even Harder Than We Thought," keynote speech at the *38th Design Automation Conference*, Las Vegas, NV, June 2001.
[52] Scott, K., and Keutzer, K., "Improving Cell Libraries for Synthesis," *Proceedings of the Custom Integrated Circuits Conference*, 1994, pp. 128-131.
[53] Stojanovic, V. and Oklobdzija, V., "Comparative Analysis of Master-Slave Latches and Flip-Flops for High-Performance and Low-Power Systems," *IEEE Journal of Solid-State Circuits*, vol. 34, no. 4, April 1999, pp. 536-548.
[54] Sylvester, D. and Keutzer, K., "Getting to the Bottom of Deep Sub-micron," *Proceedings of the International Conference on Computer Aided Design*, November 1998, pp. 203-211.
[55] Sylvester, D., Jiang, W., and Keutzer, K. BACPAC – Berkeley Advanced Chip Performance Calculator. 2000. http://www-device.eecs.berkeley.edu/~dennis/bacpac/
[56] Kim, T. and Um, J., "A Practical Approach to the Synthesis of Arithmetic Circuits using Carry-Save-Adders," *IEEE Transactions on Computer Aided Design of Integrated Circuits and Systems*, vol. 19, no. 5, May 2000, pp. 615-624,
[57] Taiwan Semiconductor Corporation. TSMC Corporate Brochure.
[58] Texas Instruments, Inc. GS30 0.15um CMOS Standard Cell Commercial Product Information Sheet.
[59] Thompson, S. et al., "Enhanced 130nm Generation Logic Technology Featuring 60nm Transistors Optimized for High Performance and Low Power at 0.7-1.4V," *International Electron Devices Meeting*, 2001.
[60] Tyagi, S. et al., "A 130nm Generation Logic Technology Featuring 70nm Transistors, Dual Vth Transistors and 6 Layers of Cu Interconnect," *International Electron Devices Meeting*, 2000, talk slides.
[61] United Microelectronics Corporation, Foundry Services Guide.
[62] Weste, N.H. and Eshraghian, K., *Principles of CMOS VLSI Design: A Systems Perspective*, 2nd ed. Addison-Wesley, Reading, MA, 1992.
[63] Williams, J. and O'Neill, J., "The Implementation of Two Multiprocessor DSP's: A Design Methodology Case Study," *IEEE International Solid-State Circuits Conference*, 2001.
[64] Yang, S. et al., "A High Performance 180nm Generation Logic Technology," *International Electron Devices Meeting*, 1998.

[65] Zhong, G., Koh, C.-K., and Roy, K., "A Twisted-Bundle Layout Structure for Minimizing Inductive Coupling Noise," *Proceedings of the International Conference on Computer Aided Design*, 2000, pp. 406-411.

[66] Zimmerman, R. and Fichtner, W., "Low-Power Logic Styles: CMOS Versus Pass-Transistor Logic," *IEEE Journal of Solid-State Circuits*, vol. 32, no. 27, July 1997, pp. 1079-1090.

Chapter 5

Finding Peak Performance in a Process
Process Variation and Operating Conditions

David Chinnery, Kurt Keutzer
Department of Electrical Engineering and Computer Sciences,
University of California at Berkeley

One of the traditional advantages of using an ASIC design methodology was that the ASIC designer was insulated from semiconductor processing issues. By restricting the circuit design to use only the cells offered in the ASIC library, and by using the "golden" tool flow provided by the ASIC vendor in the ASIC Design Kit, a designer could be assured that if the tools said that the IC would run at a certain speed then the manufactured part would as well. As the delay of integrated circuits began to be impacted by wiring delay, the simplicity of this approach was compromised. It became impossible to determine the final delay of the circuit without layout information. A number of different approaches were taken to address this problem. Many ASIC designers have taken over the responsibility for physical design as well.

In our investigation of the sources of performance differences between ASIC and custom designs, semiconductor processing proved to be one of the most significant factors. ASIC designers pay a high price in circuit performance for the luxury of ignoring processing details. This chapter gives a tutorial introduction to the many sources of performance variation in semiconductor processing. This theme is continued at a more advanced level in Chapter 14.

Semiconductor companies have different implementations of what is generally known as the same process generation, such as 0.18 micron, depending on: whether aluminum or copper is used for the metal interconnect; the dielectric constant of the insulator to reduce crosstalk between interconnect; the number of metal layers; gate oxide thickness; and silicon-on-insulator or bulk CMOS. There can be differences of 40%, or more, between the speeds of the fastest chips produced in one foundry versus the fastest chips produced in another (see Section 4). Even a single semiconductor company may have several different versions of the same process technology.

Much performance variation can also be attributed to random process variation around the nominal point. The highest clock frequency chips produced in a fabrication plant may be 20% to 40% faster than the slowest chips produced. (This estimate was based on the spread between fast and slow process conditions reported in [1] and communications with Costas Spanos, Chenming Hu and Michael Orshansky at UC Berkeley.)

Process variation can be caused by a number of factors discussed in Section 2. The behavior of the circuitry also depends on the operating conditions. Based on the variation in performance of chips produced within a process technology and a set of operating conditions, a foundry releases models for worst-case, typical, and high-speed performance. The models used by designers determine the achievable circuit performance, and it is important to understand the limitations and conservatism of these models.

There are multiple sources for performance variation of IC designs arising from processing-related issues. Section 1 provides an introduction to the process and operating conditions for which chips are designed. The cost for better operating conditions is discussed. The speed variation of ASICs due to process and operating conditions is examined.

Section 2 discusses statistical process variation and its impact on chip performance, Section 3 looks at the impact of continuous process improvements on the achievable chip speed, and Section 4 compares different process implementations of the same technology generation. The differences between custom and ASIC chips due to process and operating conditions is summarized in Section 5. Section 6 concludes with how ASICs can improve performance compared to worst case process and worst case operating conditions.

1. PROCESS AND OPERATING CONDITIONS

Standard cell libraries for a given foundry are characterized for several different process and operating conditions. The operating conditions are defined by the nominal supply voltage and operating temperature. The process conditions correspond to the speed of chips fabricated based on models provided by the foundry: slow (worst case), typical, and fast.

ASICs synthesized for slow process conditions will give a high yield at the post-layout clock frequency. There will be less chips of a higher clock frequency corresponding to fast process conditions. ASIC vendors are ultimately paid for packaged parts that work at the required speed, so ASIC designers must design for process conditions that will ensure a high yield. Thus ASIC designers typically assume worst case process conditions.

If faster chips can be sold at a premium, it is advantageous to *speed bin* chips. Chips are tested at several clock frequencies and then speed binned into groups of chips that have the same valid operating frequency. Through speed-binning, chips can be sold at clock frequencies corresponding to fast, typical and slow process conditions.

The specification for the operating conditions varies. Slow operating conditions are at a lower supply voltage and higher temperature, typically 100°C or 125°C. Sometimes 90% of the nominal supply voltage V_{DD} is used for the worst case supply voltage. Note that 90% of V_{DD} may be observed in normal operation due to variations in supply voltage (e.g. due to ground bounce) [Andrew Chang, personal communication]. Similarly, 100°C to 125°C are commonly used as worst case operating temperatures for some embedded applications [9]. Thus slow operating conditions may be quite realistic and not pessimistic – it depends on the application. Typical operating conditions are often specified at the nominal supply voltage and room temperature (25°C). Fast operating conditions are at a higher supply voltage and below room temperature.

Whether operating conditions are optimistic or pessimistic depends on the actual range of conditions within which the chips will be used.

The conditions for worst-case power consumption are different to worst case speed conditions (worst case process and worst case operating conditions). The worst-case dynamic power consumption is at the highest operating clock frequency (e.g. at fast process, high supply voltage and low temperature conditions). The worst-case leakage power is at higher temperature and higher supply voltage.

1.1 A Higher Supply Voltage Increases Drive Strength

Chapter 4, Section 5, has a more detailed discussion of the energy dissipated by gates. We present the essential information here to discuss the supply voltage operating condition.

Consider the impact of supply voltage on the drive current. The saturation drive current I_{DSAT} is given by [49]

(1) $\quad I_{DSAT} = \dfrac{W}{L} B (V_{DD} - V_{TH})^{\alpha}$

where W is the effective channel width and L is the effective channel length of an equivalent inverter. V_{DD} is the supply voltage to the gate, and V_{TH} is the threshold voltage of the transistors. B and α are empirically determined constants, where α is between 1 and 2 depending on the technology. The constant B is determined by transistor characteristics [44]:

(2) $\quad B \propto \mu \varepsilon_{ox} / t_{ox}$

where μ is the carrier mobility, which is determined by the temperature, electric field strength, and impurity concentration. The electric permittivity ε_{ox} is the permittivity of the gate oxide, and t_{ox} is the thickness of the gate oxide. Reducing the gate oxide thickness is one method of increasing the speed of a process, because it increases the drive strength. Increasing the gate oxide's electrical permittivity can also increase the speed. Intel increased ε_{ox} by 10% in their 0.13um (60nm effective gate length) process [61].

A larger drive current (dis)charges load capacitances faster, increasing the speed. From (1), the drive current can be increased by increasing the width of transistors, the supply voltage, or reducing the threshold voltage. The speed can also be increased by reducing the load capacitance (e.g. using smaller gates off the critical path).

Gates with wider transistors and higher drive strength are provided in a standard cell library, and are utilized in synthesis and layout to reduce the delay on critical paths (Section 3.5 of Chapter 4 discusses standard cell libraries). While increasing the drive strength reduces the delay, the power consumption increases substantially – preceding gates must also have their size increased to drive the increased capacitance of the wider transistors. Section 5 of Chapter 4 discusses the additional energy consumption.

Increasing the supply voltage increases the drive current, but the dynamic power dissipation. The dynamic power dissipation $P_{dynamic}$ of discharging and charging a capacitance C, switching at frequency f_{toggle}, is [44]

(3) $\quad P_{dynamic} = CV_{DD}^2 f_{toggle}$

where the switching frequency is for a cycle of the capacitance's voltage from V_{DD} to ground then back to V_{DD}.

When static CMOS gates switch, there is also dynamic power consumption due to short-circuit current. When the gate switches, there is a period of time when both the pull-up and pull-down networks are conducting and there is current from the supply to ground through the gate. Short-circuit current is not significant if the input rise (fall) time is close to the output fall (rise) time [14], and is less than 10% of the dynamic power consumption if the input and output transition times are similar in a well-designed process technology [16]. Therefore the dynamic power is dominated by the capacitance charging and discharging power consumption for a well-designed circuit.

Higher supply voltage increases the drive current, speeding up the circuit, at the expense of increased dynamic power consumption. The increase in drive current with supply voltage is at best quadratic. The increase in speed with voltage is roughly linear or worse, because the charge stored on a capacitance C is CV_{DD}. For typical high-speed operating conditions (not low voltage for low power) in 0.13um and 0.18um, the dependence of the circuit speed on voltage is approximately linear [23][60][65]. From (3), the increase in power consumption is quadratic.

The transistor threshold voltage can also be decreased, but there is static power consumption due to the subthreshold leakage current I_{off}, which depends exponentially on the threshold voltage [57]:

(4) $\quad I_{off} = k \times 10^{-V_{TH}/V_{sub}}$

where k is the reverse saturation current [67], and V_{sub} is the subthreshold slope in mV/decade. It is common for high-speed designs to take advantage of low threshold voltage on critical paths. The exponential dependence of leakage current on threshold voltage limits reduction in the threshold voltage, and static power dissipation is becoming of increasing concern [57].

The transistor threshold voltages are controlled by the threshold-adjust implants in processing. Generally ASIC designers have only a few choices of standard cell library threshold voltage (e.g. a designated low threshold voltage library corresponding to high speed and high leakage, and a designated slower library with higher threshold and lower leakage). By modifying the threshold adjust implantation in the process, custom designers can optimize the transistor threshold voltages for the design [57]. The maximum supply voltage for a process is specified to ensure device reliability, primarily, to prevent breakdown of the dielectric. Still, ASIC designers can choose a fairly wide range of supply voltages for a given standard cell library, to trade off performance versus power.

1.2 A Higher Temperature Reduces Circuit Speed

A higher operating temperature increases wire resistance. It also changes the transistor threshold voltage and decreases the carrier mobility, increasing the resistance of transistors and decreasing the drive current [15]. Thus higher temperatures reduce circuit speed.

The theoretical limit for the subthreshold slope V_{sub} is

(1) $\quad V_{sub} = \log_e 10 \times kT/q = 60$ mv/decade at room temperature

where k is Boltzmann's constant, T is the temperature, and q is the charge of an electron. In general, the subthreshold slope is proportional to the temperature. Thus increasing the temperature also increases static power consumption, because the leakage increases as the subthreshold slope is proportional to the temperature [44]. The leakage current can increase by a factor of a hundred as the temperature increases from 25°C to 125°C [22].

While the chip temperature may be lower than worst case operating conditions, 125°C is a conservative estimate for modeling standard cell library delays, which allows for spot temperatures that are higher than the average temperature on the chip.

Operating temperature is limited by the ambient temperature and available cooling methods [9]. The ambient temperature near a car engine is 165°C. Many embedded applications are limited by low cost or battery lifetime (if portable) and cannot afford to use cooling devices, operating at up to 125°C with passive cooling. A higher power chip will have higher operating temperature, as more heat is dissipated.

1.3 Custom versus ASIC Operating Conditions

It is difficult to infer the impact of increasing supply voltage on the speed of PC processors. The faster processors will be able to run at the clock speed of the lower speed bins with lower supply voltage, but some of the chips in the lower speed bins will need a higher supply voltage to operate correctly. For example, when process variation causes transistors to be slower, and hence the design operates at lower clock frequency for the same supply voltage. Increasing the supply voltage is a standard technique used when over-clocking chips [68]. However, a higher supply voltage increases the power consumption, and this reduces the battery lifetime.

Temperature is the other important operating condition. Increasing the temperature of an integrated circuit requires increasing the capability of the package to withstand and dissipate heat. With an ambient room temperature (nominally 25°C) and appropriate fans and heat sinks, AMD chips have run at operating temperatures in the range of 65°C to 95°C [38], and Intel chips have run at 55°C to 100°C [39]. More exotic cooling devices can be used to lower the temperature and improve performance. KryoTech sells systems with Freon-based refrigeration, and has shown speed improvements in K6, K6-2, Athlon, Pentium Pro, Pentium II and Alpha 21164 chips by 25% to 43%, by cooling to –40°C and increasing the core supply voltage [5][20][33][53].

Table 1 lists a variety of devices for cooling PC processors. Peltier thermoelectric conductors conduct heat better from the chip, but there can be condensation. The Peltier thermoelectric conductors consume a substantial amount of power, and the hot side of the conductor is much hotter than when using water as a coolant [48]. Water and Freon coolants can dissipate substantially more heat than air-based cooling approaches, but are very expensive, requiring pumps and a method of cooling the recirculated coolant. Water is cooled via radiators exchanging heat with the surrounding air. Freon is cooled by the same technique as used in refrigerators, expanding the Freon from a liquid to a gas to cool it. Condensation causes device failure, and solutions to avoid condensation are expensive, thus cheaper water and Freon cooling methods operate at room temperature.

Cooling Method	Cost	Coolant	Power Consumption (W)	Ambient Temperature (°C)	Chip Temperature (°C)
heatsink and fan	$10 to $50	air	1 to 4	25	40 to 80
Peltier thermoelectric conductor, heatsink and fan	$50 to $100	air	50 to 80	25	0 to 10
recirculated water cooling	$150 to $230	water	15	25	40
freon vapor phase cooling at room temperature	$350 to $400	freon	140	25	30
freon vapor phase cooling	$1,000 to $2,000	freon	160	-50	-40

Table 1. Costs and temperature ranges of some available cooling devices for Intel and AMD processors [6][42][45][47][48][53][54][56][59][62][70]. The power consumption of the water cooler was estimated based on the radiator and pump components.

Exotic cooling techniques, such as Freon refrigeration, have regularly been used only in very expensive mainframes, such as the Cray [17]. The performance improvement does not generally justify the cost for use of refrigeration even in high-performance CPUs, as a 40% speed improvement is out-paced by improvements in modern processor speeds within a year. If the refrigerated processor can be upgraded without replacing the refrigeration system, then the cost of purchasing the refrigeration system can be amortized over a longer time period. However, Peltier devices and refrigeration systems also have high power consumption, which must be factored into the cost.

Both heat sinks and fans significantly increase the overall system cost by around $10 to $50 in the current market. This cost increase will only be acceptable to a few ASIC designs, as many ASICs sell for as little as several dollars. Cost-conscious ASIC designs will be limited to plastic packaging, with only passive power dissipation, dissipating around 2 W for hand-held appliances [9].

The packaging can also affect the speed. In the 1GHz Alpha microprocessor, using flip chip packaging instead of wire bond allowed about a 10% increase in clock frequency [10]. Wire bond packaging connects pins to the perimeter of the die, whereas flip chip packaging mounts pins over the surface of the chip. Wire bond packaging was worse because the Alpha can draw more than 40A of current, which results in a detrimental IR (voltage) drop in wire bonds delivering current to the center of the chip [10].

	Frequency (MHz)	Voltage (V)	Max. Power (W)	Pin Count
Mobile Celeron	1133	1.45	23.8	478 pins
Ultra Low Voltage Mobile Celeron	650	1.10	7.0	479 balls

Table 2. Characteristics of two Intel Celeron microprocessors. Both chips have dimensions of 11.2mm by 7.2mm, an area of 80mm^2, and use micro-FCPGA packages [39].

The next example illustrates the speed reduction imposed by power limitations of ASIC packaging.

1.4 Performance Penalty Imposed by the Tight Power Constraints of ASIC Designs (contributed by Michael Orshansky)

A factor that limits the maximum clock frequency achievable in ASIC designs is a lack of cheap packaging solutions comparable in their performance to the packages used for the high-end custom designs. For example, the mobile Intel Celeron processors utilize the advanced socketable micro-FCPGA and surface mount micro-FCPGA packages. These packages have a very high thermal conductivity and are capable of effectively cooling the chip. This permits a significantly higher level of power dissipation than the cheaper packages available to ASIC designers. The situation is further aggravated by the external cooling solutions available to more expensive custom chips. These solutions are too costly for ASICs, and further limit the choice of packaging solutions, and therefore, maximum power levels.

It is instructive to estimate how much performance penalty is imposed onto the ASIC chips by the requirement of lower power consumption levels, compared to the custom chips. To estimate the power-related performance loss, we calculated at what speed the fastest custom processor would run, if it had to conform to the power limitations typically encountered by the ASIC designs using typical commercial packaging solutions.

Specifically, we looked at Intel's Tualatin Pentium III microprocessors, manufactured in the advanced 0.13um process. Some of the Tualatin chips are sold as low power Celeron chips [27]. Table 2 lists the lowest power Celerons at low and high clock frequencies.

We then identified the best packages with the roughly comparable area and pin count properties available from the packaging solutions leader Amkor, as shown in Table 3. The maximum power that can be dissipated by these packages is estimated assuming an ambient temperature of 45°C, which is a typical civilian specification for the maximum ambient temperature. The maximum chip temperature is 100°C [39]. Then the maximum power dissipation is given by

(5) $P_{max} = (T_{chip} - T_{amb})/\theta_{th}$

For the TEPBGA, the maximum power consumption is about 4W. For PBGA the maximum power consumption is 2.8W.

We now estimate the speed at which the 28 million transistor Celeron would run, if it had to be placed in a cheap package, such as the packages identified above. Using the numbers in Table 2, we construct a simple semi-empirical model to predict the scaling of power and clock frequencies with V_{DD}. We assume $V_{TH} = 0.35V$. We use the model to predict the values of clock frequency and power for different values of V_{DD}.

The analysis shows (Table 4) that if placed in ASIC packaging, the Celeron processor would be limited to a speed of about 600MHz at 4W maximum dissipation, and 530MHz at 2.8W maximum power dissipation. In comparison, the fastest Tualatin model runs at 1400MHz in 0.13um with 31W of power consumption [39]. This speed difference is a factor of 2.6×. The factor is due to using cheaper ASIC packaging, and the absence of additional external cooling (such as heat sinks and fans), which limit the chip's power dissipation.

Package Type	Ball Count	Die Size (mm^2)	Thermal Resistivity (C/W)	Power (W)
TEPBGA	452	10.2×10.2	14.5	4.0
PBGA	456	10.2×10.2	20.9	2.8

Table 3. Properties of typical ASIC packages [4]. The die area can be up to 104mm^2, which fits the Celeron with a small change in aspect ratio.

Vdd (V)	Frequency (MHz)	Power (W)
1.45	1130	23.8
1.10	650	7.0
0.94	600	4.0
0.87	531	2.8
0.72	382	1.0

Table 4. The values for the first two rows are known from Table 2. We constructed a simple empirical model to project the scaling of power and clock frequency with supply voltage V_{DD} to evaluate the projected numbers.

ASIC	Frequency (GHz)	Technology (um)	Voltage (V)	Temperature (°C)	Operating Conditions	Process Conditions	Increase in Clock Frequnecy
Tensilica Xtensa (Base)	0.150	0.25	2.30	125	worst	worst	
	0.175	0.25	2.30	125	worst	typical	17%
	0.250	0.25	2.50	25	typical	typical	43%
Tensilica Xtensa (Base)	0.200	0.18	1.60	125	worst	worst	
	0.250	0.18	1.60	125	worst	typical	25%
	0.320	0.18	1.80	25	typical	typical	28%
Lexra LX4380	0.360	0.13	*1.20	125	worst	worst	
	0.420	0.13	*1.20	125	worst	typical	17%
Lexra LX5280	0.180	0.18	1.62	125	worst	worst	
	0.230	0.18	1.62	125	worst	typical	28%
Lexra LX5280	0.266	0.13	0.90	125	worst	worst	
	0.330	0.13	0.90	125	worst	typical	24%

Table 5. Simulated ASIC clock frequency variation with process and operating conditions as specified [8][34][35]. *The worst case operating voltage of the LX4380 was not known, this is the nominal voltage.

1.5 Comparison of ASIC Process and Operating Conditions

Table 5 lists the clock frequency for some high speed ASIC CPUs, for different technologies, and different process and operating conditions. The results reported for Lexra and Tensilica show 17% to 28% difference in clock frequency comparing worst case and typical process conditions. For typical process conditions of the Tensilica Xtensa Base configuration, the difference between worst case operating conditions and typical operating conditions is 28% and 43%.

Some HSPICE simulations show that fast process conditions can be up to 24% faster than slow process conditions. There is about an additional 10% improvement when the temperature is decreased from 100°C to 25°C [Andrew Chang, personal communication].

After testing at several different clock frequencies, chips can be speed binned into groups of chips that have the same valid operating frequency. The cost overhead for testing chips at different clock frequencies is worthwhile if you can sell the faster chips at higher prices. Generally, ASIC chips are sold for fairly low prices per chip. ASIC vendors typically can't afford the additional cost per chip of testing at different clock frequencies.

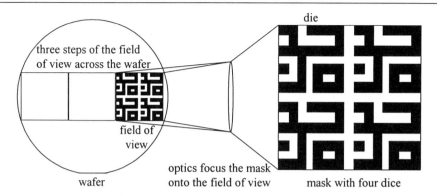

Figure 1. A simplified diagram of the mask, field of view, and die. The field of view is stepped across the wafer by stepper motors.

We can further quantify the difference between process conditions by examining variation in clock frequencies from foundries for more expensive custom processors.

2. CHIP SPEED VARIATION DUE TO STATISTICAL PROCESS VARIATION

The semiconductor process cannot be perfectly controlled, which leads to statistical variation of many process variables. Several types of process variations can occur: line-to-line, batch-to-batch, wafer-to-wafer, die-to-die, and intra-die (intra-chip).

To process a wafer, light is shone through a mask then focused by optics onto the wafer to expose selected areas of photoresist, as shown in Figure 1. Each *die* on a mask has the complete pattern of photoresist to be exposed corresponding to an individual chip. Within the *field* of view through the optics, the area of several die is exposed simultaneously. *Steppers* then move to the next location to expose the wafer with the same mask. The sequence is repeated to process the next layer [69].

Process variation can occur for multiple reasons, including non-uniform ion implantation or photoresist exposure, or lack of uniformity in oxidation, diffusion, or polishing. Variation between the production lines can occur because of differences in masks, steppers, and the optics at each line. There can be variations between each exposed field on a wafer, because of slight changes in the distance between the wafer and the optics, due to the steppers or the wafer not being perfectly flat. Varying illumination and lens aberrations can lead to a large intra-field variation of effective channel length and speed [72]. The intra-field variation can cause a 15% to 20% difference in ring oscillator speeds across the field [40]. Optical proximity effects will also cause identical patterns on the mask to be printed differently, because of optical interference from different neighboring patterns [72]. Chapter 14 discusses the impact of process variation on circuit performance in greater detail.

These process variations cause the delays of wires and gates within a chip to vary, and chips are produced with a range of working speeds. As a result, some chips may only operate correctly at slower speeds. We can estimate the total variation in clock speed caused by these process factors, by looking at the range of clock frequencies of custom processors.

Table 6 lists Intel and AMD x86 chips at the time of introduction of an architecture or technology from 0.35um to 0.13um. The difference between the slowest and fastest clock frequencies available at the time of introduction is up to 47%. Disregarding the four bold cases where higher supply voltages were used to achieve higher clock frequency, the difference between slowest and faster frequencies is up to 38%.

Intel	Date of Introduction	Technology (um)	Slowest Clock Frequency (GHz)	Core Voltage (V)	Fastest Clock Frequency (GHz)	Core Voltage (V)	Difference in Frequnacy
Pentium II (Klamath)	1997/05/07	0.35	0.233	2.80	0.300	2.80	29%
Pentium II (Deschutes)	1998/01/26	0.25	0.333	2.00	0.333	2.00	0%
Pentium III (Katmai)	1999/02/26	0.25	0.450	2.00	0.500	2.00	11%
Pentium III (Coppermine)	1999/10/25	0.18	0.500	**1.60**	0.733	**1.65**	47%
Pentium III (Tualatin)	2001/07/18	0.13	1.000	1.48	1.200	1.48	20%
Pentium 4 (Willamette)	2000/11/20	0.18	1.300	1.70	1.500	1.70	15%
Pentium 4 (Northwood)	2002/01/07	0.13	1.600	1.50	2.200	1.50	38%
AMD							
K6 (Model 6)	1997/05/07	0.35	0.166	**2.90**	0.233	**3.20**	40%
K6 (Little Foot - Model 7)	1997/01/06	0.25	0.233	2.20	0.266	2.20	14%
K6-2 (Chompers - Model 8)	1998/05/28	0.25	0.266	2.20	0.333	2.20	25%
K6-2 (Chompers - Model 8 CXT)	1998/11/16	0.25	0.300	2.20	0.400	2.20	33%
K6-3 (Sharptooth - Model 9)	1999/02/22	0.25	0.400	2.40	0.450	2.40	13%
Athlon (K7)	1999/07/09	0.25	0.500	1.60	0.650	1.60	30%
Athlon (K75)	1999/11/29	0.18	0.750	1.60	0.750	1.60	0%
Athlon (Thunderbird)	2000/06/05	0.18	0.750	**1.70**	1.000	**1.75**	33%
Athlon (Palomino)	2001/06/05	0.18	1.000	1.75	1.200	1.75	20%

Table 6. Clock frequencies of custom chips at the official introduction date of a new model, in different technologies [27][28][38][39]. The first version of each architecture is indicated in bold. If the core voltages of slow and fast versions differ, those voltages are indicated in bold.

Process Technology (um)	ITRS Printed Gate Length (um)	ITRS Physical Gate Length (um)	Intel Gate Length (um)	AMD Gate Length (um)
0.25	0.200	0.200	0.180	0.160
0.18	0.140	0.120	0.100	0.100
0.13	0.090	0.065	0.060	0.080

Table 7. Comparison of reported Intel and AMD gate lengths with the International Technology Roadmap for Semiconductors [19][23][24][43][51][50][52][60].

For marketing purposes, AMD and Intel will sometimes bin processors to slower clock frequencies than necessary for correct operation, and these chips are easily over clocked [68]. Hence the range of maximum operating clock frequencies may be less. The variation also decreases as the process matures, as factors causing sub-optimal performance are more tightly controlled. The Pentium II (Deschutes) was sold at clock frequencies of 266MHz (25% slower!) and 300MHz to meet demand, after initially being released with at 333MHz [55]. The Athlon (K75) was initially available only at 750MHz [2], and was sold at a slower frequency later. The Pentium 4 (Northwood) 1.6A has been over-clocked from its nominal clock frequency of 1.6 GHz to 2.2 GHz [26], which is expected given Intel's claims of 0.13um circuits being up to 65% faster than in 0.18um [32].

Trying to remove any bias caused by speed binning to lower speeds than necessary, we can look at the first processor introduced in each architecture, listed in bold in Table 6 (the K6-2 and K6-3 may or may not be considered sufficiently different from the K6 to include in this group, but the range is not any larger by including these). Then disregarding the K6 (Model 6) where a higher supply voltage was used to achieve higher clock frequency, the difference between slowest and faster frequencies is up to 30%.

3. CONTINUOUS PROCESS IMPROVEMENT

Top semiconductor manufacturers continuously improve the process technology and the designs within a given process generation. The National and International Technology Roadmaps for Semiconductors predicted printed gate lengths for a given process technology, but Intel and AMD have aggressively scaled gate lengths beyond this, as shown in Table 7. To correspond with the effective gate lengths from aggressive scaling, the roadmap has been updated to include the more aggressive scaling of the effective gate length.

Intel Process	Process Technology (um)	Original Gate Length (um)	Improved Gate Length (um)	Speed Improvement	Drive Current Improvement
P856 to P856.5	0.25	0.18		18%	21%
P858 to P858.5	0.18	0.14	0.10	10%	11%
P860	0.13	0.07	0.06		10%

Table 8. Intel's process speed and drive current improvements with reductions in gate width [11][12][23][60][65][71]. The 0.13um processes will show further improvements beyond what is detailed here as the technology matures. The 11% improvement reported in 0.18um was for the speed of a ring oscillator [23]

Table 8 details publications of Intel's technology scaling within each generation, beyond the initial gate length in the process [18]. In Intel's 0.25um P856 process the dimensions were shrunk by 5%, and along with other modifications, this gave a speed improvement of 18% in the Pentium II [12]. There was also a 5% linear shrink in the P860 process and the effective gate length was reduced from 70nm to 60nm [60].

AMD has released less information about its process technology. AMD's 0.25um process had 0.16um effective gate length [24], and reportedly AMD has been aggressively scaling drawn gate length from 0.12um to 0.10um in its 0.18um process technology [19]. AMD has produced 0.18um Athlon (Thunderbird) chips with aluminum interconnect at Fab 25 in Austin, and faster copper interconnect at Fab 30 in Dresden [3]. Thus AMD also has substantial variation within a process generation.

Table 9 illustrates the total range between slowest and fastest clock frequencies of processors sold by Intel and AMD. There is up to 126% difference between the slowest and fastest chips sold for a specific design within the same process generation, but this is not indicative of only process improvements. This difference is due to several reasons: some chips being binned to slower clock frequencies; higher supply voltage; process improvements, such as using lower threshold voltages to increase gate drive strength, or reducing the effective gate length; and some of the improvement is from changes in the design to improve critical paths.

Finding Peak Performance in a Process 159

Intel	Technology (um)	Slowest Clock Frequency (GHz)	Core Voltage (V)	Date Fastest Introduced	Fastest Clock Frequency (GHz)	Core Voltage (V)	Difference in Frequnecy
Pentium II (Klamath)	0.35	0.233	2.80	1997/05/07	0.300	2.80	29%
Pentium II (Deschutes)	0.25	0.333	2.00	1998/08/24	0.450	2.00	35%
Pentium III (Katmai)	0.25	0.450	2.00	1999/08/02	0.600	2.05	33%
Pentium III (Coppermine)	0.18	0.500	1.60	2001/07/30	1.130	1.80	126%
Pentium III (Tualatin)	0.13	1.000	1.48	2002/01/08	1.400	1.45	40%
Pentium 4 (Willamette)	0.18	1.300	1.70	2001/08/27	2.000	1.75	54%
Pentium 4 (Northwood)	0.13	1.600	1.50	2002/01/07	2.200	1.50	38%
AMD							
K6 (Model 6)	0.35	0.166	2.90	1997/05/07	0.233	3.20	40%
K6 (Little Foot - Model 7)	0.25	0.233	2.20	1998/04/07	0.300	2.20	29%
K6-2 (Chompers - Model 8)	0.25	0.266	2.20	1998/08/27	0.350	2.20	32%
K6-2 (Chompers - Model 8 CXT)	0.25	0.300	2.20	2000/02/22	0.550	2.30	83%
K6-3 (Sharptooth - Model 9)	0.25	0.400	2.40	1999/02/22	0.450	2.40	13%
Athlon (K7)	0.25	0.500	1.60	1999/10/04	0.700	1.60	40%
Athlon (K75)	0.18	0.750	1.60	2000/03/06	1.000	1.80	33%
Athlon (Thunderbird)	0.18	0.750	1.70	2001/06/06	1.400	1.75	87%
Athlon (Palomino)	0.18	1.000	1.75	2002/01/07	1.773	1.75	77%

Table 9. A comparison of the slowest clock frequencies of custom chip models versus the fastest produced during their commercial lifetime [27][28][38][39]. The 0.13um models will have further speed improvements.

4. SPEED DIFFERENCES DUE TO ALTERNATIVE PROCESS IMPLEMENTATIONS

In 0.18um, AMD used copper wiring whereas Intel used aluminum. Aluminum has higher resistance than copper and performance is somewhat worse as a result [25]. In 0.13um both companies' processes use copper interconnect. Intel's 0.13um process is bulk CMOS with 0.06um gate length and dielectric with relative permittivity of 3.6 [60]. AMD and Motorola in conjunction have a 0.13um silicon-on-insulator process with 0.08um gate length and dielectric with relative permittivity of less than 3.0 [43]. A dielectric with smaller permittivity reduces the capacitance between wires, and thus decreases crosstalk. According to IBM, silicon-on-insulator is 20% to 25% faster, and it is also lower power [31]. Also, smaller gate lengths reduce the resistance of a transistor. All these factors affect the speed and differentiate foundries' processes in the same process technology generation.

Intel	Technology (um)	Physical Gate Length (um)	Core Voltage (V)	n-channel MOSFET Threshold Voltage (V)	Metal Layers	Metal	Dielectric	Ring Oscillator Delay/Stage (ps)	Worst Case Ioff (na/um)
P858.5	0.18	0.10	1.5	0.40	6	Al	FSG (3.6)	10	10
P860	0.13	0.06	1.4	0.27	6	Cu	FSG (3.6)	6	10
AMD									
HiP7	0.13	0.08	1.2		9	Cu	low-k (< 3.0)		10
TSMC									
CL018LV	0.18	0.13	1.5	low	6	AlCu	low-k	25	1
CL015HS	0.15	0.11	1.5	low	7	Cu	low-k	14	17
CL013HS+	0.13	0.08	1.2	low	8	Cu	FSG/low-k	10	15
UMC									
L180	0.18		1.8		6	Al 4, Cu 2		27	0.01
L150 MPU	0.15	0.10	1.5	0.44	7	Al 5, Cu 2		14	7
L130 MPU	0.13	0.08	1.2	0.36	8	Cu	SiLK (2.7)	11	< 10

Table 10. A comparison of fastest custom processes from Intel and AMD, with fastest ASIC foundry processes reported from TSMC, UMC and IBM. Blanks appear where data was unavailable. [23][30][60][64][65][66][71]. HiP7 is a silicon-on-insulator process developed by AMD and Motorola [43]; the other processes are bulk CMOS.

Table 10 illustrates process differences between ASIC and custom foundries. A ring of inverters oscillating, with their output into the next inverter, has been used to characterize the speed of most of these technologies. However, the delay of a ring oscillator is independent of device sizing and has negligible interconnect [58], so it is not a particular good metric for comparing technologies. The difference between aluminum and copper interconnect, and the choice of dielectric, will not be apparent from the ring oscillator stage delay. Furthermore, the process conditions and operating conditions affect the delay, but what the process and operating conditions are is not stated.

Comparing Intel's reported ring oscillator stage delays with TSMC and UMC, Intel's processes have been significantly faster. The custom foundries in Table 10 have not reported intermediate 0.15um process generations, whereas the ASIC foundries more regularly make technologies available to customers – as another faster or lower power choice within a generation, or as an intermediate process generation. Comparing contemporary ASIC and custom processes in 1998, IBM's CMOS7S 0.22um ASIC process and Intel's P856.5 0.25um custom process both had ring oscillator stage delays of 22ps [21]. There are ASIC foundries with processes of comparable speed to fabrication plants used for custom chips.

Accurate comparisons are difficult because the ring oscillator delays reported do not account for interconnect delay. We can infer that the fastest processes used may be up to 20% to 40% faster than the fastest process available from some ASIC foundries, by considering the differences due to gate length, dielectric, silicon-on-insulator, and use of copper. Given the variety of techniques being considered and developed for future high-speed processes, we can expect to see a fairly wide range of performance..

Slower processes will generally be cheaper. ASIC foundries offer several processes within a process generation: slower, low power applications; high density processes; and high-speed alternatives.

4.1 The Effects of Process and Tuning (contributed by Andrew Chang and William J. Dally)

Advances in process technology combined with additional tuning can measurably increase the performance of a given design. Table 11 summarizes the benefit of a wide selection of the possible options. The listed options are equally applicable to ASICs and custom chips, though they have first appeared in custom designs, which can afford to use more expensive processes. For the average ASIC (60 FO4 delays), a 10% improvement in performance is equivalent to a 6 FO4 reduction in delay.

5. PROCESS TECHNOLOGY FOR ASICS

The design rules for an ASIC process must be fixed for standard cell library design. If there are process improvements, then the library must be characterized again and possibly redesigned (if the design rules change) to take advantage of these changes. If the old version of the library is used, then potentially as much as a 20% possible improvement in speed is lost.

Technique	Benefit	Reference/Notes
L_{eff} shrinks	10%	Hurd [29]
Low V_{TH} devices	10%	Buchholtz [13]
SOI	5-25%	Anderson [7]
		15% for static
		5% for dynamic
		>20% for pass gates
Copper + low K dielectric	<5%	Rohrer [46]
Tuning the design to the process	10%	Hurd [29]
Bulk biasing	0-10%	Tschanz [63]
Yield improvement		Standard deviation can be halved. Tschanz [63]

Table 11. Potential improvements due to process technology and tuning.

Fabrication plants won't offer ASIC customers the top chip speed off the production line, as they cannot guarantee a sufficiently high yield for this to be profitable. The fabrication plant guarantees that they can produce an ASIC chip with a certain speed. This speed is limited by the worst speeds off the production line, but chips capable of faster speeds are produced.

ASIC designers may not have access to the best fabrication plants in a particular technology generation for production of their chips.

If a specific speed is required, then the chip needs to be designed to operate at this clock frequency in worst case process conditions to ensure sufficient yield. The clock frequency needed may be set by standards or application requirements. Even if the fabricated ICs are faster than worst case, there may be no additional benefit to running faster than the target speed. Then the speed difference between typical process and worst case process conditions is lost. Both ASIC and custom designs may be limited in this manner.

5.1 Migrating ASIC Process Technology

To take advantage of process improvements, the library must be updated with the new process rules. To mitigate the overhead of redesigning the library, there are software tools that help migrate standard cell libraries. ASIC foundries can regularly update their library for the latest improvement, and release libraries for a new process sooner. The fastest processes available from ASIC foundries are fairly comparable to their contemporary custom counterparts.

Thus synthesizable ASIC designs should not lag custom designs because of process improvements and differences in process technology, if the fastest process available with standard cell libraries and regular updates is affordable. This depends on how much the chip can be sold for. Often it is more profitable to sell a slower chip that is produced in a substantially cheaper process.

It is much easier to retarget a standard cell ASIC to a different library when a new technology becomes available. To migrate to a different process generation, custom designs require substantial work from a large team of engineers as they cannot simply target the synthesis tools to a different standard cell library.

In custom designs, individual transistors will need resizing and circuits may need to be redesigned to account for the design rules, supply voltage, current, and power not scaling linearly with the process shrink. However, custom optimizations can take full advantage of a process, whereas standard cell libraries may be more conservative.

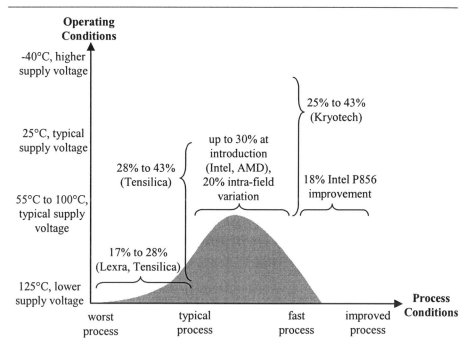

Figure 2. Diagram of speed improvement with better process and operating conditions for a given process technology generation. An approximate process yield curve is shaded at the bottom: typical process conditions correspond to acceptable yield, maybe only 10% of processors are fast enough for the fastest speed bin; and a few non-defective chips will only work correctly under the worst case process conditions. The temperatures listed are the operating temperature on the chip, not the ambient surrounding temperature.

5.2 Summary

Figure 2 gives an overview of the speed differences due to operating and process conditions. Speeds at best case operating conditions can be a factor of ×1.6 faster than speeds at worst case operating conditions. The small percentage of chips that are produced at fast process conditions may be a factor of ×1.4 faster than the speed of chips estimated for worst case process conditions. Thus fast process and fast operating conditions can be more than a factor of ×2.2 faster than worst case process and worst case operating conditions.

In addition to the speed variation of chips produced from a fabrication plant, there are process technology improvements and differences between foundries. There may be a 20% improvement within a process generation, and 20% to 40% difference between foundries. Overall due to process and operating conditions, the fastest custom processor could be a factor of ×3 faster than an ASIC produced in an older, slower process, for worst case operating and process conditions.

This is pessimistic, as operating temperature does not contribute to the clock frequency difference between custom and ASIC designs, if they are designed for the same temperature. Comparing only process conditions, the fastest custom processor could be a factor of ×2 faster than an ASIC produced in an older, slower process, for worst case process conditions. Careful manual design can create custom designs that are lower power than ASICs, and hence may be able to use a higher supply voltage to increase performance compared to an ASIC at the same power consumption and operating temperature.

Many embedded designs are used in a wide range of operating conditions and are limited by the worst operating conditions for which they are expected to operate correctly. Thus when compared to designs that are optimized for a less pessimistic operating point, they will not perform as well. An advantage of synthesizable designs is that they can be easily optimized for different operating conditions, by redoing the synthesis and layout – the price is fabricating different chips for different operating conditions, which may be too costly.

Fabrication plants won't offer ASIC customers the top chip speed off the production line, as they cannot guarantee a sufficiently high yield for this to be profitable. The fabrication plant guarantees that they can produce an ASIC chip with a certain speed. This speed is limited by the worst speeds off the production line, but chips capable of faster speeds are produced. Also, ASIC designers may not have access to the best fabrication plants in a particular technology generation for production of their chips.

6. POTENTIAL IMPROVEMENTS FOR ASICS

To improve ASIC performance, designers can test chips to see if performance is better than worst case process conditions, and higher speed are available at lower yields. ASIC designs are not limited by the worst speeds off the production line.

Foundries generally require sign off at worst case process conditions, so that they can guarantee a certain level of performance. However, produced chips can be tested to verify correct operation at higher speeds in typical process conditions. This may allow a 20% to 30% improvement in speed over worst-case speeds (see Table 5). This requires discussion with the semiconductor foundry and exploration of the yield curve for speeds of the given design – this is considered further in Chapter 14. If there is a sufficient premium for faster chips, it can be worthwhile to test ASIC chips at two clock frequencies and speed bin them.

If operating temperature and power consumption are not limited, then the supply voltage can also be increased to speed up the chip. Otherwise, it may not be possible to improve the chip's performance by allowing it to operate at a higher temperature, because cooling devices become increasingly expensive.

7. REFERENCES

[1] Allen, D. H., et al., "Custom circuit design as a driver of microprocessor performance," *IBM Journal of Research and Development*, vol. 44, no. 6, November 2000.
[2] AMD, AMD Athlon™ Processor Sets New Performance Leadership Mark At 750MHz Built On Advanced 0.18 Micron Technology, November 29, 1999. http://www.amd.com/us-en/Corporate/VirtualPressRoom/0,,51_104_543_553~713,00.html
[3] AMD, AMD Showcases High-Powered Workstations At Solidworks World 2001. February 12, 2001. http://www.amd.com/us-en/Corporate/VirtualPressRoom/0,,51_104_543_4493~640,00.html
[4] Amkor, PBGA data sheet, 2002. http://www.amkor.com/Products/all_datasheets/pbga.pdf
[5] AnandTech, Kryotech Super G2 Athlon 1.86 GHz. August 2001. http://www.anandtech.com/showdoc.html?i=1515&p=1
[6] AnandTech. MC1000 & MC2000 Peltier Coolers. August 1999. http://www.anandtech.com/cooling/showdoc.html?i=1012&p=1
[7] Anderson, C, et al., "Physical Design of A Fourth-Generation POWER GHz Microprocessor," *IEEE International Solid-State Circuits Conference*, 2001.
[8] Archide, R. (ed.). *Xtensa Application Specific Microprocessor Solutions Overview Handbook, a Summary of the Xtensa Data Sheet*. February 2000. http://www.tensilica.com/dl/handbook.pdf
[9] Bar-Cohen, A. "Thermal Packaging for the 21st Century: Challenges and Options," *International Workshop Thermal Investigations of ICs and Systems*, Rome, Italy, October, 1999.
[10] Benschneider, B., et al., "A 1 GHz Alpha Microprocessor," *Digest of Technical Papers of the International Solid-State Circuits Conference*, 2000, pp. 86-87.
[11] Bohr, M., et al., "A high performance 0.25um logic technology optimized for 1.8 V operation," *Technical Digest of the International Electron Devices Meeting*, 1996, pp. 847-850, 960.
[12] Brand, A., et al., "Intel's 0.25 Micron, 2.0 Volts Logic Process Technology," *Intel Technology Journal*, Q3 1998. http://developer.intel.com/technology/itj/q31998/pdf/p856.pdf.
[13] Buchholtz, T. et al., "A 660MHz 64b SOI Processor with Cu Interconnects," *IEEE International Solid-State Circuits Conference*, 2000.
[14] Burd, T. "Low-Power CMOS Library Design Methodology," M.S. Report, University of California, Berkeley, UCB/ERL M94/89, 1994. http://bwrc.eecs.berkeley.edu/Publications/theses/low.power.CMOS.library.MS/
[15] Chain, K., et al., "A MOSFET Electron Mobility Model of Wide Temperature Range (77-400 K) for IC Simulation," *Semiconductor Science and Technology*, vol. 12, no. 4, April 1997, pp. 355-358.

[16] Chatterjee, A., et al., "An Investigation of the Impact of Technology Scaling on Power Wasted as Short-Circuit Current in Low Voltage Static CMOS Circuits," *International Symposium on Low Power Electronics and Design*, 1996, pp. 145-150.
[17] Cray. Cray Inc. History. 2002. http://www.cray.com/company/history.html
[18] Chau, R., Marcyk, G. Intel Technology Briefing. December 11, 2000. http://www.intel.com/research/silicon/30nmpress.pdf
[19] De Gelas, J. AMD's Roadmap. February 28, 2000. http://www.aceshardware.com/Spades/read.php?article_id=119
[20] Dicarlo, L. "Alpha Chills to 767," in *PC Week*, vol. 15, no. 15, April 13, 1998. http://www.zdnet.com/eweek/acrobat/1998/98pcweek/apr13mb.pdf
[21] Diefendoff, K., "The Race to Point One Eight," Microprocessor Report, vol. 12, no. 12, September 1998, pp. 10-22.
[22] Europractice, UMC 0.18µ 1P6M Logic process with VST libraries. August 2000. http://www.europractice.imec.be/europractice/on-line-docs/prototyping/ti/ti_VST_UMC18_logic.html
[23] Ghani, T., et al., "100 nm Gate Length High Performance / Low Power CMOS Transistor Structure," *Technical Digest of the International Electron Devices Meeting*, 1999, pp. 415-418.
[24] Golden, M., et al., "A Seventh-Generation x86 Microprocessor," *IEEE Journal of Solid-State Circuits*, vol. 34, no. 11, November 1999, pp. 1466-1477.
[25] Green, P.K., "A GHz IA-32 Architecture Microprocessor Implemented on 0.18um Technology with Aluminum Interconnect," *Digest of Technical Papers of the International Solid-State Circuits Conference*, 2000, pp. 98-99.
[26] [H]ard|OCP, Intel Pentium 4 Northwood 1.6A CPU Overclock, February 1, 2002. http://www.hardocp.com/reviews/cpus/intel/p416a/
[27] Hare, C. 586/686 Processors Chart. February 2002. http://users.erols.com/chare/586.htm
[28] Hare, C. 786 Processors Chart. February 2002. http://users.erols.com/chare/786.htm
[29] Hurd, K., "A 600MHz 64b PA-RISC Microprocessor," *IEEE International Solid-State Circuits Conference*, 2000.
[30] IBM, IBM ASIC Standard Cell/Gate Array Products, January 2002. http://www-3.ibm.com/chips/products/asics/products/stdcell.html
[31] IBM, SOI Technology: IBM's Next Advance In Chip Design, January 2002. http://www-3.ibm.com/chips/bluelogic/showcase/soi/soipaper.pdf
[32] Intel, Intel Completes 0.13 Micron Process Technology Development, November 7, 2000. http://www.intel.com/pressroom/archive/releases/cn110700.htm
[33] Kryotech, Kryotech Announces 1.86GHz SuperG2™ PC Platform, August 21, 2001. http://www.kryotech.com/news_release.html
[34] Lexra, Lexra LX4380 Product Brief. June 2001. http://www.lexra.com/LX4380_PB.pdf
[35] Lexra, Lexra LX5180, LX5280 Product Brief. December 2002. http://www.lexra.com/LEX_ProductBrief_LX5K.pdf
[36] Liou, F., "We See the Future and It's Copper," *Fabless Semiconductor Association Fabless Forum*, September 1999. http://www.fsa.org/fablessforum/0999/liou.pdf
[37] McDonald, C., "The Evolution of Intel's Copy Exactly! Technology Transfer Method," *Intel Technology Journal*, Q4 1998. http://developer.intel.com/technology/itj/q41998/pdf/copyexactly.pdf.
[38] MTEK Computer Consulting, AMD CPU Roster, January 2002. http://www.cpuscorecard.com/cpuprices/head_amd.htm
[39] MTEK Computer Consulting, Intel CPU Roster, January 2002. http://www.cpuscorecard.com/cpuprices/head_intel.htm

[40] Orshansky, M., et al., "Intra-Field Gate CD Variability and Its Impact on Circuit Performance," *International Electron Devices Meeting*, 1999, pp. 479-82.
[41] Orshansky, M., et al., "Characterization of spatial CD variability, spatial mask-level correction, and improvement of circuit performance," *Proceedings of the International Society for Optical Engineering*, vol. 4000, pt.1-2, 2000, pp. 602-611.
[42] Overclockers Australia. SocketA Cooler Roundup. May 2001. http://www.overclockers.com.au/techstuff/r_socketa/index.shtml
[43] Perera, A.H., et al., "A versatile 0.13um CMOS Platform Technology supporting High Performance and Low Power Applications," *Technical Digest of the International Electron Devices Meeting*, 2000, pp. 571-574.
[44] Rabaey, J.M., *Digital Integrated Circuits*, Prentice-Hall, 1996.
[45] Rolfe, J. Chipset Cooler Roundup. February 2001. http://www.overclockers.com.au/techstuff/r_gpu_cool/index.shtml
[46] Rohrer, N., et al., "A 480 MHz RISC Microprocessor in 0.12um Leff CMOS Technology with Copper Interconnects," *IEEE International Solid-State Circuits Conference*, 1998.
[47] Rutter, D. CPU coolers compared! May 2001. http://www.dansdata.com/coolercomp.htm
[48] Rutter, D. Picking a Peltier. March 2001. http://www.dansdata.com/pelt.htm
[49] Sakurai, T., and Newton, R., "Delay Analysis of Series-Connected MOSFET Circuits," *Journal of Solid-State Circuits*, vol. 26, no. 2, February 1991, pp. 122-131.
[50] Semiconductor Industry Association, International Technology Roadmap for Semiconductors 2000 Update: Overall Roadmap Technology Characteristics. December 2000. http://public.itrs.net/Files/2000UpdateFinal/ORTC2000final.pdf
[51] Semiconductor Industry Association, International Technology Roadmap for Semiconductors 2001 Edition: Executive Summary. 2001. http://public.itrs.net/Files/2001ITRS/Home.htm
[52] Semiconductor Industry Association, "Technology Needs," in the *National Technology Roadmap for Semiconductors*, 1997 Edition, December 1997.
[53] Steinbrecher, T. The Heatsink Guide: KryoTech Athlon-900 review. May 2000. http://www.heatsink-guide.com/kryoathlon.htm
[54] Steinbrecher, T. The Heatsink Guide: KryoTech Renegade review. May 1999. http://www.heatsink-guide.com/renegade.htm
[55] Stiller, A., "Processor Talk: Of Tumbling Markets and Survival Strategies," translation by E. Wolfram from *c't magazine*, vol. 18/98, September 15, 1998. http://www.heise.de/ct/english/98/18/024/
[56] Swiftech. Swiftech Home Page. September 2001. http://www.swiftnets.com/
[57] Sylvester, D., and Kaul, H., "Power-Driven Challenges In Nanometer Design," *IEEE Design and Test of Computers*, vol. 18, no. 6, 2001, pp. 12-21.
[58] Sylvester, D., and Keutzer, K., "Getting to the Bottom of Deep Submicron," presentation at the *International Conference on Computer-Aided Design*, November 8-12, 1998. http://www-device.eecs.berkeley.edu/~dennis/talks/iccad.pdf
[59] The Card Cooler. CPU Coolers. October 2001. http://www.thecardcooler.com/shopcart/CPU_Cooling/cpu_cooling.html
[60] Thompson, S., et al., "An Enhanced 130 nm Generation Logic Technology Featuring 60 nm Transistors Optimized for High Performance and Low Power at 0.7 – 1.4 V," *Technical Digest of the International Electron Devices Meeting*, 2001.
[61] Thompson, S., et al., "An Enhanced 130 nm Generation Logic Technology Featuring 60 nm Transistors Optimized for High Performance and Low Power at 0.7 – 1.4 V," talk slides from the *International Electron Devices Meeting*, 2001.

[62] Tom's Hardware Guide. Water Coolers: Four Power Kits Starting at 200 Dollars. January 2002. http://www6.tomshardware.com/cpu/02q1/020102/

[63] Tschanz, J. et al., "Adaptive Body Bias for Reducing the Impacts of Die-to-Die and Within-Die Parameter Variations on Microprocessor Frequency and Leakage," *IEEE International Solid-State Circuits Conference*, 2002.

[64] TSMC, Technology and Manufacturing, January 2002. http://www.tsmc.com/technology/

[65] Tyagi, S., et al., "A 130 nm generation logic technology featuring 70 nm transistors, dual Vt transistors and 6 layers of Cu interconnects," *Technical Digest of the International Electron Devices Meeting*, 2000, pp. 567-570.

[66] United Microelectronics Corporation, Foundry Service Guide, 2001. http://www.umc.com/english/pdf/1_Technology.pdf

[67] Weste, N., and Eshraghian, K., *Principles of CMOS VLSI Design*, Addison-Wesley, 1992.

[68] Wojnarowicz, J. Overclocking Basics Guide, March 30, 2000. http://firingsquad.gamers.com/guides/overclockingbasicsver4/default.asp

[69] Wolf, S., and Tauber, R. N., *Silicon Processing for the VLSI Era, Volume 1: Process Technology*, Lattice Press, 1986.

[70] Wyntec, Coolers for CPU, Hard Disks, Systems, Notebooks, Monitors and Chipsets. January 2002. http://www.wyntec.com.au/coolers.htm

[71] Yang, S., et al., "A High Performance 180 nm Generation Logic Technology," *Technical Digest of the International Electron Devices Meeting*, 1998, pp. 197-200.

[72] Yu, C., et al., "Use of Short-Loop Electrical Measurements for Yield Improvement," *IEEE Transactions On Semiconductor Manufacturing*, vol. 8, no. 2, May 1995, pp. 150-159.

Chapter 6

Physical Prototyping Plans for High Performance
Early Planning and Analysis for Area, Timing, Routability, Clocking, Power and Signal Integrity

Michel Courtoy, Pinhong Chen, Xiaoping Tang, Chin-Chi Teng, Yuji Kukimoto
Silicon Perspective, A Cadence Company,
3211 Scott Blvd, Suite 100,
Santa Clara, CA 95054

A design process requires four major steps: planning, synthesis/optimization, analysis and verification. For example, to meet a timing target, a chip design has to be timing budgeted first, synthesized using building blocks, analyzed by a delay calculator, and finally verified against the timing constraints. In addition to timing, these four steps exist extensively in the other chip design objectives such as area, routability, clocking, power, and signal integrity. However, current physical implementation tools addressing only local optimization are unable to organize a global plan and predict the chip performance or design trade-off in advance. As a result, the tools leave performance on the table or cannot close the design.

This chapter describes a new physical design planning methodology: physical prototyping, which improves the traditional physical implementation approach by pulling in early design evaluation and performance estimation using fast and correlated synthesis and analysis.

1. INTRODUCTION

To successfully design and implement future generations of high-performance, multi-million gates, deep submicron (DSM) system-on-chip (SoC) integrated circuits (ICs), the front-end logic design and back-end physical design should be linked in a more collaborative manner. However, current EDA point tools are typically geared for use exclusively by one group or the other. Front-end logic designers resist engaging in physical design issues. Back-end physical designers in an ASIC flow rarely have sufficient design perspective to make high-level micro-architectural changes. The methodology relies on an iterative process between the front-end and the back-end. Moreover, in an interconnect-dominated design, simply iterating between existing tools can no longer ensure that the design process will converge to a physical implementation that meets both schedule and performance constraints. Performance constraints include timing, area, signal integrity and power requirements. Schedule constraints mean the time schedules demanded by the competitive time-to-market pressures. As a result, designers either resort to overly pessimistic designs (with wider than necessary guard banding), or suffer from the pain of many iterations between synthesis and layout. Each back-end iteration currently takes several days in order to discover if the chip can meet its constraints. Therefore, it is crucial to establish a more efficient and faster way to check design feasibility and allow trial-and-error analysis to a successful physical implementation.

On the other hand, given the size and complexity of today's chips, it is virtually inconceivable to flat-place and route all the layout objects. Thus a hierarchical design method has to be applied to manage the design complexity while preserving optimization quality comparable with the flat design approach.

The problem with traditional floorplanning is its lack of accurate physical information. One approach to address this drawback is a faster physical implementation for better prediction of the performance given the area, timing, routing resource constraints. Based on the fast physical implementation, the circuit can be partitioned more effectively into a set of macro blocks, each of which is either a set of standard cells or an IP block (design reuse). In this chapter, we will introduce a new approach to floorplanning and physical implementation, called *physical prototyping* [2][5]. Physical prototyping is a fast implementation of physical design to evaluate design trade-offs and create a realistic implementation plan.

2. FLOORPLANNING

Given a circuit netlist, traditional floorplanning tries to determine the following design configurations for a chip.
1. Block location,
2. Block shape and orientation,
3. Pin assignment,
4. Block timing budget,
5. Power routing, and
6. Clock tree.

The objectives are to meet the chip area, timing, power, and routing resource constraints. All of these problems have been studied for more than a decade (see [18] for detailed discussion), but the results still leave room for improvement due to the huge search space. For example, packing N rectangular blocks in a chip may need to search $(N!)^2 \times 2^N$ possibilities based on sequence-pair representation to find the optimum solution. This complexity is one of the reasons why the ASIC approach cannot achieve the same quality as a full-custom design. Many approaches have been reported for floorplanning in the last few decades [15][18]. They can be classified into three categories:

1. Constructive method: builds a floorplan based on a set of constraints. A partial floorplan is gradually refined until a final floorplan is obtained. Belonging to this class are cluster growth, partitioning, min-cut, mathematical programming, and rectangular dualization.
2. Knowledge-based method: derives a floorplan by applying generation rules in a floorplan knowledge database. Semi-custom floorplan generation would fall into the class.
3. Stochastic method: starts from an initial floorplan. It goes through a sequence of perturbations to search an optimal floorplan. Typical stochastic techniques are simulated annealing and genetic algorithm. The stochastic approaches have been widely used for floorplanning. The implementation of a stochastic scheme relies on a floorplan representation, where a neighboring solution is generated and examined by perturbing the representation (called a 'move'). Many researchers have explored this area [3][8][10][11][12][13][14][16][20].

Traditional floorplanners minimize the area of a bounding box, or use min-area as one of the objectives. In many cases, floorplanning is confined on a fixed-die size. It is suggested that a fixed-frame floorplanning utilize zero dead space. Zero dead space usually requires arbitrary shaped blocks. Previous rectangular or rectilinear shaped blocks are required to be non-overlapping. Such rigidity inevitably leads to wasted die area and impaired routability.

Many existing floorplanners minimize total wire length. In deep submicron design, interconnect delay and routability become more unpredictable. Furthermore, other issues such as signal integrity need to be addressed in early floorplanning stage. Thus, minimizing total wire length is no longer sufficient to meet the design requirements.

3. PHYSICAL PROTOTYPING

3.1 Early Physical Effect Evaluation

Today's advanced process technologies shrink the chip size dramatically, but at the same time they also bring many tough design challenges. Complex physical effects [1][19] on a chip such as coupling, inductance, IR-drop, electromigration, and process variation [1] make it difficult to make accurate estimates in the early stage of chip design. Conventional approaches to physical design rely on a time-consuming and error-prone detailed implementation to verify the feasibility of a design. Iterations are thus required to fix any design or electrical rule violation after a full detailed physical implementation. These physical effects, however, cannot be considered in the conventional frond-end design due to the lack of implementation details. Therefore, a fast implementation and analysis to evaluate these effects is highly valuable for the frond-end design decisions.

It is getting more difficult in a full custom design to consider all of these effects since any full chip simulation to catch the physical effects would be impractical. As a result, a full custom design is typically limited to a special IP block. An ASIC design style with some full custom IP blocks using the physical prototyping approach can easily address the performance issue and reduce the lengthy design process.

3.2 Low Cost and Realistic "What-If" Analysis

The basis of the new methodology is the creation of a physical prototype of the design. This physical prototype is used to validate the physical feasibility of the netlist to ensure a realizable implementation. The *physical prototyping* method discussed in this chapter compresses the physical feasibility check down to a few hours. Thus the long turn-around time associated with the traditional back-end loop is avoided. Many implementations of a chip can be evaluated and used to improve the top-level architecture design efficiently. For example, the placement of power pads, trunks and strips to reduce IR-drop can be determined in a "what-if" manner. Since all the instances inside each block are placed, the power consumption calculation is more accurate and realistic. Similarly, the creation of the prototype allows to generate realistic timing budgets for all sections of the chip. As a result, a physical implementation with a guaranteed design closure can be obtained without iterations. The physical prototype also provides critical information such as the chip size and aspect ratio.

Physical Prototyping Plans for High Performance 173

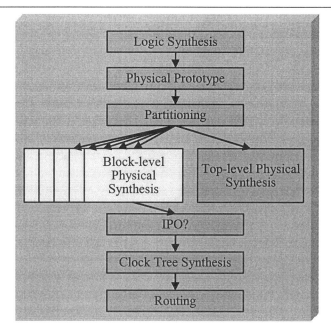

Figure 1. Design flow with physical prototyping.

3.3 A New Approach to Design Flow

Physical prototyping can be used early in a design cycle to guide the partitioning of a whole chip. The proposed new design flow is shown in Figure 1.

3.3.1 Quick Logic Synthesis

The first step in this design methodology is to perform a quick logic synthesis of a netlist. The assumption at this stage is that the netlist is functionally clean, but that the timing is not necessarily accurate. Simple wire load models (WLM) can be used at this stage.

The gate-level netlist generated through this initial quick synthesis stage and the timing constraints form the inputs to the physical prototyping stage.

3.3.2 Floorplanning

The creation of the physical prototype starts with the floorplanning of the chip. This includes floorplanning tasks such as I/O placement, macro placement and power topology. Given the increasing importance of design reuse and the large number of macros being used, a combination of interactive and automatic floorplanning is required. The best results are achieved by a manual placement of the major design elements driven by the engineer's knowledge of the chip architecture, followed by an automatic placement of the remaining elements.

For the first pass at floorplanning, a netlist, physical libraries, corresponding synthesis libraries (.lib), top-level constraints, and a technology file (process description) are created and then imported. Interactive or constraint-driven placement of I/O pads is then used to meet the chip's specification. Placement guides or constraints are usually created for the major modules and used to 'guide' the placement engine where to roughly place the module cells.

3.3.3 Quick Physical Implementation

Next the remaining standard cells are placed using a timing-driven algorithm. The placement includes a routing stage ('Trial Route') that ensures the elimination of major congestion issues. The design is then RC extracted and timing analyzed, followed by In-Place Optimization (IPO) and clock tree synthesis. All of the complex physical effects are then analyzed and the performance objectives are verified. This prototype thus serves as a design basis.

3.3.4 Realistic Planning using Physical Prototype

The prototype is used for the creation of a netlist plus physical and timing design constraints for each block. The block-level constraints are used for design refinement of each block through logic or physical synthesis. The refined blocks can be assembled with other blocks to check timing and physical closure. This process is repeated until the designers achieve their design objectives. The results can be taken back through the backend design flow. Any IPO should be done within the prototyping tool to finalize timing closure.

Note that every cell is placed during the creation of a full-chip prototype, and a trial routing that closely approximates the final routing is also performed. This guarantees that the prototype is physically feasible. Because the partitioning is based on a flat full-chip view (Figure 2), it benefits from the optimization of a flat design while supporting hierarchical design methodology. The advantage of the hierarchical approach is that a design is decomposed into modules of manageable size, which are then refined in parallel by multiple teams. Controlling the block size is the key to the productivity of the design teams because back-end physical implementation tools such as detailed place-and-route and verification tend to be capacity-limited.

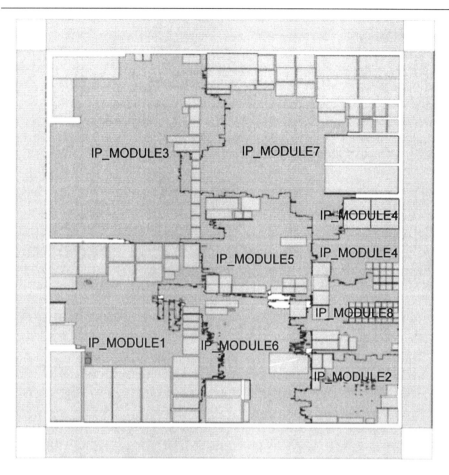

Figure 2. Flat full-chip physical prototype

3.3.5 Timing Budget Generation and Refinement

The flow from the prototype to a final implementation needs to ensure predictable design convergence at all stages of the design flow. During the early phases of the design process, most of the elements of the design are still 'black boxes' or RTL code. At this early stage, the initial timing budgets created by the prototype are rough estimations. As more portions of the design get completed, a larger fraction of the design netlist becomes available in gate-level format. During this process, the full-chip physical prototype is built on a daily basis and the timing budgets are continuously refined based on the latest prototype. When the gate-level netlist is completed, final optimized timing budgets are generated. The key to this implementation process is that the most accurate view of the design is available at all stages through the quick generation of the prototype.

The faster turn-around time of the tool will help front-end engineers effectively develop chip-level timing constraints by taking into account physical design data. The prototyping tool will also generate hierarchical timing and physical constraints for each block automatically from the chip-level constraints. This is a labor-intensive and error-prone process if done manually. The prototype's fast turn-around time and strong correlation with the back-end implementation simplify the above two tasks.

3.4 Shorter Turn-Around Time Implication

Physical prototyping must show its value by improving turn-around-time. The prototyping system should provide an order-of-magnitude speed-up in comparison to existing physical design tools. The turn-around time has significant impact on designers' productivity and design schedules. The turn-around time from a netlist through placement, routing, RC extraction, delay calculation and timing analysis within a day for a multi-million-gate design is possible with today's leading-edge prototyping tools such as First Encounter [2].

Table 1 shows data on three designs of increasing complexity from a networking company. The top section of the table shows the physical prototyping runtime using First Encounter physical prototyping tool running on a desktop workstation. All the times shown are the total times for the iterations actually required (4, 6 and 8 for design A, B and C respectively.) This demonstrates how the quick turn-around time of prototyping makes it possible to optimize the design through multiple iterations without delaying a design schedule. Timing reports are shown before and after In-Place Optimization (IPO) indicating that the prototype is a key tool to achieve timing closure; all or most timing problems are solved at this stage.

The bottom section of the table summarizes the back-end implementation steps starting from the floorplan and placement produced by the prototyping tool. The tool set includes Avant! Apollo router and Star-RCXT extractor, Synopsys PrimeTime timing analyzer and Mentor Graphics Calibre physical verification tool.

Comparison of the routing, extraction and timing analysis times between the prototyping environment and the traditional back-end implementation tools shows that the productivity gain with quick prototyping in the design cycle is significant.

Design	Design A	Design B	Design C
Prototyping Tool			
Gate count	73K	330K	1142K
Components count	49K (+ 1 memory)	86K (+ 6 memories)	145K (+ 62 memories)
Floorplan (initial creation + manual adjustment)	30min + 3min	1.0hr + 5min	1.5hr + 5min
Placement	30min (non timing driven)	1.0hr (timing driven)	1.2hr (timing driven)
Clock tree synthesis	20min	30min	40min
Trial route	2min	5min	10min
Extraction	1min	2min	3min
Timing analysis	10min	15min	20min
Timing report	-10ns setup, 9ns transition (clock 6.5ns)	-5.5ns setup, 10ns transition (clock 6.7ns)	-1.9ns setup, 17ns transition (clock 6.7ns)
IPO	1.0hr	1.5hr	2.0hrs ~ 30min
Timing report	-2ns setup, 3ns transition	-1ns setup, 3ns transition	-1.2ns setup, 3ns transition
Number of iterations	4	6	8
Traditional Back-End			
Apollo routing + antenna (non timing driven)	2.5hr	4.0hr	5.0hr
Star-RCXT extraction	50min	1.5hr	1.8hr
PrimeTime timing analysis	45min	1.5hr	2.0hr
Timing results (clock 7.5ns)	0 violation, +0.67ns setup, 2.3ns transition	no setup & hold, +0.21ns setup, 2.7ns transition (clock 7ns)	no hold, -0.15ns setup
Calibre DRC	1.5hr	2.0hr	NA
Calibre antenna check	1.0hr	1.5hr	NA
Calibre LVS	1.0hr	1.5hr	NA

Table 1. Performance data for the First Encounter prototyping tool

3.5 Hierarchical Design Flow Implication

The effective use of hierarchy is crucial in the implementation of large SoC designs. Hierarchical methodologies, however, can result in suboptimal results due to the lack of a global view. On the other hand, the flat approach is known to be effective for global optimization. To combine the benefits of both approaches, hierarchical design should start with a global view of the entire design. This global perspective enables designers to experiment, using fast placement, with optimizations such as determining the most desirable aspect ratio for each block and the optimum location of the block relative to other blocks. Using fast routing and timing analysis, designers can also get very realistic pin assignments and timing budgets for each block before proceeding to the detailed design of the block. The prototype tools provide a high-capacity environment to build a flat chip model, from which the optimal physical hierarchy can be created. Figure 3 shows the hierarchical design flow used in First Encounter.

During the creation of the hierarchy, the aspect ratio, the pin assignment and the timing budgets are generated for each design module. Again, because they are based on the aforementioned physical prototype, the elements created are guaranteed to result in physical feasibility of the chip in the back-end. Optimal pin assignment is critical to reduce the complexity of the routing between the modules. This directly translates to narrower channel width, which results in smaller die size.

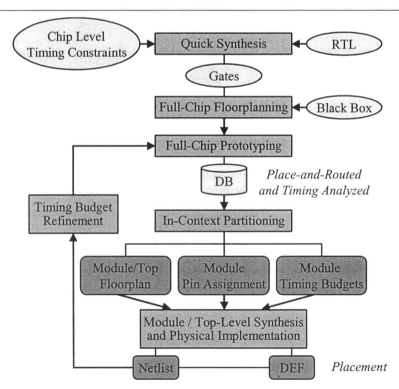

Figure 3. First Encounter hierarchical design flow.

Design size	2.8 million cells or 11 million gates
Design import and flatten	5 minutes
Placement	9 hours 52 minutes
Trial route	1 hour 29 minutes
Full-chip RC extraction, delay calculation and timing analysis	30 minutes
Maximum memory requirements	2.59 GBytes

Table 2. First Encounter capacity data

The allocation of the optimal timing budgets between the partitions is also critical to guarantee the timing closure of the chip. Without the full-chip prototype, the timing budgets are assigned arbitrarily to the modules. This leads to the situation where some modules will be impossible to implement while some other modules will have timing to spare. The result is an unnecessary iteration on under-budgeted modules, which delays the tape-out of the whole chip.

3.6 Larger Tool Capacity Implication

Traditional tools suffer from one major drawback: their internal data representations are not compact enough, limiting the maximum design size they can handle. Their processing speed is typically slow, making it impossible to bridge the front-end and the back-end effectively. The new tools need to offer lightweight efficiency to handle multi-million-gate designs with ease.

Table 2 demonstrates how First Encounter goes from a netlist to a timing-analyzed full-chip physical prototype for a multi-million-gate design in under 12 hours on a desktop workstation with a 500MHz processor. Due to the efficient memory usage, the tool can be used even on low-end computers while many of the traditional EDA tools require high-end servers.

3.7 Correlation of Timing, Routability, Power and Signal Integrity Analysis

Correlation of the timing results (RC parasitics, delay and timing reports) between a physical prototype and tape-out-quality back-end tools is critical. The correlation can be verified by feeding a placement produced by a prototyping tool into a back-end flow and comparing the final timing information against the timing estimate at the prototyping. We have confirmed a tight correlation between the two for First Encounter. For a 200MHz design using a 0.18-micron process, engineers at Infineon concluded that the timing difference is only about 50-picosecond between the First Encounter prototyping tool and PrimeTime with Star-RCXT data [6]. Further analysis confirmed that the critical paths identified by the two are identical.

The physical prototype's timing calculations are based on the placement and trial routing produced by its fast engines. The trial route data approximate detail routing very closely. Therefore the extracted RC parasitics are also closely correlated [7][17].

Correlation between the prototype and the final tape-out version of the chip is required for power and signal integrity analysis as well. The prototype is used for power network planning, and noise prevention and repair. The accurate power analysis performed at the prototyping stage eliminates the need of the over-design of power and ground networks, which is practiced commonly in the traditional design flow without prototyping.

4. TECHNIQUES IN PHYSICAL PROTOTYPING

4.1 Floorplan and Placement Techniques

In conventional placement, design modules may be confined to some "fenced" regions. Such regions are required to be rectangular or, at most, rectilinear. New techniques are now available that allow the clustering of data in a flexibly-contoured "container", a malleable region of any shape enclosing all sub-modules or circuit cells in a functional block. Such clusters are used hierarchically all the way from top to bottom, matching the logical hierarchy. A functional block containing next-level sub-blocks will contain the clusters for these sub-blocks. The user can zoom into (or out of) such a hierarchical block, with the next-level clusters shown nested, in situ, inside the parent block.

The Amoeba algorithm of First Encounter uses such an approach. Amoeba takes a unified approach to floorplanning and placement. The floorplan and the placement are simply different views of the same layout. Using the same timing engine, both aim to optimize the same set of logical, timing, and physical attributes such as wire lengths, die area, critical path timing and signal integrity. An Amoeba display of a placement can in fact be looked at as a "pseudo-floorplan" (see Figure 2).

Conventional approaches to timing driven layout typically rely on slack-based, constraint-driven algorithms. These are "explicit" timing driven methods, in that they depend on pre-specified timing constraints to direct the layout operation. They essentially treat a design as if it is flat and, using the constraints specified, try to find a solution to the extracted layout timing graph. Such techniques have been shown to work reasonably well when the designs are small. With any design of reasonable complexity, the algorithms do not perform well because the performance deteriorates very rapidly when the algorithms are over-constrained. Not only are they very hard to use and slow to run, they often achieve timing closure at the expense of layout quality such as wire length, die size, routability and signal integrity.

In contrast, Amoeba takes an "implicit" approach to timing control. Throughout the entire floorplanning and placement process it tries to exploit the natural clustering characteristic that is intrinsic in a design's logical hierarchy. Thus, instead of depending exclusively on externally imposed constraints, as traditional approaches would do, Amoeba naturally blends logical hierarchy and physical locality in its core algorithmic formulation. By exploiting the logical hierarchy, physical locality is achieved naturally, in tandem with optimizing wire length, timing performance, chip die size, and layout routability.

By maintaining the levels of logical hierarchy in the physical layout, the physical implementation is more intuitive for the logic designers. This makes it easier to ensure an effective communication between the logical and physical teams. Lack of communication between these teams is a known bottleneck in the traditional design flows.

Amoeba applies this hierarchical-locality-based approach to the unified floorplanning and placement task. At any given level of a typical hierarchical chip design, intra-module signals account for 95% or more of all the signal nets, leaving fewer than 5% that make up the inter-module signals. The physical locality implicitly leads to shorter wire lengths for these intra-module signals and they also generally require smaller drivers. Exploiting this characteristic allows Amoeba to focus on the inter-module signals, which are greatly reduced in number and also tend to be more critical. This is where Amoeba applies the timing driven strategies. Techniques such as net-weighting, soft and hard planning guides, grouping/regrouping and re-hierarchy, and hard placement fences are employed, interwoven with other features like power/ground strips and obstructions.

By applying the physical locality and intra-/inter-module-signal distribution hierarchically and incrementally, Amoeba only needs to deal with a drastically reduced number of signal paths at any given stage of the placement process. This not only greatly enhances its computational speed, but also allows Amoeba to more thoroughly explore the possible solution space to come up with an optimum solution, considering all the factors of timing, area, and power consumption.

Another differentiating characteristic of Amoeba is in the "intelligent fencing" strategy that it uses to place circuit cells. Conventional placement approaches typically demand that cells in a design module be confined to a "fenced" region that is either rectangular or rectilinear. These fences are required to be non-overlapping. Such rigidity inevitably leads to wasted die area and greatly impaired routability. In adopting the intelligent fencing strategy, Amoeba, in addition to its flexibly-contoured shapes, allows two clusters to overlap when necessary. This generally leads to more efficient die area usage, shorter wire lengths, and much higher routability.

Using the physical locality derived from the logical hierarchy and applying timing-driven strategies, the results of Amoeba are highly controllable and predictable. The integrity of the design can be maintained at each iteration of the netlist by bounding the changes to a specific Amoeba region. Traditionally place and route tools react unpredictably to relatively minor changes in the design. The ability to bound changes is very important to incorporate flows with the introduction of specially designed cells as described in Chapters 9, 10, and 12. The generation of specially designed cells is useless if the final layout is dramatically different from the layout for which the cells were originally generated.

4.2 Pin Assignment Technique

The goal of pin assignment is to decide the exact pin locations on macroblocks. The objective functions of pin assignment are typically wire length and routing congestion minimization. Traditional pin assignment uses analogical models to map pins to the available locations on a macro block. However, many important factors are neglected if pin assignment is performed without carrying out the actual routing step since interconnect is hard to estimate. Thus, it is suggested that pin assignment be integrated with global routing [4]. A global view of net information and routing resource information is critical for pin assignment and routing.

First Encounter's pin assignment is based on routing information given by Trial Route. Routing congestion is taken into account accurately this way. A pin refinement step is followed to align pins and reduce the number of crossovers, thereby improving net delay and routing complexity. This is done by building a channel graph to utilize the free space among macro blocks. A channel graph is a more efficient representation than a grid graph since there is no granularity issue in the former. As a result, pin refinement runs very fast and improves timing and routability greatly.

4.3 Routing Technique

Unlike the final implementation, the routing technology (called Trial Route) used in prototyping does not have to generate DRC-clean routing. However, it has to be very efficient, fast, and consider all the detail physical information, such as congestion, power/ground distribution, blockage, etc. The routing result from Trial Route provides a reliable basis of timing feasibility and routability analysis. Trial Route can also pass its results to a detailed router to make the final routing conform to layout design rules. The routing pattern is thus preserved and the performance estimate in the physical prototype is achievable with high correlation.

Typically, a commercial router uses a variant of the grid-based maze routing algorithm, which is usually very time-consuming and becomes a bottleneck for all of the physical design operations. In contrast, the algorithm used in First Encounter is a River-Routing algorithm [9]. First, Trial Route does global routing to plan routing resources and resolve all the congestion problems and blockages. Then, it maps the track assignment problem into a two-terminal river route problem to quickly obtain a reasonable track assignment solution. Finally, it refines the routing by ripping up and rerouting some of the nets to make the trial route result even more similar to the final routing result.

4.4 Timing Correlation Technique

Well-correlated timing relies heavily on three aspects: RC extraction, delay calculation and static timing analysis. A calibration procedure is required to fine-tune the RC extraction and the delay calculation. Actually, a simple constant scaling approach is very effective to centralize the error distribution [17]. Centralization of each delay stage's error distribution plays an important role to reduce the total path delay error, because the delay stages in a timing path may compensate each other. In addition, the speed can be dramatically improved by adaptively choosing different algorithms depending on the importance of each individual delay stage in the entire timing path.

4.5 Power Planning Technique

In the DSM design, power/ground (P/G) network design is important to prevent reliability failures due to IR drop and electromigration. Power planning is typically performed before doing any other physical implementation, since the placement and routing are all dependent on the P/G network design. Any post-layout IR drop violation can cause another iteration and lengthen the total design cycle.

A fully automatic power planning tool is almost infeasible, because the solution of design space is quite huge. Most of prototyping tools provide semi-automatic power planning capability as shown in Figure 4.

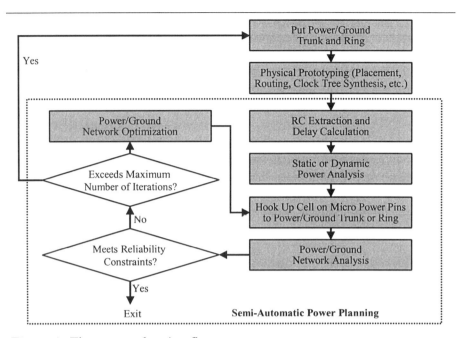

Figure 4. The power planning flow.

Typically, the designer first has to estimate the requirement of the P/G network design and place a proper number of power trunks and power rings in the design. In addition, the width of each power trunk and ring are also specified. Next, the design is placed, routed and clock tree synthesized for the purpose of prototyping, which will be the input to the power planning tool. Note that doing clock tree synthesis before power estimation is a necessary step, since an inaccurate estimate of clock network can lead to inaccurate chip power estimation and incorrect power planning.

Before doing a full-chip power analysis, RC parasitics extraction and delay calculation are necessary to obtain the loading and input slew rates for each gate. A full-chip power analysis can then be done with a static or dynamic approach. Next, P/G bus analysis follows. If the P/G bus design could not meet the reliability constraints, the power planning tool would optimize P/G network to improve the P/G bus. Typically, the optimization can only adjust the width of P/G trunks and rings. Since the P/G network is huge, too sophisticated optimization algorithm can be very time-consuming.

After several iterations of optimization, if sometimes the power planning tool still can not meet the constraints, then it will exit and feedback the necessary information to the designers. The designers have to redesign their power trunk and ring, and re-run the power planning again. The procedure above can be repeated until the design constraints are met and the P/G network design is converged.

5. CONCLUSIONS

Unlike the traditional floorplanning approaches that are blind to the underlying complex DSM physical effects, physical prototyping provides a playground where any design trade-off and constraints can be constantly monitored and verified. Moreover, this prototype can be refined using a detail implementation tool without losing consistency of design quality. Therefore, design closure can be achieved with high efficiency and predictability.

Physical prototyping tools also help to find better partitioning into manageable blocks, resulting in a hierarchical design methodology. Because the partitioning is based on the physically feasible prototype, the block-level timing budgets created are realistic and lead to a much easier task for traditional logic synthesis. The prototyping tool thus can become the hub of the design environment, covering partitioning, generation of block-level constraints, top-level design closure, clock tree synthesis, and power grid design.

6. REFERENCES

[1] Bernstein, K., et al., *High Speed CMOS Design Styles*, Kluwer Academic Publishers, 1998.
[2] Cadence, *First Encounter User Manual*, version 2002.1, March 2002.
[3] Chang, Y.C., et al., "B*-trees: a new representation for non-slicing floorplans," *Proceedings of the 37th Design Automation Conference*, 2000, pp. 458-463.
[4] Cong, J., "Pin assignment with global routing for general cell design," *IEEE Transactions on Computer-Aided Design*, vol. 10, no. 11, 1991, pp. 1401-1412.
[5] Dai. W-J., "Hierarchical Design Methodology for Multi-Million Gates ASICs," *DesignCon*, 2001.
[6] Gerousis, V., and Dai, W., "Frond-end physical design solutions provides hierarchical methodology", Design Automation and Test in Europe Conference, 2002.
[7] Goering, R., "Silicon Perspective's First Encounter helps AMD reach timing closure," *EEdesign*, May 2000. http://www.eedesign.com/story/OEG20000511S0057
[8] Guo, P.N., Cheng, C.K., and Yoshimura, T., "An O-tree representation of non-slicing floorplans and its applications", *Proceedings of the 36th Design Automation Conference*, 1999, pp. 268-273.
[9] Healey, S.T., "An improved model for solving the optimal placement for river-routing problem," *IEEE Transactions on Computer-Aided Design of Integrated Circuits and Systems*, vol. 12, no. 10, 1993, pp. 1473-1480.
[10] Hong, X., et al., "Corner block list: an effective and efficient topological representation of non-slicing floorplan," *Proceedings of the International Conference on Computer-Aided Design*, 2000, pp. 8-12.
[11] Murata, H., et al., "VLSI module placement based on rectangle-packing by the sequence pair," *IEEE Transaction on Computer Aided Design of Integrated Circuits and Systems*, vol. 15, no. 12, 1996, pp. 1518-1524.
[12] Nakatake, S., et al., "Module placement on BSG-structure and IC layout applications," *Proceedings of the International Conference on Computer-Aided Design*, 1996, pp. 484-491.

[13] Ohtsuki, T., Sugiyama, N., and Kawanishi, H., "An optimization technique for integrated circuit layout design," *International Conference on Composite Science and Technology*, Kyoto, Japan, 1970, pp. 67-68.

[14] Otten, R., "Automatic floorplan design," *Proceedings of the Design Automation Conference*, 1982, pp. 261-267.

[15] Sait, S.M., and Youssef, H., *VLSI Physical Design Automation, Theory and Practice*, IEEE press, 1995.

[16] Sakanushi, K., and Kajitani, Y., "The quarter-state sequence (Q-sequence) to represent the floorplan and applications to layout optimization," *IEEE Asia-Pacific Conference on Circuits and System*, 2000, pp. 829-832.

[17] Selzer, M., Birkl, B., and Lenke, F., "Silicon virtual prototyping using First Encounter," Evaluation Report, Motorola, Feb. 2002.

[18] Sherwani, N., *Algorithms for VLSI Physical Design Automation*, 3rd Ed., Kluwer Academic Publishers, 1999.

[19] Singh, R., *Signal Integrity Effects in Custom IC and ASIC Designs*, Wiley-IEEE Press, 2001.

[20] Wong, D.F., and Liu, C.L., "A new algorithm for floorplan design", *Proceedings of the Design Automation Conference*, 1986, pp. 101-107.

Chapter 7

Automatic Replacement of Flip-Flops by Latches in ASICs

David Chinnery, Kurt Keutzer
Department of Electrical Engineering and Computer Sciences,
University of California at Berkeley

Jagesh Sanghavi, Earl Killian, Kaushik Sheth
Tensilica Incorporated,
3255-6 Scott Boulevard,
Santa Clara, CA 95054, USA
{sanghavi, earl, ksheth}@tensilica.com

This chapter presents a novel algorithmic approach to replace flip-flops by latches automatically in an ASIC gate-level net list of the Xtensa microprocessor. The algorithm is implemented using scripts that transform a gate-level flip-flop-based net list into an equivalent latch-based design. The algorithm has been applied to several configurations of Xtensa embedded processors in a state-of-the-art CAD tool flow. The experimental results show that latch-based designs are 5% to 19% faster than corresponding flop-based designs, for only a small increase in area.

1. INTRODUCTION

Level-sensitive latches are faster than edge-triggered flip-flops as they provide some degree of immunity to clock skew and allow slack passing and time borrowing between pipeline stages. However, latches are not traditionally used in ASIC designs, despite it being possible to specify latches in the RTL. Many designers are less familiar with latch-based design methodology, and it is easier to ensure correct timing behavior in flip-flop based designs. Latches are more commonly used in custom designs, but have been identified as a design approach to increase the speed of ASICs, as discussed in Chapter 3.

Other than in register files, latches are seldom used in ASICs, because synthesis tools have provided only limited support for latch-based designs. There are no commercial synthesis tools supporting retiming of level sensitive latches. Prior to the DC98 release of Design Compiler, time borrowing was not handled correctly [12].

Ensuring correct timing behavior is more difficult than for designs using flip-flops. To determine the critical path with latches, the sequential circuitry must be analyzed to find critical loops that limit the clock period, because slack may be passed between latch pipeline stages. EDA static timing analysis tools do support this, but care is required to correctly determine the clock period with a sequential path through latches. Reported slack on each clock edge is not necessarily additive, as different paths can have worse slack on different clock edges – it is the worst sequential path that limits the clock period, not the sum of two portions of different paths! Increasing the clock period increases the amount of slack that can be shared between pipeline stages (as there is a larger window within which the latches are transparent), thus the clock period is not necessarily limited by the negative slack of the worst pipeline stage.

Latch-based designs are also more difficult to verify. One approach is to make the latches opaque and treat two latch stages (with an active high latch and active low latch) as combinational logic. As a latch's output must be valid at the clock edge causing it to go opaque, that can be used as a hard clock edge boundary for formal verification tools. Unfortunately, ASIC verification tools don't support this methodology, and alternatives such as gate-level simulation must be used for verification.

1.1 Motivation for Latch-Based Designs

Throughout this discussion, we will refer to rising edge D-type flip-flops, and active-high and active-low D-type transparent latches. Falling edge flip-flops behave in a similar manner with respect to the falling clock edge, rather than the rising clock edge.

With flip-flops the minimum clock period, T, is [7]

(1) $\quad T_{flip-flops} = t_{comb} + t_{sk} + t_j + t_{CQ} + t_{su}$

where t_{comb} is the critical path delay through the combinational logic; t_{sk} is the clock skew, the maximum difference in arrival of the same clock edge at different portions of the chip; t_j is the clock jitter, the difference between the arrivals of consecutive clock edges from the expected time; t_{CQ} is the delay from the rising clock edge arriving at the flip-flop to the output becoming valid; and t_{su} is the setup time, the length of time that the data must be stable before the arrival of the clock edge.

As discussed in detail in Chapter 3, Sections 1.3.4 and 1.3.5, with latches the minimum clock period is

(2) $\quad T_{latches} = t_{comb} + t_{DQ,L} + t_{DQ,H} + t_j$

where t_{comb} is the total critical path delay through the combinational logic between active-low and active-high latches, and the active-high and active-low latches. $t_{DQ,L}$ and $t_{DQ,H}$ are the propagation delays, from the arrival at the latch's D input to the latch's Q output, when the active low and active high latches respectively are transparent. t_j is the jitter between consecutive rising (or falling) clock edges at the same point on the chip.

Equation (2) is valid if the latch input arrives when it is transparent, and $t_{sk} + t_j + t_{su}$ before the latch becomes opaque (i.e. before the required setup time of the latch accounting for clock skew and clock jitter).

Comparing Equation (2) to equation (1), latches are less subject to clock skew and the setup time and hence latch-based designs can have a smaller clock period.

Note that latches are also subject to duty cycle jitter, variation in the length of time that the clock pulse is high and low, which must be taken into account. Flip-flops with a single-phase clocking scheme are not subject to duty cycle jitter.

Latches allow extra time (slack) not used in computing the combinational logic delay in one pipeline stage to be passed to the next stage when the latch is transparent. With latches, a pipeline stage can borrow time from the previous and next pipeline stages to finish computation. Hence a latch-based design does not need to have the delay of pipeline stages as carefully balanced, to minimize the clock period. Flip-flops present hard boundaries between pipeline stages and if one stage computes in less time, this slack cannot be passed onto other stages to allow longer computation times [7].

The Texas Instruments SP4140 disk drive read channel used latches to reduce the timing overhead, helping to achieve high clock frequencies. The RTL code for the SP4140 specified some modules to use latches [Borivoje Nikolić, personal communication]. Chapter 15 describes the use of latches in this design. However, many ASIC designs use almost solely use flip-flops and there has been no methodology for translation of these designs to latch-based netlists. For example, Lexra's LX4380 uses single-phase clocking with edge-triggered flip-flops and latches are only used in the register file [10].

Retiming for flip-flop based designs [5] is available in commercial synthesis tools. Previous research has examined retiming latch-based designs [3][6][8][9]. However, there are no EDA tools that support latch retiming, or migration from flip-flop based designs to latch-based designs. Replacing flip-flops with latches has been proposed to reduce power, by replacing a flip-flop by two back-to-back level-clocked latches, fixing the position of the latches on one clock phase and relocating latches on the second clock phase [4]. Our research is novel in presenting an algorithmic approach to generating a latch-based design from a design with flip-flops, using flip-flop retiming to locate latches of both clock phases optimally, *in an ASIC flow*.

1.2 An Automated Approach

Given the number of ASIC designs using flip-flops, we wanted to create an automated methodology for generating latch-based designs from flip-flop based designs in an ASIC flow. We did this using synthesis tools and additional scripts.

Our approach is essentially as follows:
- Start with a flip-flop based design of clock period T
- Identify flip-flops to be replaced by latches
- Replace each of those flip-flops by two flip-flops
- Perform retiming to achieve a clock period of $T/2$ – balancing the delay of pipeline stages by moving the registers, preserving functional behavior
- Replace the retimed flip-flops with latches at clock period T
- Resize gates to size the latches correctly and take advantage of the additional time savings

In Section 2, theorems show the validity of using flip-flops at a clock period of $T/2$, to allow retiming with existing synthesis tools, and we discuss how to handle input and output constraints correctly with the different clocking regime. Section 3 further details the algorithm for replacing flip-flops by latches, and Section 4 discusses the results of this approach. Then Section 5 details our conclusions from this research.

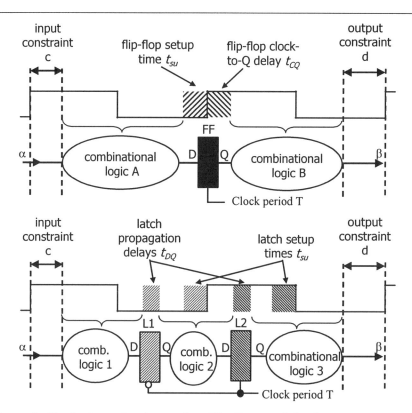

Figure 1. Equivalent circuitry using flip-flops and latches on a sequential path, from α to β. L1 is an active-low latch, L2 is active-high. FF is a rising edge flip-flop. Input α arrives after the rising clock edge with delay c. The output β must arrive at least d before the rising clock edge.

2. THEORY

2.1 Equivalence of latches and flip-flops

Theorem 1. A sequential path with a rising edge flip-flop and a sequential path with an active low latch and an active high latch are equivalent.

Given the rising edge flip-flop based sequential logic, shown at the top of Figure 1, it is possible to construct timing equivalent (with respect to the clock edges) and functionally equivalent sequential logic with active low and active high latches. Input α arrives delay c after the clock's rising edge, and output β must arrive by d before the clock's rising edge.

Proof:

To ensure functional equivalence, the output β must be the same for the latch-based circuit for all inputs α. Designate the functions performed by the combinational logic as $f_A, f_B, f_1, f_2,$ and f_3, for combinational logic A, B, 1, 2 and 3 respectively. This requirement is that

(3) $\quad \beta = f_B(f_A(\alpha)) = f_3(f_2(f_1(\alpha)))$

We can ensure this relation holds by preserving functional equivalence when positioning the latches, using retiming of flip-flops which preserves functional equivalence, as discussed later.

To ensure timing equivalence, we must satisfy relations (4) to (12) below. c_{max} is the largest input arrival time c, and c_{min} is the smallest. d_{max} is the largest output constraint d, which is effectively a setup constraint at the output.

(4) $\quad c_{max} + t_{comb1,max} < T - t_{su} - t_{sk} - t_j$

(5) $\quad c_{min} + t_{comb1,min} > t_h + t_{sk}$

where the min and max subscripts denote the minimum and maximum values of the input and output constraint delays, and minimum and maximum delays of the combinational paths.

If the input to the active low latch arrives while it is transparent:

(6) $\quad c_{max} + t_{comb1,max} + t_{DQ,L} + t_{comb2,max} < \dfrac{3T}{2} - t_{su} - t_{sk} - t_j$

If the input to the active low latch arrives while it is opaque:

(7) $\quad \dfrac{T}{2} + t_{CQ,L} + t_{comb2,max} < \dfrac{3T}{2} - t_{su} - t_{sk} - t_j$

where $t_{CQ,L}$ is the clock-to-Q propagation delay of the active low latch L1, if the input arrives when it is opaque. Correspondingly, $t_{CQ,H}$ is the clock-to-Q propagation delay of the active high latch L2.

(8) $\quad t_{CQ,L} + t_{comb2,min} > t_h + t_{sk}$

And if the inputs to both latches arrive while they are transparent:

(9) $\quad \begin{aligned} & c_{max} + t_{comb1,max} + t_{DQ,L} + t_{comb2,max} + t_{DQ,H} + t_{comb3,max} \\ & < 2T - t_{d,max} - t_{sk} - t_j \end{aligned}$

Or if the input to the active high latch arrives while it is transparent, and the input to the active low latch arrived when it was opaque:

(10) $\dfrac{T}{2} + t_{CQ,L} + t_{comb2,max} + t_{DQ,H} + t_{comb3,max}$
$< 2T - t_{d,max} - t_{sk} - t_j$

Or if the input to the active high latch arrives while it is opaque:

(11) $T + t_{CQ,H} + t_{comb3,max} < 2T - t_{d,max} - t_{sk} - t_j$

(12) $t_{CQ,H} + t_{comb3,min} > t_{h,\beta} + t_{sk}$

where $t_{h,\beta}$ is the hold time at β. Constraints (4) to (12) ensure that all the hold times and setup times are met.

Relations (5), (8), and (12) ensure hold times at latches L1 and L2, and output β, respectively, are not violated. These equations can be satisfied by adding buffers to the combinational logic to increase $t_{combi,min}$.

Relation (4) ensures the setup time is not violated at latch L1. Constraints (6) and (7) require that the setup time is not violated at latch L2. Constraints (9), (10), and (11) ensure the setup time is not violated at output β, as specified by the output constraint d.

The clock period T can be increased to ensure the setup time constraints (4), (6), (7), (9), (10), and (11) are not violated.

We need to check timing equivalence through the latches versus the flip-flop. Counting from $t = 0$ when α arrives, the output $\beta = f_B(f_A(\alpha))$ is stable by $t = 2T$ in the flip-flop circuit. As the setup and hold time constraints of the latches and output are not violated, we can conclude:

- The input to L1, $f_1(\alpha)$ is stored at the rising clock edge at time $t = T$. The output of L1 cannot change again until time $t = 3T/2$ when active low latch L1 becomes transparent on the falling clock edge.
- The next input α' cannot change the output $f_1(\alpha)$ from L1 again until $t = 3T/2$, as the hold time of L1 is not violated. Hence the output $f_1(\alpha)$ of L1 to combinational logic 2 is stable between $t = T$ and $t = 3T/2$.
- The output of L1, propagates through to L2, giving $f_2(f_1(\alpha))$ and is stored at the falling clock edge at time $t = 3T/2$. The output of L2 cannot change again until time $t = 2T$ when active high latch L2 becomes transparent on the rising clock edge.
- The output of L1 cannot change the output $f_2(f_1(\alpha))$ from L2 until after $t = 2T$, as the hold time of L2 is not violated. Hence the output $f_2(f_1(\alpha))$ of L2 is stable between $t = 3T/2$ and $t = 2T$.
  - Now as the setup and hold time constraints at β are not violated, the output to β is $f_3(f_2(f_1(\alpha)))$ at $t = 2T$, and then from functional equivalence $\beta = f_B(f_A(\alpha)) = f_3(f_2(f_1(\alpha)))$.

Thus the sequential logic with latches is functionally and timing equivalent to the sequential logic with latches.

In some cases, it can be difficult to place latches L1 and L2 appropriately. For example, using latches instead of flip-flops in a Viterbi carry-select-add unit would require placing latches within the adder (about 8 FO4 delays for adder, 2 FO4 delays for MUX, and 4 FO4 delays for comparator roughly from memory from discussion with Bora). Furthermore, to take advantage of the early arrival of less significant bits of the sum, the comparator needs to be modified – the slowest output of the adder will be the most significant bit. An analog Viterbi detector was implemented in this sort of fashion with latches, which enabled high speeds to be achieved [2].

If there is single-cycle recursion, that is a sequential critical loop of one cycle, then there is no possibility of slack passing, and impact of the setup time and clock skew are not reduced. In such cases, high-speed flip-flops give a smaller clock period than by using latches, as the delay of two latches ($2t_{DQ}$) is more than the delay of the high-speed flip-flop (t_{CQ}). Section 4 of Chapter 3 gives an example where latches are slower.

Corollary 1. A sequential path with n flip-flops is equivalent to a sequential path with $2n$ latches (active low latches followed by active high latches).
Proof:
This follows directly from Theorem 1, where input α may come from a rising edge flip-flop or active high latch, and output β may go to a rising edge flip-flop or active low latch. Considering each segment with a rising edge flip-flop separately, the flip-flop can be replaced by one active low latch and one active high latch.

Placing latches for functional equivalence at the rising clock edge, with the output of transparent low latches occurring in the same position as the output of rising edge flip-flop based circuitry is not usually the optimal placement for the latches. It is best if the latch inputs arrive during the window while the latches are transparent – not at the start or end of this window (which is when the clock edge arrives).

The latch positions can be retimed for better immunity to clock skew, preserving functional equivalence at the circuit's outputs.

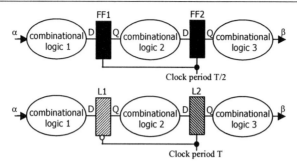

Figure 2. Equivalent circuitry using flip-flops and latches. L1 is an active-low latch, L2 is active-high. FF1 and FF2 are rising edge flip-flops. Inputs and outputs of this sequential logic are stored at a clock edge of the clock with period T.

Theorem 2. Replacing two $T/2$ flip-flops by latches.

A pipeline with two sets of rising edge flip-flops clocked at period $T/2$ is functionally and timing equivalent to a pipeline stage with a set of transparent low latches followed by a set of transparent high latches (providing setup and hold times are not violated), if the inputs and outputs are clocked at clock period T in the following manner.

The inputs to the logic before the first stage come from rising edge flip-flops, or transparent high latches, clocked with period T. The outputs of the logic after the second stage go to rising edge flip-flops, or transparent low latches, clocked with period T.

This theorem holds in general for $2n$ pipeline stages of rising edge flip-flops clocked at period $T/2$ being replaced by $2n$ successive stages registered by transparent low latches then transparent high latches alternately for each stage, providing the inputs and outputs are clocked as described above.

Proof:

As the input α is from a transparent high latch or rising edge flip-flop clocked at period T, it only changes after every rising clock edge T and is held constant at least while the T clock is low. As the output β goes to a transparent low latch or rising edge flip-flop clocked at period T, it is stored at every rising clock edge T.

The flip-flops in Figure 2 are clocked at period $T/2$. As the input α only changes after the rising T clock edge, and is held constant while the clock is low:

- The flip-flop FF1 will store the results of α propagating through combinational logic 1 at the next rising T clock edge.

- Then flip-flop FF2 stores the results of FF1 output Q propagating through combinational logic 2 at the next falling T clock edge (when there is a rising $T/2$ clock edge).
- The output of FF2 propagated through combinational logic 3 is then stored at the next rising T clock edge.

Compare this to what happens with the latches in Figure 2. As discussed in the proof of Theorem 1, the inputs of the latches are valid at the clock edge on which they go opaque, providing their hold times are not violated and the data arrives before $t_{sk} + t_j + t_{su}$. Hence the latches store the input when they become opaque. As the input α only changes after the rising T clock edge, and is held constant while the clock is low:

- The active-low latch L1 will store the results of α propagating through combinational logic 1 at the next rising T clock edge.
- Then active-high latch L2 stores the results of FF1 output Q propagating through combinational logic 2 at the next falling T clock edge.
- The output of L2 propagated through combinational logic 3 is then stored at the next rising T clock edge.

This is equivalent timing behavior at the input α and output β.

Lemma 1. Doubling flip-flops preserves correct function

Sequential circuitry with rising edge D flip-flops at period T after k clock pulses, is functionally equivalent to sequential circuitry with each flip-flop replaced by two connected rising edge D flip-flops at period T after $2k$ clock pulses (ignoring hold time violations).

Proof:

By doubling the flip-flops, every second flip-flop just provides a delay of one cycle after the first flip-flop. Hence the functional behavior does not change, but computation takes twice the number of clock cycles.

Lemma 1 provides a mechanism for replacing flip-flops at clock period T by two flip-flops at clock period T. These flip-flops can then be retimed, which preserves functional correctness [5], achieving a clock period close to $T/2$. By Theorem 2, we can replace these flip-flops at clock period $T/2$ by latches at clock period T, and the functionality is unchanged. Thus we have provided an algorithmic approach to change from flip-flops to latches in the manner proven by Theorem 1 – buffers can be inserted to fix hold time violations and the clock period can be increased to fix setup time violations at each step as needed. (These violations can be fixed in post-processing, once we have generated a latch-based design from the flip-flop based design.)

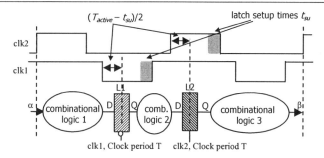

Figure 3. Optimal latch positions with respect to the clock edges, considering the setup time, with a 37.5% duty cycle for clk1 and clk2.

2.2 Handling Input and Output Constraints to Ensure Optimal Latch Placement

For simplicity, we assume that the clock skew includes the jitter. This simplifies Equation (2) to:

(13) $T_{latches} = t_{comb} + t_{DQ,L} + t_{DQ,H}$

Section 3 of Chapter 3 discusses the optimal latch positions without this simplication. More generally, the following approach could be used by assuming a clock period of $T_{latches}' = T_{latches} - t_j$.

To minimize the impact of clock skew and jitter, the optimal position for latches is when the latch input arrives in the middle of the interval when the latch is transparent (active), before the setup time t_{su} of the flip-flop. With a duty cycle of T_{active}/T, where the latch is active for duration T_{active}, the optimal placement is when the latch input arrives $(T_{active} - (t_{su} + t_{sk}))/2$ after the clock edge when it becomes transparent, and active high and active low latches are spaced $T/2$ apart (assuming equal setup times). This is shown in Figure 3.

198   Chapter 7

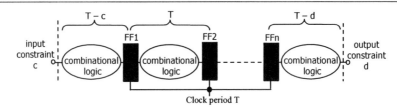

Figure 4. A flip-flop pipelined sequential path with input constraint c and output constraint d.

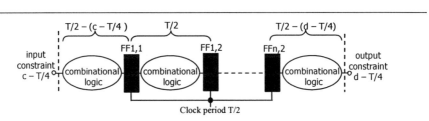

Figure 5. After replacing each flip-flop to be converted to latches with two flip-flops and retiming the flip-flops to achieve a clock period of T/2 with modified input constraints, these are the flip-flop positions on the sequential path. Flip-flop FFj in Figure 4 has been replaced by two flip-flops FFj,1 and FFj,2.

With n flip-flops separating pipe-line stages, there are $n+1$ flip-flop pipeline stages at clock period T. Consider a particular sequential path shown in Figure 4, with input constraint c and output constraint d on the path, and n flip-flops on the sequential path. The total sequential path delay is $(n+1)T$.

Our approach is then to double the number of flip-flops and retime. Retiming the $2n$ flip-flops optimally will split the path delay into $(2n+1)$ equal segments, so the delay of each segment is $(n+1)T/(2n+1)$. However, we have shown that this is not the optimal placement for the latches. When retiming, what can we set the input and output constraints to, in order to ensure that the placement of the doubled flip-flops after retiming corresponds to the optimal latch positions?

Suppose we decrease the input and output constraints, when retiming, by a total of δ, so that the total sequential path delay is reduced to $(n+1)T-\delta$. Retiming splits this into equal segments of length:

(14) $$\frac{(n+1)T-\delta}{2n+1} = T/2 + \frac{T/2-\delta}{2n+1}$$

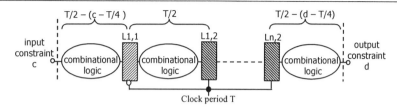

Figure 6. The latch positions from retiming the doubled flip-flops with modified input and output constraints. Latch L1,1 is active low, latches L1,2 and Ln,2 are active high. Flip-flop FFj in Figure 4 has been replaced by two latches active low latch Lj,1 and active high latch Lj,2.

Equation (14) shows that active high and active low latches can be spaced $T/2$ apart, independent of the number of flip-flops n on the sequential path, if $\delta = T/2$.

Suppose the setup, hold, and propagation times of the flip-flops and latches are zero, and the duty cycle is 50%, thus the optimal latch positions are at $T/4$ after the edge at which they become transparent. Then when retiming, $\delta = T/2$ should be split equally between input constraints c and output constraints d, the constraints become respectively $c' = c - T/4$ and $d' = d - T/4$. After retiming shown in Figure 5, this gives optimal latch positions shown in Figure 6.

More generally, the setup, hold, and propagation times for the flip-flops and latches must be considered, along with the duty cycle. The apportioning of $\delta = T/2$ to the input and output constraints can be adjusted to ensure that the retimed positions of the flip-flops are the optimal latch positions.

For our purposes, we assigned $c' = c - T/4$ and $d' = d - T/4$. The retiming software will position the $T/2$ flip-flops so that the setup constraints are not violated, which takes into account $t_{su}+t_{sk}$.

3. ALGORITHM

The user can identify critical paths or portions of the design where flip-flops are to be replaced by latches. The retiming software tools available may limit which flip-flops can be retimed. For example, the tool we used for retiming cannot retime flip-flops with gated clocks. Hence, it can be necessary to modify the design to enable flip-flops to be replaced by latches. We used implementations that only had gated clocks for the register file, and flip-flops with gated clocks were not be replaced by latches.

Having identified the region Ω where flip-flops are to be replaced by latches, the sequential inputs to Ω and sequential outputs of Ω are treated as inputs and outputs along with any primary inputs and primary outputs to Ω.

The input and output constraints to the region Ω can be characterized from the complete design.

Then each flip-flop within region Ω is replaced by two connected flip-flops. We then wish to do retiming of Ω to reduce the clock frequency from T to $T/2$, which should be close to achievable with the doubled flip-flops, as we have doubled the number of flip-flops on each sequential path in Ω.

As combinational paths from inputs of Ω to outputs may be of path delay T, they must be omitted from Ω, otherwise they will limit retiming achieving a clock period of $T/2$. Similarly, the flip-flops that are not being replaced by latches will have combinational path delays of up to T, and thus these flip-flops and paths have to be removed from region Ω. An alternative solution would be to specify two $T/2$ cycles for computation on these paths, but the retiming tools available did not handle multi-cycle paths or different clocking regimes (with clock period T for the flip-flops not being replaced, and clock period $T/2$ for the doubled flip-flops) within the region for retiming.

Input and output constraints were then modified according to the discussion in Section 2.2, to ensure optimal placement of latches. Unfortunately, the retiming software tools set negative input and output constraints to zero and don't handle multi-cycle paths, so there are no simple mechanisms for accepting constraints $c' = c - T/4 < 0$ and $d' = d - T/4 < 0$ (i.e. where the original input and output constraints are $T/4$ or less). Instead, $T/4$ was subtracted from the input and output constraints, and constraints of less than zero were set to zero.

Additionally to avoid long paths from large input constraints or large output constraints limiting retiming, the input and output constraints can be reduced to some maximum value (e.g. $T/2 - t_{sk}$) if they are too large, with the hope that time borrowing allowed with latches will help meet these input and output constraints. This was helpful when these constraints limited retiming.

Then retiming was performed on the region Ω with the modified input and output constraints, and a clock period of close to $T/2$ is achieved. The flip-flops in Ω are then replaced by latches, by traversing breadth-first sequentially from the inputs replacing flip-flops with active low latches first, then alternately active high and active low latches, until all flip-flops are replaced.

When a flip-flop was replaced by two flip-flops, but the positions of these two flip-flops were not moved in retiming (because they are on sequential paths with delay of $T/2$ between each segment), the original flip-flop should be left in that position, rather than replacing it with two directly connected "back-to-back" latches (as two latch cells have somewhat larger area and more delay than a single compact flip-flop cell).

Experiments on the T1030 Base Xtensa showed about 1,000 cases of pairs of the replacement flip-flops not being moved in retiming. Leaving these as a single flip-flop helped reduce the area of the final latch-based design, without penalizing performance. These results indicate about a third of the T1030 Base Xtensa flip-flops were on paths of delay less than $T/2$, which indicates a good opportunity for slack passing. The other 2/3 of the flip-flops were replaced by latches which were moved from this position by retiming. Of course, slack passing depends on the early arrival of all the fanins to the gates in the next pipeline stage – with transparent latches, the critical path is considered over multiple clock cycles.

Having modified Ω to a latch-based region from using flip-flops, the design is then flattened, to include Ω with the other parts of the design that remain unchanged, and compilation is performed to optimize the sizes of gates in the latch-based design. Buffers are automatically inserted to fix hold time violations.

The algorithm is illustrated for a simple example in Figure 7.

As the region Ω contains sequential loops, it is not possible to perform functional equivalence checking, because the latches are moved with respect to the original flip-flop positions. It would be viable to compare outputs of the region for given input sequences. For the experiments reported in the next section, we performed gate level simulations on the entire design to check that the latch-based gate netlist was correct.

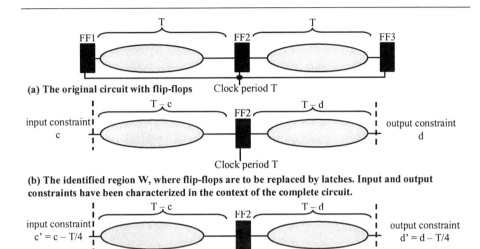

(a) The original circuit with flip-flops

(b) The identified region W, where flip-flops are to be replaced by latches. Input and output constraints have been characterized in the context of the complete circuit.

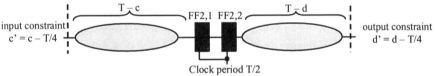

(c) Modified input and output constraints to ensure flip-flop positions after retiming will correspond to the optimal latch positions.

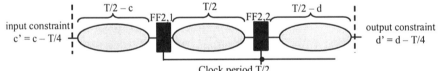

(d) Each flip-flop in region W is replaced by two flip-flops.

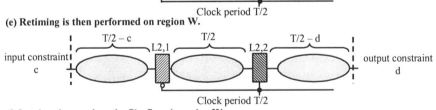

(e) Retiming is then performed on region W.

(f) Latches then replace the flip-flops in region W.

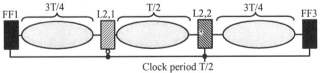

(g) Latch-based region W in the context of the complete circuit.

Figure 7. This series illustrates the algorithmic approach to replace only the flip-flop FF2. FF2 is replaced by active low latch L2,1 and active high latch L2,2. Combinational logic is shown in light gray.

4. RESULTS

Designs were mapped to the high performance, low threshold voltage Artisan 0.13 um library. The ASIC flow tools used were Synopsys DesignCompiler 2000.11-SP1 (Linux or SparcOS5) for synthesis; Synopsys PhysicalCompiler 2000.11-SP1 (SparcOS5) for global routing and floor planning; and Cadence Silicon Ensemble version 5.3.125 for detailed routing and cell placement. Scripts for the algorithm were written in Perl and DesignCompiler scripts. PhysicalCompiler took noticeably longer to floor plan and route latch-based designs (about four times the run time required for flip-flop-based versions), but the correlation between post-synthesis and post-floor planning results was good for both latch and flip-flop versions of designs.

For simplicity, we used a 50% duty cycle, with buffers inserted to fix hold time violations. It was assumed that the clock skew of 0.2ns included both the edge jitter and duty cycle jitter. The actual clock skews after clock tree synthesis and layout were reported to be about 0.13ns for the latch-based designs, which gives a 4% margin to account for jitter. From discussion in Chapter 3 of duty cycle jitter, the duty cycle jitter may be around 5% of the clock period for ASIC designs. Thus a 4% margin is a slight underestimate of the total jitter, but providing the latch inputs arrive during the appropriate time window when they are transparent, this will not affect the latch results. Future experiments will model duty cycle jitter separately.

The static timing analysis tools correctly handle latches; the static timing analysis tools were PrimeTime with Synopsys tools [11], and Pearl with Cadence [1]. These tools gave very similar critical path analyses, after back-annotation of capacitances from final cell placement.

Flip-flop based designs in the typical ASIC flow with the above tools, showed good correlation (within 10%) between synthesis, globally routed with floor plan, and final placement results. However, while latch-based designs showed speed improvements over the flip-flop versions at synthesis and floor-planned stages, the final placement results were worse than the flip-flop based designs. Given good correlation at this step on flip-flop based designs, we suspect the discrepancy for latch-based designs is a performance issue having to do with detailed routing and cell placement of latch-based designs (not something for which the tools are optimally designed).

The flip-flop based designs used for latch replacement did not have gated clocks (used to reduce power), as the synthesis tools did not support retiming with gated clocks. It would be possible to perform latch replacement in an isolated region where the only clock is a gated clock, by making the synthesis tools treat the gated clock signal as a normal clock signal within that region. A mix of different clock signals, or gated clocks, would not allow latch replacement, because of the limitations of the synthesis tools. It is expected that future versions will allow gated clocks to be retimed, resolving this issue.

The latch-based versions of designs passed all gate level simulation tests. This checked that the latch-based netlists gave the right outputs for about twenty sets of test input sequences. Functional verification to compare the latch-based designs with the original flip-flop based designs was not possible, because of the retiming needed to place latches in speed-optimal positions – the position of latches and flip-flops does not correspond, and so functional verification cannot be done with a direct comparison.

The flip-flop gate-level net lists were synthesized from RTL. These net lists of up to 30,000 gates were converted to predominantly latch-based net lists by the algorithm discussed. The run times on these gate-level net lists for the algorithm were typically around 8 hours on Pentium IV 1.7 GHz machines with the Linux operating system (for the T1030 and T1040 Xtensa Base configurations), including retiming and compilation after retiming for area recovery. After the algorithm generates the latch-based design, compilation steps to optimize gate sizes can take several hours.

The synthesis command line instructions are quite slow for some netlist manipulations. If Verilog net list manipulations were coded directly (e.g. in C), instead of via scripts, the run time spent on net list manipulation (around 75% of the 8 hours for the Base configurations) would be reduced substantially. The run times were fast enough to provide experimental proof of the viability of this approach, and scripts greatly reduced the time taken to write this prototype of the algorithm software.

The T1030 and T1040 Base Xtensa is a synthesizable 32-bit embedded processor, with a single-issue five-stage pipeline [13]. Table 1 presents the synthesis results for four different configurations, and Table 2 shows the post floor-planning results for these designs. For our purposes, the primary difference between the T1030 Base Xtensa and T1040 Base Xtensa implementations was the path from the data tag memory to the cache: in T1030 Base this path was combinational and hence not included in flip-flop replacement; in T1040 Base this path has flip-flops on it, which can be replaced by latches. The T1030 MAC16 configuration includes a 16-bit multiply-accumulator with the T1030 Base Xtensa. Compared to the T1040 Base, the T1040 Base_no_loop_no_pif does not include the processor interface and some of the external interface logic, and it does not include the logic for zero-overhead loops.

Design	Number of			Area (mm²)	% Area Increase	Clock Period (ns)	% Speed Increase
	Gates	Flip-Flops	Latches				
Flip-flop T1030 Base Xtensa	24,241	2,780	1,288	0.341		2.35	
Latch T1030 Base Xtensa	24,618	1,248	5,017	0.332	-2.6%	2.20	6.8%
Flip-flop T1040 Base Xtensa	24,711	2,837	1,288	0.351		2.56	
Latch T1040 Base Xtensa	25,546	976	6,121	0.351	0.0%	2.24	14.3%
Flip-flop T1040 Base_no_loop_no_pif	17,258	1,708	1,024	0.227		2.56	
Latch T1040 Base_no_loop_no_pif	18,185	517	4,048	0.238	4.8%	2.28	12.3%
Flip-flop T1030 MAC16	27,854	3,209	1,288	0.400		2.40	
Latch T1030 MAC16	29,109	1,009	6,722	0.400	0.0%	2.28	5.3%

Table 1. Synthesis results for flip-flop and latch versions. These clock frequencies are for the worst case process and worst case operating corner.

Design	Area (mm²)	% Area Increase	Clock Period (ns)	% Speed Increase
Flip-flop T1030 Base Xtensa	0.367		2.58	
Latch T1030 Base Xtensa	0.379	3.3%	2.34	10.3%
Flip-flop T1040 Base Xtensa	0.382		2.67	
Latch T1040 Base Xtensa	0.402	5.2%	2.40	11.3%
Flip-flop T1040 Base_no_loop_no_pif	0.246		2.60	
Latch T1040 Base_no_loop_no_pif	0.272	10.6%	2.18	19.3%
Flip-flop T1030 MAC16	0.440		2.61	
Latch T1030 MAC16	0.452	2.7%	2.46	6.1%

Table 2. Post-physical compiler results for the flip-flop and latch versions from Table 1. These clock frequencies are for the worst case process and worst case operating corner.

We believe that some post-synthesis cases have achieved area reduction because of the area recovery performed after retiming. Area recovery also limits the increase in the number of latches. The area recovery limits any increase in size for the latch-based designs, and in some experiments has reduced area slightly. Note that two latches are slightly larger than a flip-flop in the standard cell library. However, as the latch-based designs have higher gate count and more wires, the area does increase due to the additional wiring after floorplanning and global routing with Physical Compiler. The target clock period was 2.0 ns in all cases.

4.1 Theoretical Limit to Latch-Based Speedup

From Equation (20) of Chapter 2, *if the pipeline stages are very carefully balanced* (to within a gate delay), the clock period with flip-flops is bounded by:

$$(15) \quad \begin{aligned} T_{flip-flops} &\geq t_{comb} + t_{sk} + t_j + t_{CQ} + t_{su} \\ T_{flip-flops} &\leq t_{comb} + t_{sk} + t_j + t_{CQ} + t_{su} + t_{gate} \end{aligned}$$

where t_{gate} is the delay of a combinational gate. Then from Equations (1) and (2), the reduction in clock period by using latches is bounded by

$$(16) \quad \begin{aligned} T_{flip-flops} - T_{latches} &\geq t_{sk} + t_j + t_{CQ} + t_{su} - t_{DQ,L} - t_{DQ,H} \\ T_{flip-flops} - T_{latches} &\leq t_{sk} + t_j + t_{CQ} + t_{su} + t_{gate} - t_{DQ,L} - t_{DQ,H} \end{aligned}$$

As stated earlier, Equation (2) assumes that the latches are placed so that the latch inputs arrive while the latches are transparent and before the setup constraint. Additionally, we must consider the impact of pipeline stages not being balanced in the flip-flop based design. If the pipelines are not well balanced, the reduction in delay by using latches is more than this, as latches allow slack passing.

Typical values observed in this technology are 0.2 ns for the clock skew (including clock jitter); 0.2 ns for the clock-to-Q delay of high-fanout flip-flops; 0.1 ns for the flip-flop setup time; 0.2 ns for the D-to-Q propagation delay of high-fanout latches; and 0.15 ns for a high-fanout combinational gate. With these values, the improvement by using latches can be estimated from (16). Note that the critical paths had high-fanout flip-flops and latches, and thus these delays were used for analysis.

For a typical clock period of the Base Xtensa of about 2.6 ns, the bounds in (16) give an expected percentage speed improvement of between 4% and 10%. The observed performance improvements of 5% to 19% are more than this because of additional slack passing – the flip-flop pipeline stages are not as well balanced as assumed in (15), and thus there is more gained by slack passing.

Note that retiming was run on the flip-flop version of the T1030 Base Xtensa, and there was no speed improvement. The pipeline stages with flip-flops were as well-balanced as they could be for this design. The Xtensa designers expect that the other designs are also well-balanced [Ricardo Gonzalez, personal communication].

The T1040 Base_no_loop_no_pif wire load models over-estimate the load, because of the removal of some of the logic (wire load models weren't separately characterized for each design). The latch-based designs can take advantage of the additional slack after placement.

Note that the largest drive strength of active-high (TLAT) and active-low (TLATN) cells from the library used was X4, whereas X8 flip-flops were available. In the latch-based designs, the latches on the critical paths have high fanout, and are fairly slow (around 0.2ns delay) because of the limited drive strength. If larger latch drive strengths were available in this technology, the D-to-Q latch propagation delay should be reduced to around 0.1ns, which would improve some results by up to 4% (estimated from Equation (16)).

5. CONCLUSION

We have overcome some of the limitations of existing ASIC tools for handling latch-based designs, providing a theoretically valid and working methodology for retiming latches by retiming flip-flops. We have demonstrated a successful approach to replacing flip-flops on critical paths by latches to speed up ASICs, providing actual speed improvements of 5% to 20% on real commercial designs.

In this chapter we outlined some of the limitations on latch-based ASIC designs. Hopefully, by showing that latches provide performance improvement over traditional flip-flop ASICs with minimal area penalty, future tools and standard cell libraries will provide more support for latch-based designs.

6. ACKNOWLEDGMENTS

Thanks to Albert Wang of Tensilica, for help with the initial exploration of this approach. Thanks to Bill Huffman of Tensilica, for discussion of constraints when retiming flip-flops to ensure optimal placement of latches. Our gratitude to the other engineers at Tensilica, who provided support for the ASIC flow. Thanks to Professor Borivoje Nikolić of UC Berkeley, for discussion of the advantages of latch-based designs and existing ASIC approaches. Thanks also to anonymous reviewers for their feedback.

7. REFERENCES

[1] Cadence, "Theory of Operation: Transparent Latches," *Pearl User Guide*, 1998, pp. 6.2–6.13.
[2] Fukahori, K., et al., "An Analog EPR4 Viterbi Detector in Read Channel IC for Magnetic Hard Disks," *Digest of Technical Papers of the International Solid-State Circuits Conference*, 1998, pp. 380-381.
[3] Ishii, A.T., Leiserson, C.E., and Papaefthymiou, M.C. "Optimizing Two-Phase, Level-Clocked Circuitry," *Journal of the Association for Computing Machinery*, January 1997, pp. 148-199.

[4] Lalgudi, K.N., Papaefthymiou, M.C. "Fixed-Phase Retiming for Low Power Design," *International Symposium on Low Power Electronics and Design*, August 1996, pp. 259-264.

[5] Leiserson, C.E., and Saxe, J.B. "Retiming Synchronous Circuitry," *Algorithmica*, 6, 1991, pp. 5-35.

[6] Lockyear, B., and Ebeling, C. "Optimal Retiming of Level-Clocked Circuits Using Symmetric Clock Schedules," *IEEE Trans. on Computer-Aided Design*, September 1994, pp. 1097-1109.

[7] Partovi, H., "Clocked storage elements," in Chandrakasan, A., Bowhill, W.J., and Fox, F. (eds.). *Design of High-Performance Microprocessor Circuits*. IEEE Press, Piscataway NJ, 2000, pp. 207-234.

[8] Sapatnekar, S. S., and Maheshwari, N. "A Practical Algorithm for Retiming Level-Clocked Circuits," *Proceedings of the IEEE International Conference on Computer Design*, 1996, pp. 440-445.

[9] Sapatnekar, S. S., and Maheshwari, N. "Optimizing Large Multiphase Level-Clocked Circuits," *IEEE Trans. on Computer-Aided Design of Integrated Circuits and Systems*, vol 18, no. 9, September 1999, pp. 1249-1264.

[10] Snyder, C.D., "Synthesizable Core Makeover: Is Lexra's Seven-Stage Pipelined Core the Speed King," *Microprocessor Report*, July 2001.

[11] Synopsys, "Optimization and Timing Analysis: 11. Timing Analysis in Latches," in *Design Compiler Reference Manual* v.2000.11, pp. 11.1-11.34.

[12] Synopsys. *Synopsys Design Compiler – Whitepaper*. February 1998. http://www.synopsys.com/products/logic/dc98_bckgr.html

[13] Tensilica, *Xtensa Microprocessor – Overview Handbook – A Summary of the Xtensa Microprocessor Databook*. August 2001. http://www.tensilica.com/dl/handbook.pdf

Chapter 8

Useful-Skew Clock Synthesis Boosts ASIC Performance

Wayne Dai
Computer Engineering Department,
University of California at Santa Cruz,
1156 High Street,
Santa Cruz, CA 95064, USA

David Staepelaere
Celestry Design Technologies Incorporated,
1982A Zanker Road,
San Jose, CA 95112, USA

1. INTRODUCTION

It is no secret that controlling clock skew is crucial to meeting overall design timing [6][7][14]. However, pursuing the traditional goal of eliminating or reducing clock skew may not yield the best design timing. Intelligently skewing clock arrivals can improve performance, achieve timing closure, reduce power, increase reliability margin, and reduce simultaneous switching noise—all by fixing just one signal. This technique is commonly referred to as either "useful" or "intentional" clock skew [15]. Useful skew has been applied manually for many years in high performance custom designs. Recently, a commercial tool has become available that can exploit useful clock skew automatically and systematically to improve timing within the standard ASIC design flow [4]. In this chapter we first explore the traditional zero skew methodology and then introduce the concept of useful skew followed by the description of an ASIC design methodology employing useful clock skew. We then discuss clock and combinational logic co-design and simultaneous clock skew optimization and gate sizing.

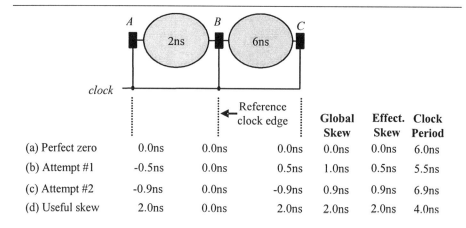

Figure 1. Skew is a local constraint. Figure (a) shows ideal zero skew solution. Figures (b) and (c) show two different implementations with different global skew values. Figure (d) is the optimum useful skew schedule.

2. IS CLOCK SKEW REALLY GLOBAL?

The goal of clock design in the standard ASIC design flow is typically either to bound or to minimize *global* clock skew. Global clock skew is the largest skew value among all pairs of flip-flops within a clock domain, which is simply the difference in arrival time between the earliest and latest arriving clock signals. However the approach of minimizing global skew may produce sub-optimal system timing. This happens because the logic and clock design steps are completely separated—the logic designer focuses on optimizing the critical paths while the clock designer attempts to control global clock skew.

Figure 1 shows an example of two pipeline stages between three sets of flip-flops. Suppose that the logic design has a critical path for each pipeline stage of 2ns and 6ns. Thus, *under the key assumption that the clock arrives simultaneously at all flip-flops*, the clock period must be at least 6ns. See Figure 1(a).

Standard clock tree design practice completely ignores logic timing and seeks to implement a tree that delivers acceptable global skew. For this example, assume one potential clock tree can deliver relative arrival times of -0.5ns, 0ns, and 0.5ns with respect to the center flip-flop. We will call such a set of arrival times a *clock skew schedule*. In this schedule the clock arrives at the left flip-flop 0.5ns before the center flip-flop, and arrives at the right flip-flop 0.5ns afterwards, yielding a global skew of 1.0ns (Figure 1(b)).

Let's consider another potential clock tree that allows the clock edge to arrive at the left and right flip-flops 0.9ns before the center flip-flop (that is, -0.9ns, 0.0ns, -0.9ns). Obviously, the schedule represents a global skew of 0.9ns. See Figure 1(c). The prevailing zero-skew clock design methodology would prefer the second implementation to the first one, since 0.9ns is a smaller skew than 1.0ns. Unfortunately this would be a mistake if the goal were to improve overall design timing.

The mistake results from ignoring the logic and focusing solely on global clock skew. Since there is no direct combinational logic path from the left flip-flop to the right flip-flop, any skew between them is totally irrelevant. Recall that the first solution, in Figure 1(b), had a global skew of 1.0ns resulting from this pair of flip-flops (0.5 minus -0.5). If we amend our interpretation of clock skew to consider only arrival time differences between sequentially adjacent pairs of flip-flops, we find that the first alternative has an effective skew of only 0.5ns, which makes it a better alternative than the second solution, in Figure 1(c), with 0.9ns of skew. So, clock tree synthesis that considers the presence or absence of logic paths can produce better timing than one that seeks only to minimize the worst case global skew.

The key observation above is that skew is not a global constraint, but rather a set of local constraints between different pairs of flip-flops connected through combinational logic. Looking beyond connectivity to the actual delay numbers can, in fact, lead to an "optimum" clock skew schedule. For this particular example, the optimum schedule is: 2ns, 0ns, 2ns, which also happens to be a very bad solution in terms of the global skew metric. The optimum clock skew schedule balances the delays of the pipeline stages, allowing a clock period of 4ns. By adjusting clock skew, it effectively transfers excess slack from the first pipeline stage to the second. We will explain this "cycle stealing" or "useful skew" concept in the next section.

3. PERMISSIBLE RANGE SKEW CONSTRAINTS

Current clock design methodology depends on the assumption that a smaller skew means a better design: A chip with zero or minimal skew either performs better or offers more robust timing. Moreover, most current clock design tools seek to minimize skew through buffer insertion or clock routing. The previous section illustrated that better overall timing can be achieved by considering the relationship between logic and clock timing than is possible by treating each separately. This section formalizes this idea through the permissible skew range concept.

Let F_i and F_j be two flip-flops connected by combinational logic. D_i and D_j are the logic arrival times and C_i and C_j are the clock arrival times to F_i

and F_j respectively (see Figure 2(a)). When data travels from F_i through combinational logic to F_j it may propagate through multiple paths, but its delay is bounded, below and above, by d_{ij}, the shortest path delay from F_i to F_j and by D_{ij} the longest path delay. For a given clock frequency, in order for the data to be latched properly, there are constraints between the propagation delay from F_i to F_j and the clock skew between C_i and C_j. These constraints are commonly referred as the *setup* and *hold* constraints, and ensure that a single active edge both launches and captures data. For the purposes of this discussion we use the following simplifying assumptions: all clock domains have a uniform clock period, T, and all path are single-cycle.

The hold check ensures that data launched from F_i does not arrive at F_j until after a library dependent hold time constraint t_h (Figure 2(b)):

(1) $\quad C_i + d_{ij} \geq C_j + t_h$

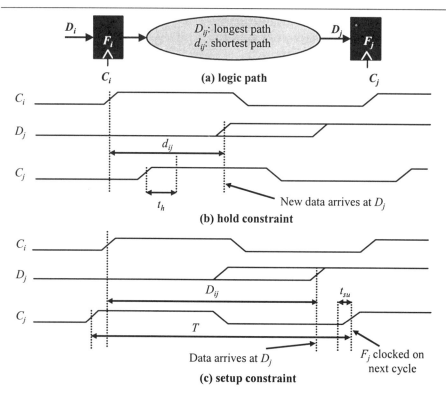

Figure 2. (b) Shortest path constraint and (c) longest path constraint for the clock skew of the circuit shown with flip-flops in (a).

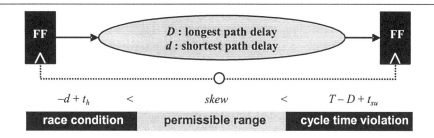

Figure 3. The permissible range for the clock skew between flip-flops.

The setup constraint ensures that data is able to propagate from F_i to F_j in one clock cycle. Data launched from F_i must arrive on the next edge at F_j before a library dependent setup time t_{su} (see Figure 2(c)).

(2) $\quad C_i + D_{ij} \leq C_j + T - t_{su}$

By combining the setup and hold constraints, we can derive both an upper and lower bound for the clock skew between F_i and F_j. We call this two-sided constraint the *permissible range* (Figure 3):

(3) $\quad t_h - d_{ij} \leq C_i - C_j \leq T - D_{ij} - t_{su}$

4. WHY CLOCK SKEW MAY BE USEFUL

Looking at Figure 4, for example, in the case of zero skew, the circuit operates correctly with a clock period of 9ns with a margin of only 0.5ns from cycle time violations. Should some capacitive coupling, IR drop effects, or process variation cause skew to vary slightly, there is only 1ns between FF1 and FF2 to prevent race conditions.

On the other hand, if we use non-zero clock skew to move the skew points further within their permissible ranges, we can achieve either of the following benefits: if we reduce the clock period to 7ns, the circuit still operates correctly but runs at a faster clock frequency, improving the performance of the circuit. Alternatively, keeping the cycle time at 9ns leaves a 2.5ns safety margin to guard against cycle time violations. In effect, the original critical path (8.5ns) has changed to a non-critical one (6.5ns). Increasing the skew to a positive value can raise the safety margin against race conditions between FF1 and FF2 to 3ns, rendering a more robust timing for the circuit.

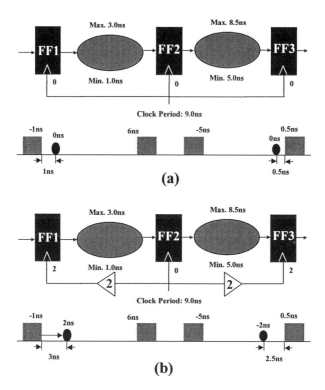

Figure 4. Useful skew creates wider safety margins.

This technique is also referred to as *cycle stealing*: it lowers the clock period by transferring the hold time slack from the slow combinational logic to the fast one and transferring the set up time slack from the fast combinational logic to the slow one. Effectively, it shifts the slacks to critical paths. See Figure 5.

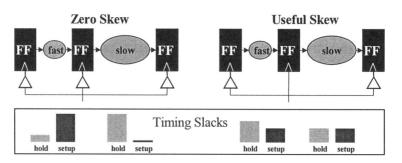

Figure 5. Shifting timing slacks to critical paths.

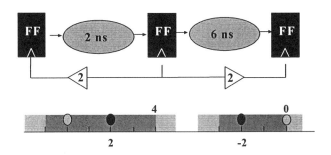

Figure 6. Useful skew can increase clock frequency.

The additional timing slack achieved through useful skew can be used to increase chip performance or simplify timing closure.

Logic designers or the actual circuit owners want their chip to run faster. In Figure 6 a useful skew schedule can reduce the clock period to 4ns, even though the longest logic delay is 6ns.

On the other hand, ASIC vendors care mostly about timing closure. If their customers finish logic design and hand over the net list for physical design, they want to be able to close the timing without iterations. In this example, let's say after physical design this path is changed to 7ns. In a zero-skew methodology this means their customer is then faced with a timing violation. With useful skew, on the other hand, we can actually close the timing even with this 7ns surprise from the physical design side (Figure 7).

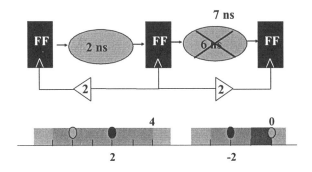

Figure 7. Useful skew can speed up design closure.

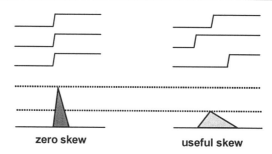

Figure 8. Useful skew can reduce peak current and simultaneous switching noise.

In a zero-skew methodology the goal is for every flip-flop to switch at the same time. The closer we get to realize this goal, the closer we get to two other stumbling blocks: simultaneous switching noise and peak current. If everything switches at exactly the same time, the current drawn from the power supply at that instant may easily be so large that even with a very small lead inductance, the *di/dt* switching noise can inadvertently cause the chip to fail. The large peak current can also harm the chip—especially when the chip contains sensitive circuits, such as analog functions. In contrast, in a useful skew methodology, the instantaneous currents are typically lower since all flip-flops do not switch simultaneously (Figure 8) [3][11].

5. USEFUL SKEW DESIGN METHODOLOGY

For custom design, it is no secret that designers make use of clock skew to optimize the design. It has been used for mainframe design, such as the IBM 360/91 [1], the Amdahl 580, the IBM 3033, and the IBM 370/168 [12], and recently for high performance microprocessors, such as the IBM S/390 [13] and the DEC Alpha [2].

For the standard cell ASIC design flow, a clock tree synthesis tool, ClockWise, is available to make use of clock skew automatically and systematically [4]. This tool uses intentionally created local skew to improve the timing on critical path.

The useful skew clock design methodology consists of five steps: permissible range generation, initial skew scheduling, clock tree topology synthesis, clock net routing, and clock timing verification [16].

Permissible range generation combines logic path delay information with system-level timing constraints to derive a set of local skew constraints between each pair of sequentially adjacent flip-flops. Note that this process is very similar to standard static timing analysis method commonly used during design.

Initial skew scheduling uses the permissible ranges to determine whether a feasible clocking schedule exists. If so, it computes a schedule consisting of required arrival times to each register. Most circuits have a number of feasible clock schedules. Initial skew scheduling selects one that produces the biggest benefits—either the best performance or the most robust timing (the fewest critical paths). Since that step chooses the best skew value from the permissible range, it's key to managing the skew.

Clock tree topology synthesis uses the clock schedule, the register placement locations, technology information, and buffer descriptions to determine a buffered clock tree topology that delivers the clock signal according to the schedule. At minimum, clock tree topology synthesis should deliver a tree that either meets the initial skew schedule, or comes as close as possible. However, as mentioned above, there may be many schedules with similar timing benefits, but some may be easier to synthesize than others depending on physical factors, such as flip-flop placement. So, in general the clock tree topology synthesis step is free to modify the schedule on the fly in order to make it easier to realize and to control the trade-offs such as area vs. timing [10].

Clock net routing takes the clock tree topology generated in the previous step and implements the layout of the clock net. This step includes routing each branch in the clock tree, connecting buffers and clock pins, and delivering the skew according to the schedule.

Clock timing verification assesses the quality of the clock tree implemented by the previous steps. It analyzes the clock tree by extracting the parasitic resistances and capacitances, and calculating the delays and skews, comparing those results with the target schedule to verify that the system is free of race conditions and can operate correctly at the required clock frequency. The verification feedback quantifies the tolerable variations in clock arrival time. For 0.25um and deeper submicron technologies, clock tree verification must also analyze coupling and IR drop effects and report potential impact on the clock signal and the timing of the chip. Finally, verification must indicate where any violations occur and how to fix them.

Useful skew is relatively new to many; however, industry IC design leaders in networking and consumer electronics applications have proven in silicon that the results are valuable. Clock generation for ASICs can be done in a mater of hours, where otherwise it would have taken weeks.

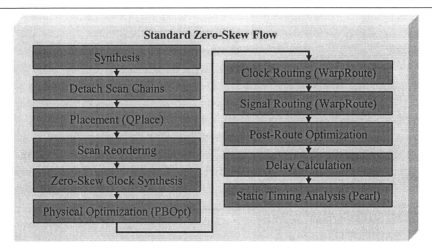

Figure 9. Conventional zero-skew design flow.

6. USEFUL SKEW CASE STUDY

The method for applying useful-skew to a physical design flow may vary according to the specific needs of the flow and the designer. The following description outlines one technique that has been applied and proven effective for using ClockWise to augment the Cadence physical design flow. It uses a two-phase approach that first applies zero skew and then useful skew. Note that this approach is general and can be applied to other design flows as well.

First the designer applies a standard zero-skew flow as illustrated in Figure 9.

After static timing analysis, the designer should have a good understanding of the design timing from both an ideal-clock and a propagated clock analysis. From the ideal clock analysis, the designer gains insight on performance limits based on traditional zero-skew methodology. From the propagated clock analysis, the designer realizes the actual impact of clock skew on critical path timing.

The designer is now ready to apply useful-skew to the design with the flow in Figure 10.

Table 1 shows some performance characteristics of useful-skew for a graphics chip design. This design uses the Artisan library and the 0.15 micron TSMC process. It has 283,000 cell instances and 322,000 nets.

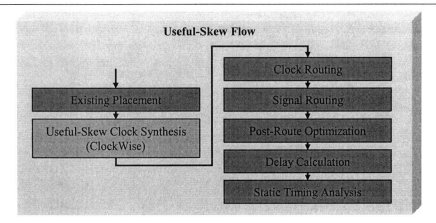

Figure 10. Useful-skew design flow.

Generally, better quality of result can be realized when the user applies the physical optimization step before applying useful-skew to the design. Some path timing may have unusually large cell delay and interconnect delay caused by nets with either high-fanouts or long wire lengths. These timing violations can be easily solved with high-fanout net synthesis and/or repeater insertion. Solving these transition-induced violations before applying useful-skew to the design will generally provide better timing results. The value of useful-skew is that it can continue to perform timing optimization through cycle-stealing when physical optimization reaches its performance limit.

The following are some general observations gained from applying useful skew to a variety of designs. Designs with large numbers of pipelined logic paths can greatly benefit from useful-skew as these designs typically have greater flexibility for time. Designs where the difference between the longest and shortest path between any two flip-flops is small gives useful-skew more flexibility as there is generally greater hold-time margin. Although the performance improvement due to useful-skew is largely a design-dependent phenomenon the slack improvement is typically in the range of 5% to 15% of the clock period.

	Flip-Flops	Period (ns)	Slack (ns)	Buffers
Zero-Skew	16,770	5.5	-0.2	1,036
Useful-Skew	16,770	5.5	0.3	1,048

Table 1. The performance improvement by using useful-skew in a graphics chip. The number of buffers increased slightly.

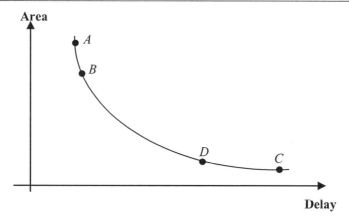

Figure 11. Trade-off in area and delay.

7. CLOCK AND LOGIC CO-DESIGN

For simplicity, in the conventional ASIC design methodology, we separate the clock design from logic design. While clock designers ignore path timing and strive only to minimize global clock skew, the logic designers assume zero-skew clocking and focus on controlling long path delays. Gate sizing is a commonly used technique for logic optimization. The general character of the area-delay trade-off for gate sizing alone without clock skew optimization is shown in Figure 11 [8]. For a loose delay constraint, the area penalty is small, but for a tighter delay constraint, the area penalty becomes very large. By applying clock skew optimization, we move timing slacks form non-critical paths to critical paths, thus create a larger timing budget and reduce the area penalty. While we achieve a big area saving from A to B on the curve for the critical paths, we pay a small area penalty from C to D for the non-critical paths. This illustrates the motivation of simultaneous clock skew optimization and gate sizing and clock and logic co-design.

8. SIMULTANEOUS CLOCK SKEW OPTIMIZATION AND GATE SIZING

The clock skew optimization problem can be formulated as a linear program problem [6]:

Minimize T
Subject to:

(4) $$C_i + d_{ij} \geq C_j + t_h$$
$$C_i + D_{ij} \leq C_j + T - t_{su}$$

If we combine clock skew optimization with gate sizing, then d_{ij} and D_{ij} are a function of transistor sizes, w_1, w_2, ..., w_n. Each constraint in the optimization problem is of the form:

(5) $\quad Posynomial(w_1, w_2, ..., w_n) + Convex(C_1, C_2, ..., C_m) \leq K$

Where K is a constant.

Since a posynomial function in x can be transformed into a convex function in z by mapping x to e^z, Substituting transistor width w_i by e^z_i in d_{ij} and D_{ij} and leave C_i along, the problem is a convex programming problem, which can be solved easily [8].

The solution may be arrived in two steps. First, a convex programming problem is solved where only long path constraints are considered and short path constraints are ignored [5][8]. Next the short path violations are reconciled by minimum padding [9]. For the first step, the following two procedures are iterated until all long path constraints are satisfied for a given clock period: a generalized PERT method is applied to detect violations of the clocking constraints and the size of the most sensitive gate on critical path is identified and sized: p-channel or n-channel width for the gate is bumped up by a small amount depends on whether the gate is undergoing a rise or fall transition. As the result, the arrival time at flip-flop outputs calculated by PERT is the optimal skew to the flip-flops, or an optimal clock schedule.

It has been reported that simultaneous clock skew optimization and gate sizing achieved significant better results compared with gate sizing alone [8]. For example, the clock period of a 20-stage pipeline with 4043 gates has been improved from 40.38ns to 14ns. While the area penalty for gate sizing alone was 38%, the area penalty for simultaneous skew optimization and gate sizing was 26.5%. There was not significant difference in computing time. In most of the cases, simultaneous skew optimisation and gate sizing achieve the clock period achievable by gate sizing alone at about half of the area penalty. In some cases, simultaneous skew optimization and gate sizing achieve significantly lower clock period that gate sizing alone.

9. CONCLUSION

Few people will question that the clock is the single most critical signal in System-on-a-Chip (SoC) design. This is the net which first encountered the deep submicron challenges: interconnect delay, coupling noise, IR drop, electro-migration, process variation, and on-chip inductance. It becomes more and more difficult, if not impossible, to achieve "exact zero skew." Even if power and area are no object, and assuming the designers do a good job and achieve almost zero skew, the large amount current drawn from the power supply will introduce simultaneous switching noise, which may inadvertently cause the chip to fail—especially if it contains sensitive analog and RF circuits. On the other hand, if we view clock skew as a manageable resource, we can employ useful skew to achieve wider safety margins, increase clock frequency, speed up design closure, and reduce peak current and simultaneous switching noise. With increasingly design margins, logic and clock can no longer be designed separately—simultaneous clock skew optimization and gate sizing will become common practice in ASIC design in the near future.

10. REFERENCES

[1] Anderson, S. F., Earl, J. G., Goldschmidt, R. E., and Powers, D. M. "The IBM system/360 model 91: floating-point execution unit," *IBM Journal*, pp. 34-53, Jan. 1967.
[2] Bailey D. W., and Benschneider, B. J., "Clocking design and analysis for a 600-MHz alpha microprocessor," *IEEE Journal of Solid-State Circuits*, pp. 1627-1633, Nov. 1998.
[3] Benini, L., et al., "Clock skew optimisation for peak current reduction," *Journal of VLSI Signal Processing Systems for Signal, Image, and Video Technology*, pp. 117-130, June/July 1997.
[4] Celestry. Clockwise: Useful-Skew Clock Synthesis Solution. 2002. http://www.celestry.com/products_clockwise.shtml
[5] Chuang, W., Sapatneker, S. S., and Hajj, I. N. "A unified algorithm for gate sizing and clock skew optimisation to minimize sequential circuit area," *Proceedings of the International Conference on Computer-Aided Design*, pp. 220-223, 1993.
[6] Fishburn, J. P. "Clock skew optimisation," *IEEE Transactions on Computers*, pp. 945-951, July 1990.
[7] Friedman, E. G., ed. *Clock Distribution Networks in VLSI Circuits and Systems*, IEEE Press, 1995.
[8] Sathyamurthy, H., Sapatnekar, S. S., and Fishburn, J. P. "Speeding up pipelined circuits through a combination of gate sizing and clock skew optimisation," *Proceedings of the International Conference on Computer-Aided Design*, pp. 467-470, Nov. 1995.
[9] Shenoy, N. V., Brayton, R. K., and Sangiovanni-Vincentelli, A. L. "Minimum padding to satisfy short path constraints," *Proceedings of the International Conference on Computer-Aided Design*, pp. 9-27, Oct. 1988.
[10] Tsao, C.-W., and Koh, C.-K. "UST/DME: A clock tree router for general skew constraints", *Proceedings of the International Conference on Computer-Aided Design*, pp. 400-405, 2000.
[11] Vittal, A., Ha, H., Brewer, F., and Marek-Sadowska, M., "Clock skew optimisation for ground bounce control," *Proceedings of the International Conference on Computer-Aided Design*, pp. 395-399, 1996.
[12] Wagner, K. D., "Clock system design," *IEEE Design and Test of Computers*, pp. 395-399, 1996.
[13] Webb, C. F. et al. "A 400MHz S/390 microprocessor," *Proceedings of the International Solid-State Circuits Conference*, pp. 168-169, 1997.
[14] Xi, J. G., and Dai, W. W-M., "Useful-skew clock routing with gate sizing for low power design," *Journal of VLSI Signal Processing Systems for Signal, Image, and Video Technology*, vol.16, no. 2-3, June-July 1997, pp. 163-79.
[15] Xi, J. G., and Dai, W. W-M., "Useful-skew clock routing with gate sizing for lower power design," *Proceedings of the 33rd Design Automation Conference*, pp. 383-388, June 1996.

[16] Xi, J. and Staepelaere D., "Using clock skew as a tool to achieve optimal timing," *Integrated System Design*, April 1999.

Chapter 9

Faster and Lower Power Cell-Based Designs with Transistor-Level Cell Sizing

Michel Côté, Philippe Hurat
Cadabra, a Numerical Technologies Company,
70 West Plumeria Drive,
San Jose, CA, 95134, USA

Much effort is going into providing optimization capabilities for standard cell-based design. While the industry is focusing mainly on enhancing design tools, the standard cell libraries that are the foundation of most designs do not get the same attention. Most designs are developed using general-purpose libraries. In general, chip designers cannot afford the time and human resources that custom design requires.

This chapter describes a Power and Performance Optimization (PPO) flow that leverages cell creation for design optimization. This flow delivers higher performance and reduces power consumption in cell-based designs

The PPO flow optimizes the transistor sizes of each cell instance in the context of the design and automatically creates the optimized cell layout, while maintaining the compatibility of the cell's footprint. The optimized cells may then be re-inserted using an ECO place-and-route flow.

1. INTRODUCTION

Today, many semiconductor products are dependent on a semi-custom design methodology based on synthesis and place-and-route tools. Demand for faster completion time makes this methodology very attractive. Engineers can focus on describing the design, and the flow ensures implementation of this design in silicon within a reasonable time.

In comparison, a full-custom design methodology is very expensive in terms of resources. It takes a large team of engineers and a long time to complete these designs. Full-custom design methodologies are still used because they provide much higher performance, and designers have more flexibility to tune performance versus power consumption. For example in the 0.18um process technology generation, microprocessor products (with a full-custom methodology) have up to 2GHz operating frequency, while the companion chipsets operate at around 350MHz.

Part of the reason for this "performance gap" is that full-custom designers have the ability to optimize their designs at the transistor level. They have access to custom macro cells, circuits using dynamic logic, etc. On the other hand, ASIC designers implement optimizations at the gate level or RTL level and, therefore, cannot reach such high performance.

To close the performance gap, much effort is going into providing optimization capabilities for cell-based design. The industry is mainly focusing on design tools to accelerate the timing convergence and increase design performance. In order to improve cell-based designs, EDA companies provide post-layout optimizations or implementation solutions that combine synthesis and place-and-route techniques.

However, the standard cell libraries that form the foundation of most designs do not always get the same attention. Most designs are developed using general-purpose libraries. Because of the time to market pressure and the lack of resources, semi-custom designers cannot afford the same level of customization that custom design methods provide.

The Power and Performance Optimization (PPO) flow is a methodology to improve cell-based design by optimizing the standard cells in the context of the design.

2. OPTIMIZED CELLS FOR BETTER POWER AND PERFORMANCE

Figure 1 illustrates how an optimized cell can improve power and performance. In this example, synthesis chose an inverter with 2X drive strength to drive the net *out*. If timing is not met after place-and-route because the *out* load capacitance is larger than anticipated, in-place optimization needs to increase the buffer size. The optimization tool would then pick the next available cell in the library such as a 3X drive strength. However, the ideal drive to meet the timing goal could be a 2.5X drive strength cell that does not exist in the library. If the critical path is on the *out* rising edge, then only the P transistor, driving a 0→1 transition, needs to have a 2.5X drive strength, while the N transistor, driving a 1→0 transition, may only need a 1.5X drive strength. Typical standard cell libraries don't have this fine granularity of drive strengths, and often don't have cells with skewed drive strengths. By sizing up the P transistor, the cell performance is increased. Downsizing the N transistor not only reduces the power, but also decreases the input capacitance of the cell, which increases the speed of the critical *in* falling transition and further reduces power consumption.

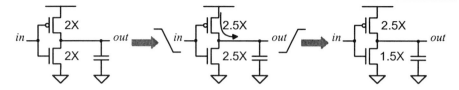

Figure 1. An example of an inverter driving a load capacitance, where the inverter can be optimized to better drive the capacitance on the critical rising output transition.

The principle demonstrated in Figure 1 can be applied to any cell in the library. For instance, Figure 2 illustrates the optimization of complex cells with several output nets. Conventional libraries have a limited set of drive strengths for each function of the library. Complex cells with multiple outputs have the same drive strength for both rising and falling transitions, and for each output. In Figure 2, the adder cell has 2 output nets, *Sum* and carry out (C_{out}) which have the same 2X drive strength: the pull-up chains of P-channel MOSFET transistors and the pull-down chains of N-channel MOSFET transistors have a driving capability of 2 units. In the final design, the net capacitances on *Sum* and C_{out} are usually different, and the timing requirement for rising and falling transitions may be different. Therefore *Sum* and C_{out} may require different drive strengths and different rising and falling drive strengths. As shown on the two adder variants of Figure 2, variant (a) needs a stronger drive for net C_{out} than for the net *Sum*. On the other hand, variant (b) requires a strong drive for the net *Sum* as well as for the falling edge of C_{out}.

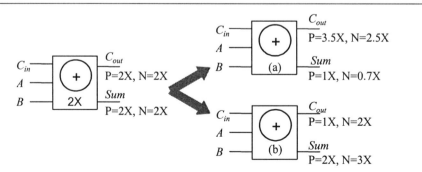

Figure 2. Optimizing an adder: variant (a) has increased drive to net C_{out}, whereas (b) has been optimized to drive only the falling edges of *Sum* and C_{out} more strongly.

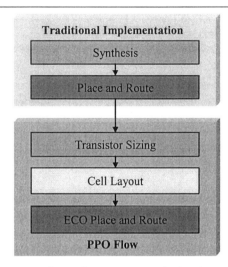

Figure 3. The Power and Performance Optimization flow, creating new drive strengths of cells after place-and-route, and then replacing cells in an incremental place-and-route (ECO) flow.

Developing and maintaining so many cell variants in a library is not practical. Library development would be too time-consuming and synthesis with a large library may be suboptimal. But limited drive strengths limit performance of traditional design implementations. By dynamically expanding the library using custom cells that are optimized for design specific contexts, the limits of using a static library are overcome.

3. PPO FLOW

As shown in Figure 3, the PPO flow starts from a design implemented in a standard cell ASIC flow. The design has been synthesized and place-and-routed using an existing standard cell library with traditional gate-level tools. The library may have been optimized for the specific synthesis and place-and-route flow, but not for the design. Beyond the optimization provided by gate-level design techniques, further power and performance improvements can be achieved by optimizing the standard cells in the context of the design. The design goes through one additional optimization phase at the transistor level. The transistor-sizing tool automatically resizes the standard cell transistors to find the combination that will best meet power and speed goals. The tool provides a list of optimized cell netlists, and then the layout of these optimized cells is created. Once the optimized cell layouts are completed, the original design is updated through ECO place-and-route.

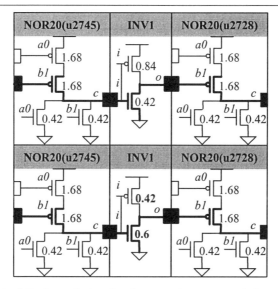

Figure 4. Critical Path optimization by transistor up and down sizing

The aim with the Engineering Change Order (ECO) place-and-route flow is to implement incremental changes in the layout without perturbing the layout too much, to avoid problems with convergence. To do this, the footprints of the new cells must be compatible with the footprint in the original layout.

3.1 Transistor Sizing Tool

A transistor level optimization tool, such as Synopsys AMPS [4], resizes the transistors of the cells depending on their context to meet or improve the timing performance while reducing the power consumption. Transistor-level sizing tools are often used in custom design, but have generally not been used in the standard cell flow, because of the historical lack of adequate standard cell generation tools.

Transistor level optimization tools read in the SPICE transistor netlists of the standard cells, the design netlist, the design parasitic data and the timing constraints. It analyzes the timing with its embedded transistor-level static timing analysis tool and then resizes the transistor sizes. Transistor level optimization tools do not usually change the circuit design, but upsize and downsize the transistors to find the best cell depending on its context.

Figure 4 illustrates how a transistor-sizing tool creates the right cell for its context. The highlighted critical path goes through the falling edge (N transistor) of the inverter. Up-sizing the N transistor to provide a drive strength of 1.5X on the falling edge can increase the speed of the signal going through this inverter. At the same time, the P transistor can be downsized to 0.5X because it does not belong to the critical path. Downsizing the P transistor also reduces the gate capacitance viewed by the gate driving the signal through the inverter, increasing the speed and reducing the power.

The transistor-sizing tool automatically defines the optimal transistor sizes. It adjusts the P/N ratio of the pull up and pull down transistor chains based on the context of each cell. It takes into account the post-layout wire capacitance, input slopes and the criticality of power and timing of these transistors (e.g. whether or not they belong to the critical path). The minimum and maximum P/N ratio is also constrained in order to preserve the cell integrity, as explained in Section 3.4.

Multiple instances of the same cell may result in different optimized versions. The transistor level optimization tool must consider the netlist hierarchy. For instance, Synopsys AMPS' hierarchy management maintains the original netlist hierarchy [5], and thus keeps the notion of individual cells. This is required to enable the cell to re-enter the gate level design flow.

To avoid the creation of too many similar cells, the transistor-sizing tool must provide a way to limit the number of cell variations it creates. The user of Synopsys AMPS can specifies a maximum number of new cells, or a maximum of cell variants for a given cell [5].

The result of transistor sizing is a new design netlist containing optimized cells that were not a part of the original cell library.

3.2 Cell Layout

To propagate the result of the optimization tool to the original library and design, the layout views of optimized cells need to be created. We used Cadabra's Automated Transistor Layout (ATL) tool to create optimized cell layouts in the same standard cell layout architecture as the original library. Starting from the optimized SPICE netlist created by the transistor level optimization tool, Cadabra tools offer two paths to implement the optimized cells: migration and creation [3]. Migration is used when the transistor sizes can be applied directly to the original cell layout. In that case, the Cadabra tool reads the optimized netlist and the original layout [2] as shown in Figure 5(a). In order to minimize the perturbation introduced by the optimized cells in the placed and routed block, the Cadabra tool derives footprint information from the original layout. As shown in Figure 5(b), footprint information such as the pin locations and porosity is extracted. When applying the new transistor sizes to the original layout, the Cadabra tool tries to preserve the footprint of the original layout, as shown in Figure 5(c).

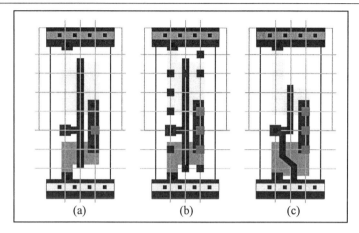

Figure 5. The cell layout of an inverter (a) before, and (c) after optimization. Cell footprint information is extracted in (b).

However, it is not always possible to accommodate the new transistor sizes in the original layout. This is particularly evident when doing performance optimization, where some transistors may grow too large to fit in the existing cell architecture without being folded. Then the optimized cell layout cannot be derived from the original layout, since transistor folding implies a different transistor placement. For these cells, we used Cadabra's unique ATL capability to automatically create layouts from SPICE netlists [2]. Once the architecture and layout style has been specified through the graphical user interface, Cadabra's tool does not require any cell specific information. It analyzes the SPICE netlist connectivity, defines the optimal transistor folding and placement, and routes the cell following the user-specified layout style preferences. Finally, Cadabra two-dimensional compaction creates the LVS and DRC correct layout of the optimized cell. ATL is the result of years of experience and research and development at Cadabra Design Automation and has been in production use at major semiconductor companies and library vendors since 1996.

It is important to note that the cell layout is significantly changed when it needs to be created from scratch instead of being derived from the original layout. In that case, the cell is likely to get larger, and the footprint compatibility is lost. A change in footprint may create a change at the block level. In order to minimize these perturbations of the block level place-and-route and reduce the cell layout runtime, it is recommended to constrain the transistor-sizing tool to maximize the use of layout migration versus creation. Section 3.4 explains how to create such constraints.

The cell layout success depends on both capabilities: creation and migration. It is critical to be able to preserve the footprint constraints and to use a powerful compaction technology to maintain the cell layout quality and density.

3.3 ECO Place and Route

Once the optimized cell layout is created, the optimized design is loaded in the place and route tool in ECO (Engineering Change Order) mode. Most place and route tools have an ECO mode, where an existing layout block is updated to reflect some netlist changes. In ECO mode, the tool preserves the block boundary (area and pin placement) and tries to preserve as much as possible the cell placement and signal routing.

But ECO place and route can only accommodate minor changes in the design. If the perturbation in the block layout is too much, rerouting may change the wire parasitic capacitances significantly and the convergence of the flow is no longer guaranteed. In the case of PPO, the design netlist and connectivity stay the same. Block routing changes may only come from an area or more generally footprint change of some cells, thus changing the cell placement. The cell area would only change if the transistor sizes changed significantly. In order to minimize that side effect, a set-up phase analyzes the cell of the original library and derives new constraints for the transistor sizing tool to limit the change in cell footprint and area, as described in the Section 3.4.

Final performance and power verification are performed at the transistor level with the final optimized design data.

3.4 Library Analysis

The PPO flow requires some preliminary set-up. This set-up phase has two main objectives:
- Provide cell specific information to the transistor optimization tools
- Constrain the transistor sizing to preserve the cell behavior and try to control the impact of sizing on the cell area in order to minimize the cell footprint and minimize the perturbation during ECO place and route after optimization

Prior to PPO, the analysis of the standard cell library used to implement the original design is mandatory in order to provide the needed data for the PPO flow. We will use the term PPO-views for this data, and a PPO-ready library refers to a library that has been analyzed and for which the PPO-views are available. This analysis is not design dependent and is run only once for a standard cell library.

The PPO-views contain commands for the transistor optimization. The PPO-views include three main types of data:
- Special cell declaration: the transistor-level static timing engine requires some special definitions, in order to properly handle special cells such as sequential elements.
- Electrical data: the sizing constraints are required in order to preserve the electrical integrity of the cell.
- Topological data: the transistor sizing is usually constrained to keep transistor widths within a certain range in order to minimize the perturbation in cell footprints during layout of the sized cells.

3.4.1 Special Cell Declaration

The transistor-level static timing engine requires special cell declarations for latches in order to identify sequential elements, and correctly analyze the timing through these elements and perform timing checks. Special cell declarations identify transistors that latch data, and feedback transistors to ignore during the transistor-level static timing analysis. If a cell declaration is not supplied for a latch, transistor optimization may treat the latch as combinational logic [5].

Some other special cells, such as bus repeater and clock buffers are not suitable for sizing. Therefore, sizing must be prevented by specific commands to the transistor-sizing tool.

3.4.2 Electrical Integrity Constraints

It is recommended that all the critical circuits of the design should be constrained in order to preserve the electrical integrity of the cell during sizing. Here are a few examples of such sizing constraints:
- Minimum and maximum P/N ratio (limiting skewing of the drive strength)
- The P/N ratio of pull up and pull down transistor chains may also be constrained to keep the same ratio.
- For sequential elements, the transistor-sizing tool is not allowed to change the transistor sizes of the core sequential elements within the latch.
- It is also not recommended to up-size other transistors of latch cells, such as clock drivers and output buffers, as it may impact the behavior of the core.

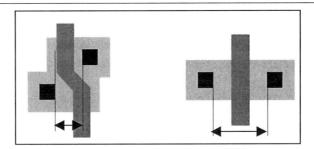

Figure 6. Removing the gate bend increases the contact-to-contact distance.

3.4.3 Topological Constraints

The objective of the topological constraints is to minimize the perturbation in cell footprints during layout of the sized cells. Thus the transistor optimizer is constrained with a minimum and a maximum size for each transistor of the cell. The ultimate minimum size is the minimum transistor size defined by the technology. However, it is not always possible to keep the same cell width even when only downsizing transistors, because of design rule constraints (DRC).

For instance, minimizing a transistor with bent gates may increase the cell width. As shown in Figure 6, when downsizing a transistor with a bent gate the transistor may become too small to allow for the gate bend and the gate bend must be removed, thus spreading the contacts apart.

Figure 7 illustrates another case where a smaller transistor can actually increase the cell size. Consider placing a transistor of minimum length, 0.18um, between two diffusion contacts. If the diffusion width is the same as the contact height, then the transistor may be placed minimum distance to the contacts, in that case 0.20um. Thus the total distance from contact to contact is:

$0.20 + 0.18 + 0.20 = 0.58$um

However, if the transistor has a smaller width than the contact height, then the design rule constraints require that the transistor be 0.05um from the diffusion that surrounds the contact and a regular diffusion extension of 0.18um must be enforced around the contact.

Hence the total distance is:

$0.18 + 0.05 + 0.18 + 0.05 + 0.18 = 0.64$um

Defining the minimum and maximum sizes of each individual transistor cannot be easily achieved without a detailed analysis of the layout topology. This is beyond the capability of a visual inspection or a simple layout analysis.

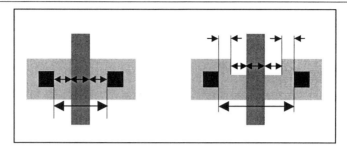

Figure 7. Diffusion-gate spacing increases the contact-to-contact distance

The topological constraints for a cell are the minimum and maximum sizes of each transistor of the cell in the format of the transistor-sizing tool.

3.5 PPO Views Creation

The special cell declarations and electrical integrity constraints are library specific. When the library designer creates the cell circuit, he can easily provide these constraints.

In comparison, the creation of topological constraints for minimum and maximum transistor sizes requires a sophisticated analysis of the layout. The goal is to find the minimum and maximum transistor sizes while maintaining a compatible footprint to the source layout. This includes achieving the same area, and keeping the same cell porosity and routability. Reusing the original layout data and the footprint information as described in Figure 5 of Section 3.2, the Cadabra tool is able to automatically derive these minimum and maximum transistor sizes and export them to the transistor-sizing tool.

In the case of power only optimization, the transistor-sizing tool can be constrained to not increase the size of any MOSFET. In that case, the maximum transistor size is the original size.

The library analysis is run only once for a standard cell library; subsequent design optimizations with the PPO flow do not require this analysis to be run again. This step is critical for the convergence of the flow, as it minimizes the ECO layout perturbation. It also controls the transistor sizing to preserve the electrical integrity of the optimized cells.

4. PPO EXAMPLES

4.1 Power Optimization

To check the PPO flow, we used an 8051 IP block using a commercially available 0.18um library of over 600 standard cells. The library included a very rich set of cells with fine drive strength granularity: the combinational cells had up to seven different drive-strengths and the sequential elements had six drive-strength variants. The circuit design of these cells was already optimized for power. Most of the transistor sizes are of the minimum contact size, except for the output drivers.

We used a traditional ASIC flow, including synthesis and place and route, to implement the ~13,000 gate design. It is important to mention that synthesis was run to optimize both timing and power. Once synthesis with Design Compiler has met timing constraints, Power Compiler was used to reduce the power consumption using cells from the standard cell library (this is pre-layout). In-place-optimization (IPO) was also run at the final stage of the place and route, as shown in Figure 8. This design implementation made good use of the rich set of library cells, as it used almost a third of the cells available in the library.

Further improvements were possible with the PPO flow by creating new cells with finer drive strength granularity than in the standard cell library, where these new cells are optimized for their layout context. The PPO flow further reduced the power consumption of the 8051 by 15%, creating 300 optimized cells. An optimization iteration, which includes transistor-level optimization, cell creation and ECO place-and-route, runs overnight. About half the time was spent running AMPS and the remaining half was spent creating the optimized cells with Cadabra tools. ECO place-and-route was very fast on this design.

4.2 Performance Optimization

We ran the PPO flow for performance on a bus controller of around 12,000 gates, which had about 70,000 transistors using 178 different cells of a 0.35um standard cell library.

The original design's worst critical path was 18.5ns in the worst case process and worst case operating conditions. After optimization, the worst critical path was 16.02ns, which was a 13.5% improvement. At the same time, power was reduced during optimization. Power dissipation analysis before and after optimization reported 18% lower power dissipation for the optimized design.

To achieve that improvement, the PPO flow generated 300 optimized cells, changing 22% of the transistors in the design. Of those, 70% of the transistors were downsized and 30% were upsized.

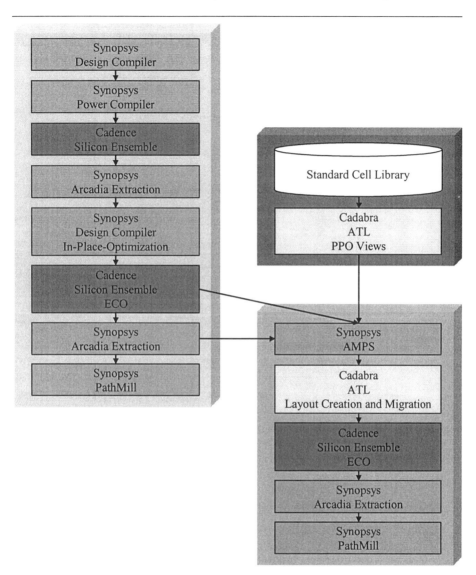

Figure 8. Detailed PPO flow used for power optimization

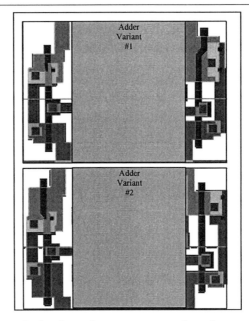

Figure 9. After optimization, there are several variants of the basic adder cell. The variants of inverters that buffer the outputs of the adder are shown.

The most critical paths were in the Arithmetic and Logic Unit (ALU) using adder cells. After optimization, the adder instances of these paths became different optimized cells with different *Sum* and C_{out} and rising and falling drive strengths as shown in Figure 9. Only the output buffers of the adder variants are shown.

Among the 300 optimized cells needed to optimize the design performance, 187 were created by migration, while 30% of the optimized cells required layout creation from the SPICE using ATL. As all the steps were automated, a single iteration of the PPO flow on this 12,000 gate design was completed in less than a day.

On this design, PPO increased performance by 13.5% and reduced power consumption by 18%

5. FLOW CHALLENGES AND ADOPTION

The PPO flow provides significant benefits, but its large-scale adoption still faces some challenges.

Current transistor sizing tools have limited capacity, on the order of 50,000 to 100,000 transistors [4], which translates to about 12-25Kgates. The design needs to be partitioned into fairly small blocks to take advantage of transistor sizing and allow power and performance optimization. This is a common practice in high performance designs [1], but would require a methodology change for most chip designers. Recent breakthroughs in the transistor-level simulation domain have shown tremendous increase in capacity, going from 100,000 transistors to several million transistors [6]. As of today, these improvements have not been applied to transistor sizing tools.

In any optimization flow, timing convergence is critical. In the PPO flow described in this chapter, the ECO place-and-route step is a critical component. When optimizing for power, the PPO flow maintains the cell footprints, therefore, optimized cells can replace original cells without impacting placement or routing. However, to maximize the benefit of performance optimization larger drives are required, thus making it impossible to guarantee the footprint compatibility. In that case, we can keep footprint compatibility for non timing-critical cells and allow transistor growth on a reduced set of critical instances. This allows us to update the placement using ECO. However, ECO routing cannot repair the routing if the modified placement creates too many violations. To increase the timing convergence, it is important to constrain the transistor sizing in order to reduce the perturbation at the block level and limit the change for ECO. It is also recommended to use more slack in place-and-route in order to increase the chance of success with ECO.

Designers are used to traditional standard cell flows, where each cell is characterized for timing and where timing models are created and used for gate level sign-off. In our optimized block, the timing information must be updated. Since the cells changed, the characterized timing data is no longer valid. New timing information must be derived from the optimized transistor netlist. Optimized cells could be characterized, but this is very computationally intensive and 100% correlation is unlikely with transistor-level timings due to guard banding. Since this block is optimized at the transistor level, timing signoff should be done at the transistor level. The new timing of this block, once verified, can be captured in a form of a timing stamp. This approach to timing analysis is common in custom IP development, though not yet in ASICs.

Limitations of capacity and block level timing models make this flow particularly well suited for hard-IP development.

6. CONCLUSIONS

Current cell-based design methodology is limited by the set of standard cells available in the library. Optimization is limited by the coarse granularity of drive strengths and lack of skewed drive strength in libraries. Including these variants would lead to a huge library set, which would very quickly become unmanageable. Moreover, gate-level timing data are more conservative than transistor level timing due to guard banding during cell characterization. Optimizing the cells in their context at the transistor level removes these limitations.

Power and Performance Optimization flow presented in this chapter combines a cell creation and transistor-sizing tool. Other PPO flows are possible with different design optimization tools, such as synthesis, and place and route tools.

By giving the design optimization the flexibility of adding or optimizing standard cells in the context of the design, the PPO flow delivers higher performance and reduces the power consumption of cell-based designs. By offering more optimization than traditional cell-based designs, PPO is especially well suited for semi-custom designs and hard IP development, where high performance and/or low power are critical, and where transistor level optimization is more frequent than in ASIC design. PPO combines the efficiency of cell-based design and the accuracy and optimization of custom design, thus reducing the performance gap between ASIC and custom design.

7. REFERENCES

[1] Northrop, G.A., and P. Lu., "A Semi-custom Design Flow in High-performance Microprocessor Design," *Proceedings of the 38th Design Automation Conference*, June 2001, pp. 426-431.

[2] Numerical Technologies. *abraMAP abraKAZAM Documentation – Cadabra Tools*, version 2001-1.

[3] Numerical Technologies. Cadabra Tools Datasheet
http://www.numeritech.com/ntproducts/?id=12

[4] Synopsys, AMPS Datasheet. November 1999.
http://www.synopsys.com/products/analysis/amps_ds.html

[5] Synopsys, *AMPS Reference Manual*, version 5.4, January 2000.

[6] Wang, S., and Deng. A., Delivering a Full-chip Hierarchical Circuit Simulation & Analysis Solution for Nanometer Designs. 2001.
http://www.nassda.com/HSIMWhitePaper.pdf

Chapter 10

Design Optimization with Automated Flex-Cell Creation
Beyond Standard-Cell Based Optimization and Synthesis

Debashis Bhattacharya, Vamsi Boppana
Zenasis Technologies Incorporated,[1]
1671 Dell Avenue
Campbell, CA 95008, USA

BACKGROUND

A standard-cell based automated design flow for digital circuits is a mixed blessing for designers. Historically, use of pre-characterized and silicon verified "standard cells" was driven by the designers' need to design and verify large digital circuits without having to verify the behavior of the circuit at the transistor level – transistor-level design and verification of a multi-million gate digital circuit is simply too resource-intensive to be commercially viable for most digital designs. Standard cells provided relatively fine-grained control over the structure of the digital circuit, and yet were amenable to manipulation using automated synthesis tools that made it possible to design complex digital integrated circuits (ICs) with a team of less than 10 engineers.

On the other hand, from its inception, "quality" of the designs created by such automated standard-cell based design flows, has been deemed "poor to barely acceptable", by almost any measure of quality, including clock speed, area/die-size, power consumption, etc. In a series of studies including a special session at the 37th Design Automation Conference (DAC 2000) [8][11][18], it has been estimated by various researchers that designs created by such automated design flows are slower by at least a factor of 6, and larger in design area by at least a factor of 10, when compared to similar designs created and/or optimized manually. An in-depth examination of

---

[1] Zenasis Technologies, Inc., is an IC design automation company developing hybrid gate- and transistor-level optimization technologies (patents pending) that bridge the performance gap between standard-cell based and custom design.

these studies' findings shows that a significant portion of such deficiency in quality – e.g. a quarter or more of the speed shortfall – can be due to the use of a fixed, pre-defined library of standard cells.

Over the years, various forms of manual intervention and manual tweaking of designs generated by automated flows, have become commonplace. Such practices include:

1. Use of specially (usually manually) crafted macro-cells, such as special memories, and special functional blocks like barrel shifters and multipliers.
2. Use of special directives in synthesis, to force use of specially crafted macro-cells for implementation of known problem constructs in register transfer level (RTL) description of digital circuits.
3. Special layout techniques for structured circuits, e.g. tiling for datapath circuits.
4. Use of tactical (design-specific) cells – which are similar in size to typical standard cells, as opposed to macro-cells which tend to be much bigger – that are created manually, based on prior experience with similar design, or experience with prior versions of the same design.

Among these, use of special macro-cells and use of tactical cells has become virtually routine for all high-performance designs created using automatic design tools, especially for high-performance designs that run into *timing closure* problems. Timing closure problem occurs due to inaccuracies in timing estimation during automated creation (synthesis) of design, usually linked to incorrect estimation of interconnect delays. According to surveys done by Collett International, a renowned market research firm, greater than 60% of all ASIC designs have timing closure problems [10].

PARADIGM SHIFT – CELL CREATION DURING OPTIMIZATION

The quest to overcome the limitations of standard-cell based design methods, leads naturally to the creation of new, design- and context-specific cells – designated *flex-cells* – during the process of optimizing a given digital design. It's well known in the design community that virtually every high-performance design project that relies on automated (synthesis-oriented) design flows, also uses design-specific *tactical cells* which are identified and created manually, and then utilized in the design via a combination of RTL coding style and synthesis directives. In fact, without the use of such tactical cells, the gap in quality between automated and handcrafted designs would be even wider than it is now.

From a superficial point of view, flex-cell based design optimization automates the creation of tactical cells, and thereby helps bridge the quality gap. However, a deeper examination of the flex-cell based optimization process makes it amply clear that the full impact of such optimization goes far beyond providing a better framework for tactical cell creation. An overall view of one possible IC design optimization process using flex-cells, is provided in Figure 1.

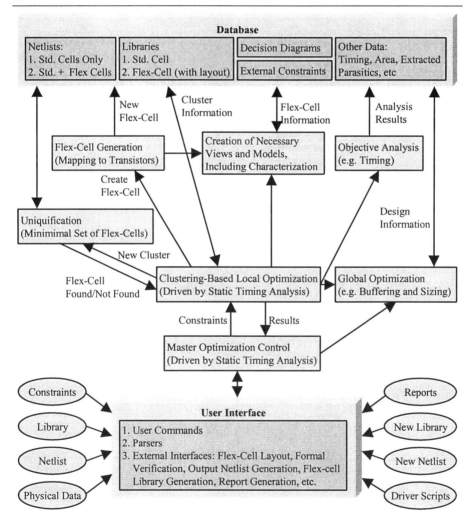

Figure 1. A possible IC design optimization process using flex-cells.

1. FLEX-CELL BASED OPTIMIZATION – OVERVIEW

The IC design optimization process of Figure 1, which is geared towards enhancing the performance – i.e. increased clock speed – of the design, is inspired by the time-tested (manual) process of *local optimization of a digital design driven by global analysis*. The global analysis, for this process, consists of accurate Static Timing Analysis (STA) [14][16], and local optimization consists of an overall control mechanism employing two key steps designated *clustering* and *mapping*, respectively.

Conceptually, the clustering process consists of identifying the best candidate regions in the design, for local optimization: this search is driven by the results of the STA, for performance optimization. The clustering process yields a set of "clusters" – groups of one or more standard cells – which need to be replaced by new flex-cells created for their respective timing contexts. The mapping process takes as inputs the clusters, and their respective timing contexts, and determines a reasonably small set of "best candidate flex-cells" which should be used to replace the clusters. The *optimization control* process searches through this set of best candidate flex-cells to determine whether replacing one or more clusters with flex-cells improve the overall timing of the given design.

The close coupling between the clustering and mapping processes is key to success of this optimization technique. Specifically, the mapping process is tailored to choose from a variety of techniques that can be used to create new flex-cells, based upon the inputs it receives from the clustering process. Such mapping techniques can include time-tested methods of gate sizing and (continuous) transistor sizing, as well as techniques typically found in manual design flows, e.g. creation of new (appropriately sized) transistor-level implementation of the function of a given cluster of standard cells.

1.1 Mapping – The Practical Choices

While details of the mapping process can be quite involved, at a minimum, this process must provide the capability for: (i) ensuring functional correctness of the resultant transistor-level design; (ii) meeting design targets, for example, performance (possibly measured using propagation delay) of the flex-cells generated, given the timing contexts for the intended use of the flex-cells; (iii) meeting other implementation constraints, such as maximum length of N- or P-transistor chains in the flex-cell, the required signal output (i.e. drive strength) for the design-specific cell, desired input capacitive load of the design-specific cell, etc.; (iv) minimizing the number of transistors in the design-specific cells, subject to the design of the IC design; and (v) sizing the transistors of the transistor-level netlists for the design-specific cells, as necessary.

Design Optimization with Automated Flex-Cell Creation

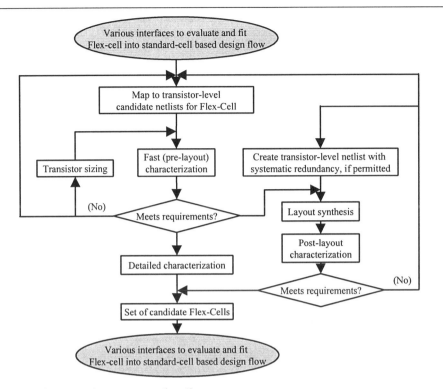

Figure 2. Mapping process details.

A more detailed view of the mapping process is shown in Figure 2. The inputs to this mapping process can include (i) a set of structural netlists composed of standard cells, otherwise designated "clusters", and (ii) a set of performance constraints for each individual cluster. As mentioned earlier, the set of clusters, and the set of performance constraints for the clusters, are identified by a clustering process that precedes the mapping process (see Figure 1). The clustering step essentially partitions the output of a conventional logic synthesis tool, either using heuristics to guide the partitioning process, or using a systematic search procedure such as a branch-and-bound search.

The important features of mapping process include: (i) creation of transistor-level netlist; (ii) fast characterization that incorporates implementation context of the flex-cells; (iii) transistor sizing; (iv) accurate (and slower) characterization of final transistor-level netlist; (v) optional layout synthesis with transistor sizing, via a layout synthesis tool; (vi) parasitic extraction and accurate post-layout characterization if layout synthesis is performed; (vii) generation of views to fit the flex-cells into a standard-cell based design flow. This is in sharp contrast with conventional, automated, transistor-level design optimization techniques that primarily derive their benefits from transistor sizing [12].

The mapping process includes four key sub-steps as shown in Figure 2: (i) transistor netlist generation, (ii) transistor netlist evaluation; (iii) transistor topology alteration, and (iv) preliminary transistor sizing. A variety of algorithms and heuristics are available in the literature for generation of a transistor netlist given the original cluster. For example, several techniques for deriving transistor netlists use the Binary Decision Diagrams (BDDs) [1] representing the function of the cluster, as starting points. A BDD is a well-known data structure based on acyclic directed graphs used to represent functions commonly encountered in digital circuits, and recently, several researchers, including Liu and Abraham [19], Gavrilov et al. [13], Newton et al., and Kanecko et al. [17], have demonstrated techniques to derive transistor netlist structures using BDDs.

Most of the existing mapping algorithms, available in published literature are geared towards working with very simple objectives, such as minimizing transistor count [2][22]. Moreover, many of the methods suffer from relatively high computational complexity. In addition, several algorithms have been proposed to synthesize pass transistor logic for use in highly custom design environments [22]. In practice, however, the optimization criteria and design requirements for the generated flex-cells, are not static, but are varied and complex even across different parts of a given IC design. Consequently, practical transistor netlist generation processes need to start with the invocation of a plurality of algorithms to generate multiple candidate flex-cells that may ultimately be used in the target design. The algorithms can use various forms of BDDs, including Reduced Ordered BDDs (ROBDDs) [5], Free BDDs (FBDDs) [3], as well as non-BDD representations of the function of given clusters [24], to derive the transistor-level structures. Free BDDs or FBDDs [3] are a variety of BDDs in which different paths traced through the structure can have input variables appearing in different orders. Another variety of BDD referred to as ordered BDD (OBDD) [1][5], imposes a rule that variables encountered during tracing any path through the structure will always follow a fixed order. ROBDD [5] is a special case of OBDDs, where there is exactly one (unique) ROBDD for each unique function (i.e. ROBDDs are canonical).

Despite the use of multiple algorithms, to start with, the transistor netlist generation process may not yield topologies that meet the constraints imposed on the cluster by its context of use. In such cases, a topology alteration process needs to be invoked to explore multiple alternative topology implementations, given the functionality of a cluster. For example topology alteration process may include using a variable reordering in the decision diagram (ROBDD, FBDD, etc.) representations of the cluster. The

Design Optimization with Automated Flex-Cell Creation 247

topology alteration process may also use multiple decomposition methods for the function, such as Boole-Shannon, Kronecker, Roth-Karp, Positive Davio, Negative Davio, and Ashenhurst techniques. The multiple transistor topologies, which result from the topology alteration process, can be evaluated using various metrics derived from the constraints mentioned earlier, to obtain a ranked list of flex-cells where a higher rank indicates greater suitability for use in the given context. For purposes of timing optimization, an appropriate mix of SPICE-like transistor-level timing analysis as well as faster (and more approximate) switch-level timing analysis techniques, e.g. static timing analysis at the switch/transistor level, is key to ranking the candidate transistor netlist topologies properly.

Some results and uses of the flex-cell generation process discussed thus far are illustrated in Figure 3. This diagram illustrates a flex-cell that results when a portion of an IC design is mapped to transistors, with the primary goal being performance optimization. The cluster in question is shown in Figure 3(a). It has only one critical input, namely input a. In this context, a critical input denotes an input such that the delay from this input to the output of the cell limits the overall performance of the cell. A candidate flex-cell generated by the mapping process, is shown in Figure 3(b), and performance improvement that would result from replacing the cluster by the flex-cell, is shown in Figure 3(c).

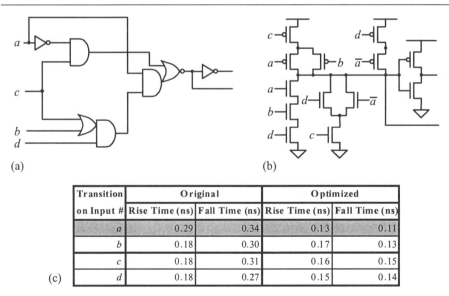

Figure 3. Flex-cell generation illustrated: (a) original cluster of standard cells; (b) flex-cell created by mapping process; (c) performance improvement resulting from replacement of cluster by flex-cell.

While most of the discussion in this chapter is focused on performance optimization of designs, flex-cells can be also used to reduce transistor count of a design – and thereby reduce area and/or power consumption of a given design. In fact, a careful study of Figure 3 shows that the number of transistors in the flex-cell is significantly smaller than the total number of transistors in the original cluster of standard cells. Additionally, flex-cell based optimization can be used to significantly reduce the total number of placeable instances in a design, as well as the total number of interconnects between instances in a design. Such reduction, in turn, benefits the final design quality in a multitude of ways, including higher quality results during place-and-route phase, and potentially improved signal integrity and noise problems due to fewer interconnects between cells.

The impact of a change in transistor topology and transistor sizing on the performance of a flex-cell, is complex. Various combinations of choices made in the above processes, may result in a large set of candidate design-specific cells. As a result, the mapping process typically includes a selection step. The selection step begins with ranking the candidate flex-cells using a sophisticated cost function, that evaluates the quality of each candidate design-specific cell, measured using various appropriate target metrics, such as input-to-output delay through the design-specific cell, number of transistors, stack-depth (i.e. length of a path through N- or P-transistors), input load capacitance, output drive strength, etc. A limited number of candidate flex-cells from the top of the rank-ordered list are then chosen for use in the overall design optimization loop shown in Figure 1. In a simplified optimization scheme, the flex-cell selection process can be greedy, or iterative in nature. Other sophisticated search schemes may be employed in flex-cell selection process, including linear programming, dynamic programming, branch-and-bound search techniques, etc., or some combination thereof, to achieve near-optimal design of the flex-cells.

Although the above description has been implicitly focused on the static CMOS family of logic circuits, the mapping process described above is broadly applicable for creation of the NMOS or the PMOS networks individually, if the target IC design implementation calls for using another family of MOS circuit design, including various forms of dynamic CMOS, some combination of static and dynamic CMOS, etc.

1.2 Optimization Control

A conceptual view of the optimization control process that drives the clustering and mapping steps, is shown in Figure 4. This process can be varied to suit specific design optimization goals. Such variations typically include techniques such as dynamic programming, simulated annealing and genetic algorithms

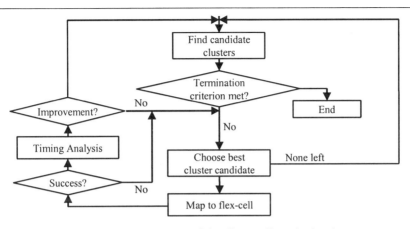

Figure 4. A conceptual illustration of the flex-cell optimization process.

For instance, a branch-and-bound search process can be used to drive the global optimization process. This involves (i) exploring alternative choices, to determine which choice is better, since such determination cannot be made a priori (i.e. branching), and (ii) ruling out some possible choices as being "obviously bad" (i.e. bounding).

2. MINIMIZING THE NUMBER OF NEW FLEX-CELLS CREATED

As shown in Figure 1, flex-cell based optimization process can invoke a *uniquification* process to minimize the number of functionally unique flex-cells created during optimization of a given design. In the context of flex-cell synthesis, uniquification focuses on identifying the minimum number of unique flex-cells required for optimization, and it is different from the "uniquify" command found in synthesis tools.

Further, since a pre-defined standard-cell library is fully expected to be available as in input to the optimization process, the uniquification process can also be used to identify near or exact matches (depending on the tolerance of the IC or flex-cell design process). In fact, as a matter of policy, if transistor-level implementations of the standard cells are available as input to the optimization process, creation of flex-cells that have equivalent or near-equivalent matches in the available standard-cell library, should be avoided within limits of practicality. Matching, especially with variations such as allowing input permutations and complementation, can be a computationally expensive process.

There exists a rich body of literature addressing the problem of matching a target functional description with the functionality of a given block of logic composed of one or more standard-cells [6][20]. Methodologies employed by typical existing approaches to perform matching are illustrated in Figure 5, which shows two potential matches in a cell library, for the given functionality. Additionally, there exist significant related work that attempt to match portions of a digital circuit (specification) with portions of another digital circuit (implementation) with an objective of comparing their functional equivalences, utilizing both structural and functional information [15][21].

In the context of flex-cell based synthesis, especially one aimed at enhancing performance of the target design, uniquification needs to take into account both functionality, and the timing contexts in which each unique cell is to be used. Thus, in addition to functionality, the target for matching is annotated with timing constraints related to its intended use in a design. In fact, while the current discussion is focused on timing, in general, constraints specified as part of the target may be related to various other metrics of interest including power, area, noise margins, slew, input/output capacitances, drive strength, footprint size, pin-placement, etc.

Typically, the addition of such constraints will decrease the potential of any matches in a specific library of cells (standard-cells, flex-cells, or mix). For example, Figure 6 shows a modification to the matching problem of Figure 5, with added constraints on the rise and fall timing on the pin-pair z-f: the worst z-f rise transition is required to occur in exactly 0.15ns and the worst z-f fall transition is required to occur in exactly 0.12ns. These constraints reduce the potential matches from two as shown in Figure 5(b), if only functional information is considered, to one, i.e. (a:y, b:z, c:x). This is the only match that satisfies the functionality and the timing constraints.

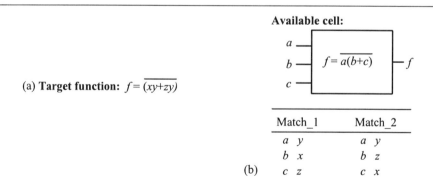

Figure 5. (a) Given logic function; (b) two functional matches in a cell library.

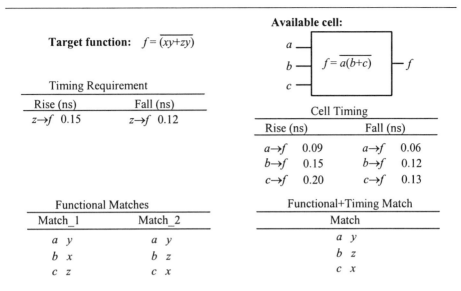

Figure 6. Matching for a given functionality, with additional constraints on pin-to-pin rise and fall times.

The timing constraints of the example of Figure 6 were exact. However, during optimization, the target constraints are often inexact. For example, it might be acceptable for a cell to match the target timing constraint within a certain bound or variance. In fact, most constraints are naturally expressed as values within a range. For example, a timing constraint may be expressed as having a rise time that should be no worse than 0.23ns.

In order to capture variances in timing, timing constraints can be expressed in terms of a nominal value and associated range. For example, consider Figure 7 that includes the concept of variance with the example of Figure 6. Each input pin to the output pin has a rise and a fall time constraint expressed in terms of a nominally required value and a variance as shown in a cell timing chart in Figure 7(b). Thus, the constraint on pin x-f can now be interpreted as having a rise time delay that is preferred to be 0.20ns, but is no worse than 0.21ns (i.e. 0.20ns + 0.01ns) and is no smaller than $-\infty$ (0.20ns $-\infty$). With this modified criterion, there is still one potential match satisfying the functionality and timing constraints, viz., that of (a:y, b:z, c:x), shown in Figure 7(c).

Since quick determination of satisfaction (or lack thereof) of constraints is key to minimizing the number of flex-cells, another possible approach that can be adopted is the computation of signatures that capture and filter out impossible matches from the existing library. Consider the example of Figure 8 that illustrates the construction and use of signatures based on constraints for an example target objective of timing shown in Figure 8(a).

Available cell:

Target function: $f = \overline{(xy+zy)}$

$f = \overline{a(b+c)}$ with inputs a, b, c and output f

Timing Requirement				Cell Timing			
Rise (ns)	Var (ns)	Fall (ns)	Var (ns)	Rise (ns)		Fall (ns)	
$x \to f$ 0.20	<-∞,0.01>	$x \to f$ 0.12	<-0.01,0.03>	$a \to f$	0.09	$a \to f$	0.06
$y \to f$ 0.08	<-∞,0.02>	$y \to f$ 0.06	<-∞,0.01>	$b \to f$	0.15	$b \to f$	0.12
$z \to f$ 0.15	<-∞,0.01>	$z \to f$ 0.16	<-∞,0.01>	$c \to f$	0.20	$c \to f$	0.13

Functional Matches		Functional+Timing Match
Match_1	Match_2	Match
a y	a y	a y
b x	b z	b z
c z	c x	c x

Figure 7. Matching for functionality, taking into account desired pin-to-pin timing including timing uncertainty.

Target function: $f = \overline{(xy+zy)}$

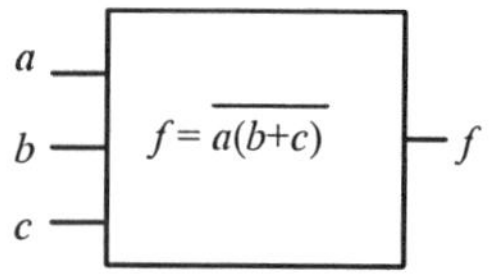

Timing Requirement

Rise (ns)	Var (ns)	Fall (ns)	Var (ns)
$x \to f$ 0.20	<-∞,0.01>	$x \to f$ 0.12	<-0.01,0.03>
$y \to f$ 0.08	<-∞,0.02>	$y \to f$ 0.06	<-∞,0.01>
$z \to f$ 0.15	<-∞,0.01>	$z \to f$ 0.16	<-∞,0.01>

Available cell:

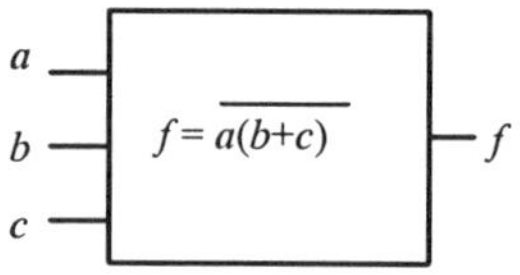

$f = \overline{a(b+c)}$ with inputs a, b, c and output f

Cell Timing

Rise (ns)		Fall (ns)	
$a \to f$	0.09	$a \to f$	0.06
$b \to f$	0.15	$b \to f$	0.12
$c \to f$	0.20	$c \to f$	0.13

MATCHING PROCESS USING RISE-TIME SIGNATURE

Sorted list of required rise-times including allowed positive variance

L1: [0.10, 0.16, 0.21]

Sorted list of available rise-times

L2: [0.09, 0.15, 0.20]

Check for feasibility between

L1: [0.10, 0.16, 0.21]
L2: [0.09, 0.15, 0.20]

POTENTIAL FOR MATCH
(In case of failure, no match possible)

Figure 8. Use of signatures for matching, with pin-to-pin timing constraints.

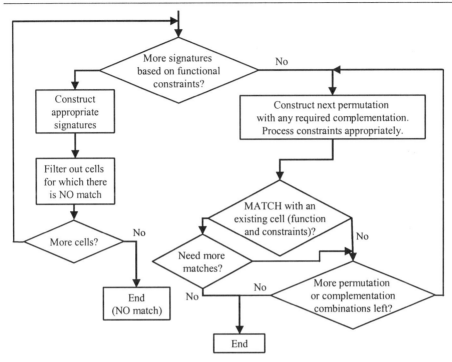

Figure 9. A possible implementation of matching process taking into account pin-to-pin timing constraints, including variances in such timing constraints.

The desired signature could consist of, for example, a sorted set of rise and fall times, since such a sorted list is independent of any permutations on input pins. It is also possible to combine rise and fall times and construct signatures comprised of sorted lists of functions of rise and fall times that also remain independent of any permutation or potential complementations at the inputs of the library cells.

In order to account for variances on constraints, such signatures can use one or more fixed corners within the range of variability, e.g. the worst possible rise time (nominal + worst tolerance), the worst possible fall time (nominal + worst tolerance), etc. Once the fixed corners are chosen, similar lists can be constructed with the sorted rise and fall times for the existing library cells, and the signatures for the library cells can be matched against the signatures for any potential feasible matches. The feasibility is checked by ensuring that for every target constraint value on a pin (e.g. rise time), there exists at least one value on an actual pin that satisfies the constraint.

The flowchart of Figure 9 shows one possible implementation of the desired matching process. This implementation allows for construction of multiple signatures based on constraints, and use of such signatures for filtering existing library cells. Once the signature checking phase is complete, exhaustive permutations and complementations of candidate cells are then generated. A determination is next made whether there exists a cell in the library that satisfies the target functionality and constraints. The permutations and complementations are exhaustively searched in an iterative loop, until a match is found, or it is determined that the signature checking phase will not yield any additional matches.

Note that matching process of Figure 9 allows multiple matches to be generated. In the presence of multiple matches, selection criteria can be utilized to select the most appropriate match. Such criteria may include, for example, preferences for certain constraints and how closely the cell matches the given constraints (based on the nominal values).

3. CELL LAYOUT SYNTHESIS IN FLEX-CELL BASED OPTIMIZATION

Automated cell layout synthesis [9] plays a key role in "closing the loop" with respect to creation of the actual layouts of the flex-cells that are designed as transistor-level netlists, during the mapping process in Figure 1. Layout synthesis takes as input, the transistor-level netlists of the flex-cells, various fabrication process technology parameters including layout design rules (DRC rules), desired standard cell architecture parameters including cell height, number of tracks, well and implant specifications, etc., and creates the detailed transistor layout – polygons that will be eventually fabricated on silicon substrate. Layout synthesis commonly includes further tuning of the sizes of the transistors in the flex-cells, especially to ensure that the timing characteristics of the flex-cells, post-layout, closely match the desired timing characteristic passed to layout synthesis, as input.

A key objective for layout synthesis step is compatibility with standard-cell library blocks such that the flex-cells created can be mixed seamlessly with the predefined standard cells that are used in the rest of design – from a layout point of view. The compatibility of the flex-cells and the standard-cells, at the layout level, enables the final IC design to be highly customized (i.e. using flex-cells) and yet stay flexible enough to use standard-cells where possible and/or desired.

It is also interesting to note that the use of carefully controlled cell-layout synthesis is a powerful process that can yield design benefits even without the use of other flex-cell related transistor transformations. Such approaches have been used to derive additional design benefits in specialized high-performance design flows [7].

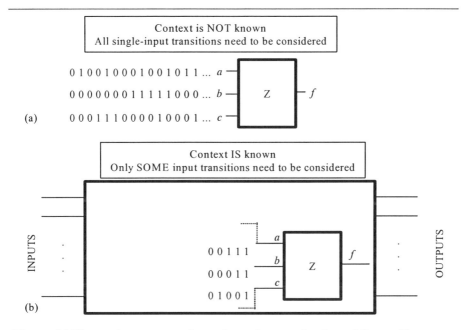

Figure 10. Illustrating context-dependent characterization of flex-cell.

4. GREATER PERFORMANCE THROUGH BETTER CHARACTERIZATION

As shown in Figure 2, a fast characterization step is used repeatedly during mapping, to obtain estimate of the timing characteristics of the flex-cells being created, since the design constraints are known and have in fact been used as the basis for generating the flex-cells. The use of appropriately chosen characterization mechanisms, throughout the mapping and optimization process, is key to the success of such flex-cell based design optimization techniques. Broadly speaking, characterization mechanisms used in flex-cell based optimization process can be divided into two categories: pre-(cell-)layout, and post-(cell-)layout.

Pre-layout characterization includes characterization of flex-cells, as they are created, taking the context of use into account. Such use of context-dependent information during characterization provides the benefits of more accurate characterization as well as savings in computational resources required for such characterization.

To illustrate the characterization process, Figure 10 shows a cell Z having three inputs, a, b, and c. In a conventional design methodology, cell Z can be instantiated in any portion of the design. Hence, a characterization method that does not have knowledge of the context in which a cell is to be used has to consider the possibility that signal transition may propagate from any input to the output, for any possible valid input combination on the other inputs.

In contrast, consider the scenario that occurs in the alternative design flows where flex-cells are generated on-the-fly. In this scenario, the design environment for a cell is known, defined by the places of instantiation of the cell in the design. Hence, certain combinations of input values may not be possible for application to cell Z, as illustrated in Figure 10. As shown in Figure 10(b), only a subset of the possible set of exhaustive input transitions, need to be considered due to the known design context of cell Z.

In fact, the pseudo-code that follows depicts a simple process of generating the minimal set of test vectors that provide complete characterization of a given flex-cell implementing a combinational function. The set of vectors, generated by the procedure above, can be refined further with knowledge of context of use of a specific combinational flex-cell, as has already been illustrated in Figure 10.

```
GenerateSmartVectors(function)   {
    // supportSize represents the number of variables
    // that function depends on
    supportSize = function.SupportSize();
    for (i = 0; i < supportSize; i++)   {

        // compute boolean difference w.r.t input i
        boolDiffFunction = function.BooleanDifference(i);

        setOfVectors = boolDiffFunction.EnumerateMinterms();
    }
}
```

The benefits of pre-layout characterization are more pronounced when applied to a group of cells (standard cells, or flex-cells), in a design. Consider, for example, the circuit shown in Figure 11, which is a simple four-stage chain of NAND gates. Table 1 compares the worst propagation delays through this chain of gates, as determined by both gate-level STA and transistor-level timing analysis for five circuits. Each row in Table 1 represents one unique circuit configuration.

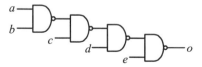

Figure 11. Obtaining more accurate timing estimate by characterizing a group of cells at the transistor-level.

Drive Strengths at Stage 1 2 3 4	Gate-Level Timing (STA) (ns)	Transistor-Level Timing (SPICE) (ns)
0 0 0 0	0.40	0.186
1 1 1 1	0.22	0.166
2 2 2 2	0.15	0.157
4 4 4 4	0.61	0.581
0 1 2 4	0.41	0.332

Table 1. Comparing gate-level STA with transistor-level characterization.

The first column of Table 1 represents the drive strengths of the gates at each stage, where zero (0) representing the smallest drive strength and four (4) representing the largest drive-strength. The second column shows the worst propagation delay through the design as determined by gate-level (STA). The third column represents the worst propagation delay through the same design as determined by transistor-level timing analysis. Note that the results are based on representative state-of-the-art process and delay models; and the same load capacitance and input slew values have been used for each of the runs.

Clearly, transistor-level timing analysis is much more accurate than gate-level timing analysis. In fact, the differences in timing analyses can be dramatically different in certain kinds of certain configurations as seen in the first circuit configuration (i.e. the first row in the Table 1) where each NAND gate is of drive strength 0.

The extent of differences vary based on factors such as the delay models being used, circuit topology, model extraction technology, transistor simulation techniques being used, fabrication process to which the design is targeted at, circuit design style and drive-strengths of the circuits under consideration. In general, gate-level STA is made inherently conservative to account for potential inaccuracies in the creation of the gate-level abstraction from the transistor-level circuit.

Due to this well-known conservative nature of gate-level STA, it is possible to improve the accuracy of analysis of performance of a design by considering clusters (i.e. partitions or groups) of standard-cells and analyzing and optimizing the clusters at the transistor-level. In fact, when members of such clusters can be physically placed in close proximity to each other, simply making a new cell out of the group, with a pre-defined shape, and characterizing the whole group as one entity at the transistor level, can provide a much more accurate estimate of timing than can be achieved via gate-level STA.

A simplified abstraction of the process of optimizing clusters in this manner is illustrated shown in Figure 12. The circuit in Figure 12(a) includes standard cells A1, ..., A12 with a critical path through the cells A1-A4-A8-

A10-A11. The circuit of Figure 12(a) is analyzed at the transistor-level in the areas of interest, on critical regions, by creating larger cells C1, C2 and C3, as shown in Figure 12(b). Cells C1, C2 and C3 are characterized at the transistor-level, and gate-level STA is performed using the new cells. This results in a more accurate (and hence more aggressive!) timing analysis of the design under consideration.

An important aspect of this process is that larger standard-cells, corresponding to C1, C2 and C3 are created, characterized and modeled to comply with cell-level design methodology. It is also important to note that the number of intra-cell wires (i.e. interconnects) is smaller in circuit of Figure 12(a) compared to those in Figure 12(b). Thus, such characterization of relatively large clusters of cells results in reduction in the number of non-characterized interconnects in the design. In fact, if computational resources were available to perform a detailed transistor-level analysis of the entire circuit under consideration, one could get the most accurate analysis of the timing characteristics of the design. However, considering the practical limitations of computational resources, it is typically feasible to characterize only a few hundred new cells having 20 to 40 transistors per cell, in a reasonable amount of time such as 1 to 2 days.

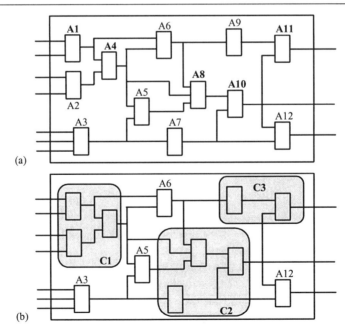

Figure 12. Better optimization through transistor-level characterization of clusters: critical path A1-A4-A8-A10-A11; timing in circuit (b) can be computed more accurately (and hence, more aggressively) than in circuit (a); circuit (b) can be optimized better as a result.

The post-layout characterization step operates closely with layout synthesis and post-layout parasitic extraction tools and flows. In general, post-layout characterization needs to employ a highly accurate device-level simulator such as SPICE or variations thereof, although it is possible that other tools, including macro-modeling tools with sufficient accuracy, may be used for this step, in future. The inputs to the post-layout characterization step are: (i) layouts generated in the layout synthesis step, (ii) models of the devices used in the layout, and (iii) optionally, the context of use for each design-specific cell in the specific design for which the design-specific cell is created.

In principle, the output of post-layout characterization step can be fed back to drive the layout synthesis step, again, with altered constraints, and the steps of layout synthesis, parasitic extraction, and post-layout characterization can be iterated until the target timing characteristics of the flex-cell(s) being designed and evaluated, are satisfied. In practice, though, such iterations are limited due to the resource-intensive nature of the steps, and post-synthesis characterization is typically done for a wide range of contexts in which the flex-cell(s) are likely to be used, much like their standard-cell counterparts.

5. PHYSICAL DESIGN AND FLEX-CELL BASED OPTIMIZATION

As minimum feature size of fabrication processes have progressed to 0.18um, and smaller, it has become virtually impossible to create designs, especially, high-performance designs, without incorporating detailed physical design information into the synthesis and optimization process. The dominant factor guiding this development, of course, is the greater role played by interconnect delays in determining the overall delay of critical paths in a design. Flex-cell based optimization is no exception to this trend. The STA in flex-cell based design must take into account actual wire delays, loads, and slew degradation differences between different parts of the same interconnect net – or good estimates thereof derived from physical design knowledge. The local optimization steps, including clustering and mapping must also take into the impact of wire delays, loads, and slew degradation profiles of "nets of interest".

Various intermediate steps can be taken, as flex-cell based optimization transitions from traditional wire-load model based load computation to physical design based load computation. A necessary step is to understand placement of the cells – standard-cells and flex-cells – and to estimate the wire lengths of individual nets, utilizing various well-known parameters like half-perimeter, number of terminals on the net, fraction of bounding box covered by cells, fraction of bounding box occupied by blockages, and the like [4]. At high utilization, or in the presence of high congestion, more detailed routing information become essential to allow accurate estimation of the delays and loads that need to be taken into account by various stages of the flex-cell based optimization tool.

Key to the incorporation of such physical design data into a flex-cell based optimization process, is the use of fast incremental placement algorithms. Such incremental placement algorithms can be based on well-known techniques like quadratic placement, force-directed placement, simulated-annealing [23], etc., or some appropriate combination of multiple placement techniques. Important issues that need careful attention, for effective use of incremental placement, are (i) execution time of the incremental placement algorithm, and (ii) quality of result – as measured by correlation to final placement that will be generated by whatever place-and-route tool is to be used for actual layout. A practical solution may need to make a variety of trade-offs to achieve the desired speed, potentially at the cost of some quality degradation. Such trade-offs may include (a) relaxing requirements to generate "legal" placements, i.e. allowing placements generated to have some design-rule violations like cell overlap, (b) invoking incremental placement after a set of optimization steps are completed as opposed to invoking incremental placement after every change made during optimization, and (c) using simpler algorithms like force-directed placement, as opposed to more sophisticated placement techniques.

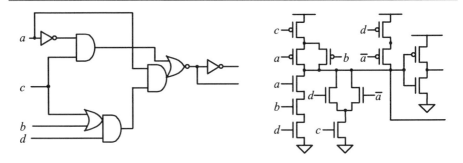

Figure 13. Structural comparison of unoptimized cells vs. optimized flex-cell.

Transition	Original		Optimized	
on Input #	Rise (ns)	Fall (ns)	Rise (ns)	Fall (ns)
a	0.29	0.34	0.13	0.11
b	0.18	0.30	0.17	0.13
c	0.18	0.31	0.16	0.15
d	0.18	0.27	0.15	0.14

Table 2. Unoptimized cells vs. an optimized flex-cell: timing comparison.

6. CASE STUDIES WITH RESULTS

Experiments studying flex-cells and their use in design optimization are presented in this section. Results from these experiments help demonstrate the substantial benefits that can be derived from custom-designing a design context by using crafted flex-cells and also provide evidence supporting the feasibility of applying the methodology in present-day design.

The first experiment presents an idea of the savings achievable by replacing a set of conventional standard-cells with a customized flex-cell implementation. These results are shown in Figure 13 and Table 2, which demonstrate the structural and timing advantages of flex-cells, respectively.

As shown in Figure 13, the conventional, standard-cell, implementation used 5 cells, 22 transistors and 9 wires, while the optimized, flex-cell, implementation reduces to a single cell consisting of a mere 13 transistors and no (zero!) global wires. There is also potential significant improvement in the timing characteristics (see Table 2). In the design context from which this example was taken, the flex-cell was created to optimize the critical path in the design between input *a* and the output. The worst-case delay for that critical path was improved from 0.31ns in the conventional implementation to a remarkable 0.13ns in the flex-cell implementation.

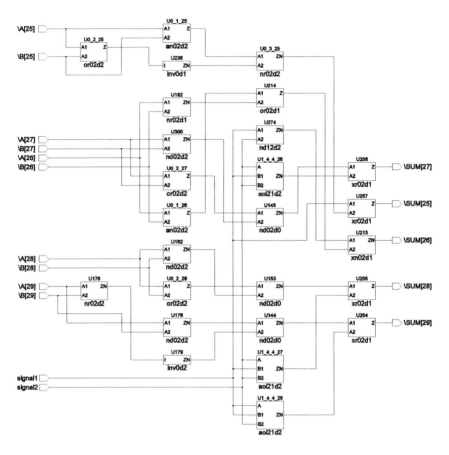

Figure 14. An adder design using 26 conventional standard-cells.

Having seen how flex-cells can achieve significant structural and timing advantages over unoptimized standard-cells, let us now turn our attention to study the benefits of flex-cells at the full design-level. We first present results on a small adder design. Figure 14 and Figure 15 present the results of using flex-cell based optimization for this adder design. Figure 14 shows the schematic diagram (Cadence schematic) of the original adder produced by a state-of-the-art commercial standard-cell synthesis tool. The worst timing critical path on this design had a delay of 0.54ns. Flex-cells were then used to optimize this design for improving its timing performance. The optimized design, as shown in Figure 15 (Cadence schematic), had a worst-case critical path delay of only 0.38ns, an improvement of nearly 30%!

Design Optimization with Automated Flex-Cell Creation 263

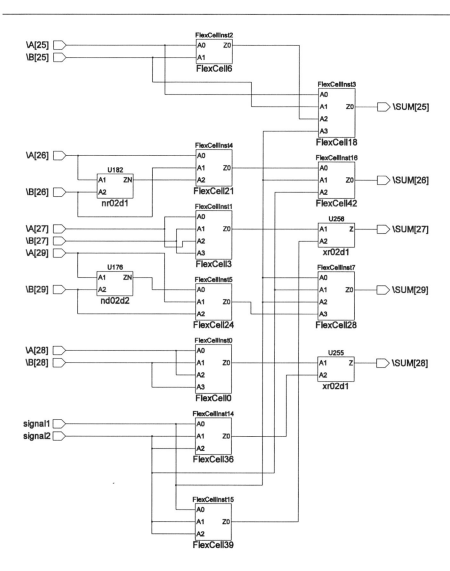

Figure 15. The adder design comprising 14 flex-cells.

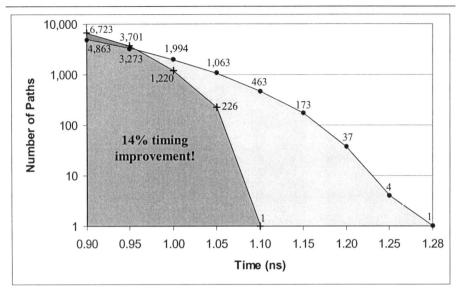

Figure 16. Improving the number of paths violating timing, with flex-cells.

It is also instructive to see how the worst critical paths in a design get altered by the introduction of flex-cell based optimization. Figure 16 demonstrates a graph that plots the number of paths that violate a specific timing constraint against the timing constraint for two versions of another adder design. The first design, represented by the outer curve, was produced by a state-of-the-art commercial standard-cell synthesis tool. The second design, represented by the inner curve, was produced after applying flex-cell based design optimization to the original design.

Figure 17 presents a pictorial view of how individual timing paths can be improved by applying flex-cell based optimization. In this example, the original critical path was running at 2.23ns while flex-cell based optimization using an adder that was optimized using flex-cells and a *single* flex-cell at the end of the path reduced the critical path delay to 1.83ns, an improvement of 19% over the combinational path delay on this set of timing paths.

Let us now look at data that demonstrates that it is indeed feasible to use flex-cells in an automatic design optimization procedure. One of the potential difficulties with the use of flex-cells is that the creation of their layout may potentially be a time consuming process that interferes with the speed of front-end optimization/synthesis tools.

Design Optimization with Automated Flex-Cell Creation

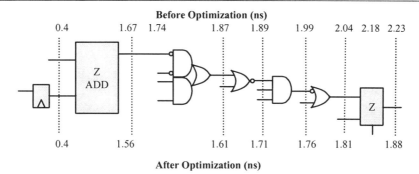

Figure 17. Flex-cell based optimization of timing critical paths.

	(Pre − Post)/Pre		How Often Pessimistic?
	Mean	Standard Deviation	
Cell Delay (Rise)	6%	5%	93%
Cell Delay (Fall)	10%	7%	96%
Output Slew (Rise)	4%	4%	94%
Output Slew (Fall)	8%	7%	88%

Table 3. Pre-layout accuracy can work for flex-cell based design.

However, as the data in Table 3 demonstrates, it is sufficient to work with pre-layout estimates of flex-cell characteristics. Even for a characteristic as sensitive as timing data, experiments on a large set of flex-cells show that pre-layout data is within about 5%-6% of post-layout data. In fact, the pre-layout data can be tuned to be slightly pessimistic in order to eliminate costly optimization and place-and-route iterations.

Finally, we summarize the results of optimizing some moderate-sized blocks from industrial circuits, using the flex-cell based optimization process. The results are shown in Table 4. It is clear that performance enhancements to the tune of 10-20% are easily achievable using this methodology. As tools based on this methodology mature, it is expected that significantly greater than 20% improvement in performance — over-and-above what traditional standard-cell based design optimization flows can deliver — will be achieved through flex-cell based design optimization.

Design Name	# Cells/Instances	Original Delay (ns)	Flex-Cell Delay (ns)
5-bit Adder	26	0.54	0.38
32-bit Adder	122	1.31	1.10
Individual Circuit Module1	~5600	4.10	3.47

Table 4. Summary of results for sample circuits.

7. CONCLUSIONS

The set of flex-cells, either alone or in combination with standard-cells, provides an optimally tuned set of building blocks for the target IC design, where optimality is measured against accepted and definable (i.e. quantifiable) metrics like clock speed, die size, power consumption, etc.. By allowing the transistor-level structures to be manipulated, flex-cells open up a new dimension in the optimization of automatically created designs. Flex-cells allow both the transistor-level structure and sizing to be manipulated freely. Such flexibility does not come for free, as is to be expected. A host of issues ranging from the "correct" choice of flex-cell mapping (to transistors) technology, to minimizing the number of flex-cells to be used in a design, to efficient characterization of flex-cells, to the interplay between flex-cell based optimization and other important issues like physical design information, need to be addressed. Preliminary results using flex-cell based optimization suggest that when employed properly, this methodology holds promise of significant benefit to the process of optimizing automatically created digital designs.

8. REFERENCES

[1] Akers, S.B., "Binary Decision Diagrams," *IEEE Trans. on Comp.*, vol. C-27, June 1978, pp. 509-516.

[2] Berkelaar, M. R. C. M., and Jess. J. A. G., "Technology Mapping for Standard-Cell Generators," *Proceedings of the International Conference on Computer-Aided Design*, 1988, pp. 470-473.

[3] Bern, J., Gregov, J., Meinel, C., and Slobodova, A., "Boolean Manipulation with Free BDD's: First Experimental Results," *Proceedings of the European Design and Test Conference*, 1994, pp. 200-207.

[4] Bodapati, S., and Najm, F.N., "Prelayout Estimation of Individual Wire Lengths," *IEEE Transactions on VLSI Systems*, vol. 9, no. 6, 2001, pp. 943-958.

[5] Bryant, R.E., "Graph-Based Algorithms for Boolean Function Manipulation," *IEEE Transactions on Computing*, vol. C-35, Aug. 1986, pp. 677-691.

[6] Burch, J. B., and Long, D. E., "Efficient Boolean Function Matching," *Proceedings of the International Conference on Computer-Aided Design*, 1992, pp. 408-411.

[7] Burns, J. L., and Feldman, J. A., "C5M – A Control-Logic Layout Synthesis System for High-Performance Microprocessors," *IEEE Transactions On Computer-Aided Design of Integrated Ciruits and Systems,* Vol. 17, No. 1, Jan. 1998.

[8] Chinnery, D.G., and Keutzer, K., "Closing the Gap Between ASIC and Custom: An ASIC Perspective," *Proceedings of the 37th Design Automation Conference*, 2000, pp. 637-642.

[9] Cirit, M., and Hurat, P., "Automated Cell Optimization," *Numerical Technologies White Paper*, 2002.

[10] Collett International, *1999 IC/ASIC Functional & Timing Verification Study*, 1999.

[11] Dally, W.J., and Chang, A., "The Role of Custom Design in ASIC Chips," *Proceedings of the 37th Design Automation Conference*, 2000, pp. 643-647.

[12] Fishburn, J., and Dunlop, A., "TILOS: A Posynomial Programming Approach to Transistor Sizing," *Proceedings of the International Conference on Computer-Aided Design*, 1985, pp. 326-328.

[13] Gavrilov, S., et al., "Library-less Synthesis for Static CMOS Combinational Logic Circuits," *Proceedings of the International Conference on Computer-Aided Design*, 1997, pp. 658-662.

[14] R.B. Hitchcock, "Timing Verification and Timing Analysis Program," *Proceedings of the 19th Design Automation Conference*, 1982, pp. 594-604.

[15] Jain, A., et al., "Testing, Verification, and Diagnosis in the Presence of Unknowns," *Proceedings of the. VLSI Test Symposium,* 2000, pp. 263-269.

[16] Jyu, H.-F., et al., "Statistical Timing Analysis of Combinatorial Logic Circuits," *IEEE Transactions on VLSI Systems*, vol. 1, no. 2, 1993, pp. 126-137.

[17] Kanecko, M., and Tian, J., "Concurrent Cell Generation and Mapping for CMOS Logic Circuits," *Proceedings of the Asia-South Pacific Design Automation Conference*, 1998, pp. 223-226.

[18] Keutzer, K., Kolwicz, K., and Lega, M., "Impact of Library Size on the Quality of Automated Synthesis," *Proceedings of the International Conference on Computer-Aided Design*, 1987, pp. 120-123.

[19] Liu, C.P., and Abraham, J.A., "Transistor-Level Synthesis for Static Combinational Circuits," *Proceedings of the 9th Great Lakes Symposium on VLSI*, 1999, pp. 172-175.

[20] Mohnke, J., and Malik, S., "Permutation and Phase Independent Boolean Comaprison," *Proceedings of the European Design Automation Conference*, 1993, pp. 86-92.

[21] Paruthi, V., and Kuehlmann, A., "Equivalence Checking Combining a Structual SAT-Solver, BDD's and Simulation," *Proceedings of the International Conference on Computer Design*, 2000, pp. 459-464.

[22] Reis, A., Reis, R., Auvergne, D., and Robert, M., "The Library Free Technology Mapping Problem," *Proceedings of the International Workshop on Logic Synthesis,* 1997.

[23] Sherwani, N., *Algorithms for VLSI Physical Design Automation*, Kluwer Academic Publishers, 1995, 2nd ed.

[24] SIS Abstract Page, University of California at Berkeley, http://www-cad.eecs.berkeley.edu/Respep/Research/sis/abstract.html

[25] Taki, K., "A Survey for Pass-Transistor Logic Technologies – Recent Researches and Developments and Future Prospects," *Proceedings of the Asia-South Pacific Design Automation Conference*, 1998, pp. 223-226.

[26] Weste, N., and Eshraghian, K., *Principles of CMOS VLSI Design*, Addison-Wesley, 1985.

Chapter 11

Exploiting Structure and Managing Wires to Increase Density and Performance

Andrew Chang, William J. Dally
Computer Systems Laboratory,
Stanford University

1. INHERENT DESIGN STRUCTURE

Judicious use of custom techniques can significantly improve the quality of ASIC designs. Custom approaches *exploit* the inherent design structure and *explicitly* manage routing. Traditional ASIC approaches *discard* a great deal of this structure and only *implicitly* manage wiring through the use of global design constraints and optimization tools. A simplified design flow for a typical ASIC is illustrated in Figure 1. The designer writes an RTL description of the function (Verilog or VHDL). A logic synthesis tool generates a gate-level netlist by mapping the design into a target standard cell library. Finally, a place-and-route (P&R) tool generates the physical design of the chip by first placing the cells, and then interconnecting them. The electrical properties of the wires are extracted from this layout. The designer checks chip timing by calculating the critical path delays using the extracted parasitic loads. There are four main problems with the traditional ASIC approach:

1. **Little is known about the performance of the design until the very end.** Power and timing are critically dependent on the resistance and capacitance of key wires. These values are only known *after* place and route. Also, as on-chip frequencies and edge-rates increase, the effect of parasitic inductance grows in importance. Yet, the detailed routing is critical to determining these values as well.

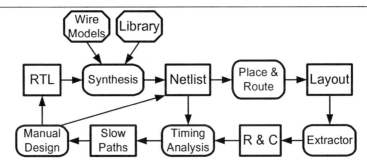

Figure 1. The traditional ASIC RTL flow.

2. **The designer has little control.** If the design fails to meet power and timing goals after P&R and after re-powering the cells along the critical paths, it is very difficult to fix the problem by changing the RTL. This is especially frustrating in the cases when the solution (structuring the wiring) is obvious, but there is no direct way to tell or force the tools to do it.

3. **The results are unstable and sensitive to small changes in the netlist.** If the designer changes the RTL later in the process to fix a logic error, the resulting netlist and placement may change drastically creating a whole new set of timing and power problems.

4. **The designer gains little insight and intuition about the design.** The designer is at the mercy of tools to identify problem areas. While timing analysis tools attempt to find and report the worst-case paths, due to process and design variations, the true offenders can be other signals within the same "bin" of uncertainty [20]. In the ASIC flow, placement and routing are authored automatically. The designer may need to *deconstruct* the results to determine both the proximate and ultimate sources of error as original intent is neither encoded nor preserved by the process. The underlying cell libraries are usually immutable, generic and supplied by a third-party; eliminating a valuable problem solving option.

A common practice to reduce the opportunities for disasters is to use and enforce conservative wire models for parasitic loads. In addition to sacrificing performance and efficiency, this technique can still fail. First, gates are oversized as the wire model intentionally sets loads several standard deviations beyond the mean so that there are fewer surprises. Second, as most models do not accurately account for resistance or inductance, they are least accurate for the longest wires. But, long nets are exactly the wires *most likely to be problematic*. Finally, for wires with significant resistance, increasing the buffer drive strength will not fix the timing problem. But, since much of the design structure has already been discarded, at this point it is difficult for either automated tools or the designer to fix problems by repartitioning the logic.

Recognition of the benefits of considering design structure earlier in the ASIC flow has spawned a generation of "physically aware" [3][23] synthesis tools. These tools attempt to automatically identify placement regions and group logic gates during synthesis. These tools attempt to simultaneously perform placement and routing and replace wire-load models with the results from test routes. The most sophisticated of these formulate the placement and routing problems as convex optimizations and allow multiple concurrent target criteria. While these tools represent a significant improvement over the traditional ASIC methodology, they still offer the designer little control or insight into problem regions. Also, as these tools neither manipulate the composition nor change the quality of the underlying cell library, they address only a subset of the possible problems.

2. SUCCESSIVE CUSTOM TECHNIQUES FOR EXPLOITING STRUCTURE

There is a continuum of custom design options ranging from simple floorplanning to handcrafting every transistor and manually routing each wire. Progressive gains in performance and density come from taking successive steps down this path at the cost of increasing effort. Almost everyone who builds high-performance ASICs today takes at least the first step by floorplanning the chip.

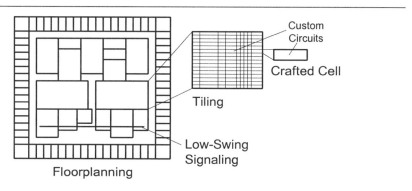

Figure 2. Customization options for the ASIC designer.

2.1 Increasing Wire Structure through Floorplanning

Custom designers structure the wiring so that they can identify key wires and their parameters at the beginning of the design process, rather than at the end. The drive and timing of these wires is then pre-planned in the design. This approach gives a stable design with early visibility to timing and power problems. It does require a "tall-thin designer" as logic designers need to understand the circuit and layout implications of their choices. With upfront knowledge, designers have much more control over the result and can exploit the structure they see within the logic.

A number of different techniques for custom design that structure wires at different levels with different amounts of design effort are illustrated in Figure 2. They form a step-by-step approach to building better ASICs with less effort. The initial steps involve no custom cells and only the last few steps involve custom circuits. The first step is to floorplan the chip into smaller than usual regions – $1,000\chi$ by $1,000\chi$ (where χ is the minimum intermediate routing metal pitch – normally M2). This corresponds to an area of approximately 15,000 gates. Signals that cross between these regions must follow specific disciplines (e.g. no combinational paths across regions). This floorplanning structures all of the inter-region wiring. The length and density of these wires is known up front and the logic and circuits can be built using this knowledge. For example, pipeline stages may be added to account for the known delay of long wires. Often this knowledge can also be exploited in the physical design by pre-routing these signals.

The next step in structuring wires is to tile the datapaths within the $1M\chi^2$ blocks. The designer partitions structured arrays into rows and columns and assigns library cells to each datapath cell. This is, floorplanning at the $20K\chi^2$ or 300-gate level. A further extreme that can yield additional benefits is to pursue the creation of regions to the sub-50-gate level [17]. In addition to structuring processor datapaths involving registers, adders, ALUs and multiplexers, this approach has also been shown to be useful in structuring arrays that exist in control logic for issue, decoding, scheduling and matching [1]. In general, there are few large logic circuits that are devoid of structure.

Exploiting Structure and Managing Wires to Increase Density and Performance 273

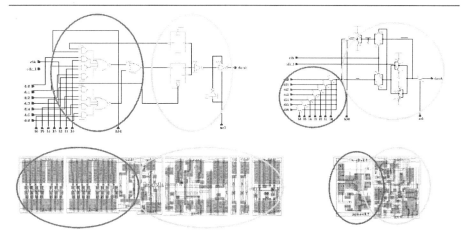

Figure 3. Comparison of standard cell versus crafted-cell implementations of a multiplexed latch with a tri-stated output.

As discussed in Chapter 4, the cells in existing libraries encompass limited functionality requiring the use of several simple gates to emulate a complex function. Figure 3 compares the standard cell (on the left) and *crafted-cell* (on the right) implementations of a 7-input multiplexed latch with a tri-state output. The gate is composed of two distinct functions: the input multiplexer and the static latch with tri-state output. The crafted-cell version requires 19 transistors organized as two gates. The standard cell design requires 40 transistors organized as seven gates and is almost a factor-of-4 larger. The crafted-cell implementation is only a factor-of-1.8 larger than a full-custom reference implementation. While the standard cell version is almost a factor-of-7 larger than the reference.

Our experience [16] shows that the level of effort to develop a crafted-cell, including generating all the 'views' needed to support the standard-cell tools, is quite low (1-10 person days depending on complexity) and that a very small number of *reusable* cells (10-20) is sufficient to realize most datapaths. In their simplest incarnation, crafted-cells are just larger standard cells with enhanced functionality matched to the application. These cells are placement and pin compatible with P&R, but they have three distinct advantages over their standard cell brethren. First, they eliminate the mixing of global and local wires by removing the set of intra-cell routes from the view of the P&R tool. These nets are kept short, stay on lower level metal layers and reduce congestion by minimizing interference with inter-cell routing. Second, by folding together several functions, redundant insurance and safety devices, input diodes, input and output buffers can be eliminated. Within a crafted-cell, the exact application and environment of the circuit is known and the insurance is no longer needed and the design margins can be recovered. Finally, the judicious use of aggressive sizing and custom circuits can be employed in these cells. Crafted-cells can be realized at different levels of ambition. Fusing several library cells together provides the simplest implementation of crafted-cells and realizes the first advantage. A slightly more aggressive implementation allows arbitrary static CMOS circuits to be employed within the cell. Finally, for very special cases, custom circuits can be employed within the cell.

Custom circuits can be used selectively within a predominantly ASIC flow to further improve the quality of designs. The guiding principle is *maximum benefit for minimum effort*. As a result, two candidate areas for application are critical datapath elements and low swing signaling on critical long nets. The simplest method for applying custom circuits to critical datapath elements is to implement the key function as a hard-macro with self-contained timing generation. This encapsulation facilitates integration into the standard ASIC tool flow. Traditionally, integrating crafted-macros into the ASIC timing flow was a very manually intensive effort. More recently, commercial tools that both create the requisite vectors and automatically script characterization simulations have become available [5], and greatly simplify the task. In our experience, crafted-macros implemented in domino logic and composed of up to 20K-transistors can be integrated into rows of static CMOS logic effectively. Figure 4 shows a portion of the datapath for a 64-bit floating-point unit [13][16]. The majority of gates are tiled static CMOS logic. However, each of the two large blocks is a half-array of the 64x64-multiplier and is implemented in a differential domino logic style. The crafted-macro for each half-array contains all necessary timing chains and interface logic.

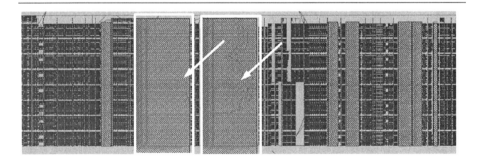

Figure 4. Application of a domino logic crafted-macro within rows of tiled static CMOS logic.

Multiplier	FO4 delays	Area ($10^6 \chi2$)	Relative Area	Transistors	Area Efficiency ($\chi2/T$)
Mitsubishi	24.3	1.53	0.82		
NEC Multiplier	27.0	2.19	1.18	135,318	16.2
MIT/Stanford	29.8	1.86	1.00	124,574	14.9
Fujitsu	32.5	1.63	0.88	82,500	19.8
Mitsubishi	35.2	2.35	1.26	78,800	29.8
Fujitsu Multiplier	41.0	1.65	0.89	60,797	27.1

Table 1. Comparison of reported multiplier implementations. The number of transistors in the Mitsubishi multiplier was not reported.

Despite the mix of implementation styles, the performance and density of the resulting multiplier array is comparable to several previously reported designs [8][9][11][15][18]. Table 1 presents a comparison of the key characteristics of these designs.

Finally, selective use of low swing signaling for critical long nets can improve performance and simultaneously reduce energy dissipation. Low-swing transceivers can employ signal pre-emphasis to increase their propagation velocity relative to full-swing signals [6].

2.2 Example Applications

Exploiting structure and managing wires has significant impact on design quality. In this section we describe three illustrative design examples: the datapath portion of a microprocessor register fetch stage implemented with crafted-cells; a 64-bit addsub [10] implemented as a crafted-macro; and a 4-input 4-output random logic function implemented as a PLA.

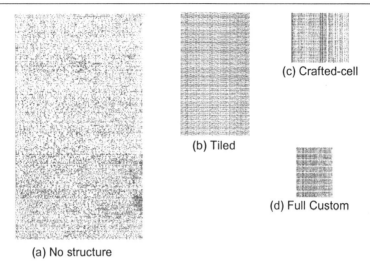

(a) No structure

Figure 5. Comparison of layouts of a 64-bit datapath in four design styles.

2.2.1 Register Fetch Stage

Datapath tiling can be exploited at several levels of customization. The impact on area and delay of different design styles is shown in Figure 5 and Table 2. These results are from four implementations of a 64-bit microprocessor register fetch stage. This datapath includes a 7-ported 75 entry 64-bit register file, a six entry reservation station, bypass multiplexers, 18-bit immediate insertion logic, and thirteen 1-bit condition code registers. The baseline custom design was implemented in a 0.5um L_{eff}, CMOS process with five levels of metal, the value of χ equal to 1.8um. The design contains 10376 cells and occupies $1036\chi \times 747\chi$ in area. Figure 5 displays the four layouts with all designs drawn to the same scale.

Parameter	Custom	Crafted	Bit-sliced	Automatic P&R
Area	1	1.64	5.25	14.5
Delay	1	1.11	2.23	3.72
Gate Load	1	1.09	2.29	2.29
M2 Length	1	1.07	4.19	34.9
M3 Length	1	1.63	2.52	7.92
Effort	1	1.13	0.06	0.06

Table 2. Comparison of four implementations of a 64-bit microprocessor register fetch stage.

The custom approach (5d) manually generated from circuit schematics, required 80 unique cells totaling 367 unique transistors. Each cell was manually placed and routed. The crafted-cell implementation (5c) was created from a structural Verilog model that was manually mapped to a basic 91-gate standard-cell library supplemented by 7-crafted-cells (totalling 48 unique transistors). Bit-slices were manually tiled and function blocks were manually ordered. This design was automatically routed after the detailed placement. The 'bit-sliced' standard cell implementation (5b) was created by the synthesis of a 1-bit slice from the crafted-cell structural Verilog source code targeting a basic 91-cell library. The individual bit-slices were ordered and assembled manually. The 'automated' standard cell implementation (5a) was created by synthesis of the full design from the crafted-cell structural Verilog source code. The resulting netlist was then automatically placed and routed. The same metal layer conventions and external pin constraints were used in all four cases. All four cases employed strictly single-ended static CMOS logic.

2.2.1.1 A Comparison of Density

The differences in quality of the four designs highlight the benefits of exploiting design specific structure, managing routing and employing crafted-cells. Structuring the design into bit-slices reduced the area by almost a factor-of-2.7 due the regularity of the wires. Next, creating and employing crafted-cells, saved another factor-of-3.2. The addition of only 7 extra cells helped to overcome the inefficiency of building multiplexed input latches and multi-ported registers from the cells traditionally available in the standard library. Also, by aggregating additional functionality, the crafted-cells converted inter-cell wiring into intra-cell routing further reducing the global wiring burden. The area advantage of full custom over the crafted-cell approach was mainly due to additional logic function compression, tuned P:N ratios, and the grid alignment penalties required to interoperate with the automated routing system.

2.2.1.2 A Comparison of Delay

The results for delay are similar. The automated design significantly lags the performance of the other three examples. There are three main causes of the additional delay. First the excess routing results in additional parasitic capacitance. Second, the base cells in the standard library poorly match the required datapath functions resulting in additional logic levels and signal inversion overhead. And, third, the limited range of available drive strengths combined with the additional wiring load exacerbated the poor matching of drives to loads. The crafted-cell design outperforms the bit-sliced design by almost exactly a factor-of-2. This advantage is gained by removing redundant levels of logic and signal inversions, and by matching drivers to internal loads. The factor-of-1.1 speed advantage of full custom design is due to the sub-optimal device sizing at the boundaries of crafted-cells and the increased parasitic wiring capacitance due to the larger area.

2.2.1.3 A Comparison of Routing

The difference in metal routing lengths confirms the importance of pre-planned structure. As expected, the custom design uses the minimal M2 and M3 resources as all contacts are made by abutment. While the crafted-cell version is auto-routed, the detailed manual placement enforces minimal routes. The increased metal usage in the bit-sliced design is due to its larger area as well as the inefficiencies that result from assembling complex cells from inexact primitives. Here, the intra-cell routing captured by crafted-cells is exposed and results in excess inter-cell wires. The metal use in the fully automated design reflects the "spaghetti" wiring that results from fully automated place and route. A key result is that without the advantage of structure, *automatic P&R makes some short wires a little bit better, and some long wires much worse.*

2.2.1.4 A Comparison of Design Effort

A comparison of the relative design efforts for the four designs is also informative. At the rate of 22 t_{unique}/wk, the custom design required 17 weeks from inception to verified layout. As the crafted-cell design required one-sixth the unique transistors, it was completed in only three weeks. Both the bit-sliced and fully automated design required one week each to import the netlists, perform logic synthesis and mapping, experiment with P&R runs, generate and verify layout. By carefully selecting a critical subset of design specific functionality to customize, the crafted-cell approach yields significant improvements in design quality with reasonable effort

2.2.1.5 A Comparison of Placed Instances

The original custom design has 10376 cells. The crafted-cell version has 10560 cells representing a 2% growth. The netlists for the bit-sliced design and the automatic P&R design are identical and were created through synthesis and mapping to a target standard cell library. The synthesis tool required 31066 gates to implement the original function.

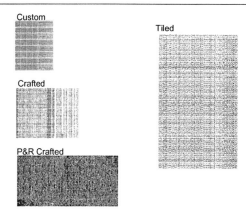

Figure 6. Comparison of P&R crafted with three preceding designs.

Parameter	Custom	Crafted	P&R Crafted	Bit-Sliced	Automatic P&R
Area	1	1.64	2.7	5.25	14.50
M3 Length	1	1.63	3.9	2.52	7.92

Table 3. Characteristics of the P&R implementation using crafted-cells.

2.2.1.6 Automatic P&R of the Crafted-Cell Netlist

The results from applying automatic P&R to the crafted-cell netlist while enforcing the crafted-cell pin constraints are shown in Figure 6 and Table 3. The area of the P&R version is a factor-of-2.7 larger than the full-custom reference and a factor-of-1.5 larger than the original crafted-cell design. While the area of this P&R design is an improvement over both the bit-sliced and other automatic P&R design, the M3 wire lengths are worse, reflecting the additional difficulties the routing tool encountered accessing the denser cells used in the crafted-cell netlist.

2.2.1.7 Automatic P&R Destroys Design Structure

Automatic P&R destroys inherent design structure and makes some short wires shorter and some long wires much longer. Clear insight into these effects is provided by the wire load distribution shown in Figure 7. This graph presents the actual distribution of wire loads for three implementations on a linear-log scale. The reported loads are proportional to the actual wire lengths. Wire load distributions are presented for the crafted-cell design, the P&R crafted-cell design and the original automatic P&R design. The X-axis of the graph represents the capacitive wire loads on a log-magnitude scale ranging from 1fF to 100pF. Wires at the same load point generally have the same length. The Y-axis of the graph represents number of nets at each load point on a linear scale.

The wire loads in the crafted-cell design occur in tight distributions around clear peaks reflecting the inherent structure of the design. The clear structure of the wire loads mirrors the inherent structure of the nets and their clear "preference" and grouping around "natural" lengths. Also, in this design only the CLK net has a wire load of greater than 10pF.

Applying P&R to the crafted-cell netlist, results in three key changes. First, the loading on a significant number of originally lightly loaded nets becomes even smaller. This reflects an improvement in the routing of short nets and demonstrates the efficacy of the optimization algorithms for reducing wire lengths in some cases. However, two negative results also occur: the loading increases on a significant number of nets reflecting longer routes and spreading occurs on the originally tight distributions reflecting the effects of mixing the routing for nets from different natural groupings and subsequent loss of inherent design structure. The resulting congestion further increases the loading and wire lengths for some originally long nets. In this design, over 350 nets have a wire load of greater than 10pF.

The wire load distribution from the original automatic P&R implementation reflects the added consequences of the increase in total number of nets due to the increase in the number of placed instances. It also continues to show the spreading of load distributions indicating the additional loss of design structure. The combination of increased wires and mixing of routes increases congestion and causes over 500 nets to have wire loads in excess of 10pF.

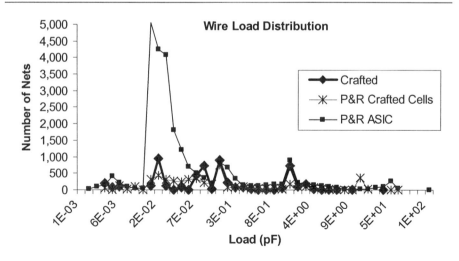

Figure 7. Comparison of wire load distributions.

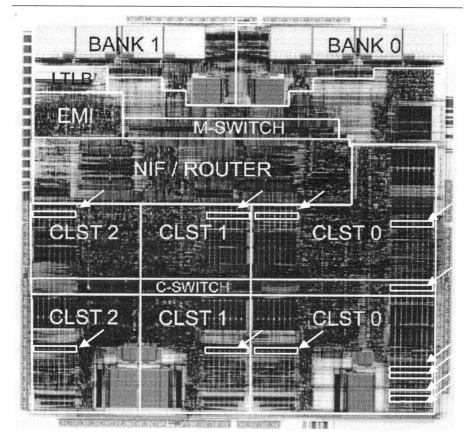

Figure 8. Twelve instances of the 64-bit addsub crafted-macro.

2.2.2 A 64-bit Addsub Crafted-macro

Both crafted-Cell and crafted-macro techniques were used to implement the Multi-ALU Processor (MAP) [16]. The MAP chip consists of 2.5-million logic transistors and 5-million total transistors. It was fabricated in 0.7um L_{dr}, 0.5um L_{eff} CMOS Bulk technology and resulted in an 18.25mm by 18.3mm die size. The design includes seven 64-bit execution pipelines (six integer units and one floating-point unit). The use of crafted-cells and crafted-macros enabled the design to be implemented from RTL to GDS with 30 man-years of effort over two calendar years. This effort included the creation of a full gate level RTL model, creation of a test infrastructure, process technology setup, design and characterization of two cell libraries totaling 200 cells, five RAM arrays, five megacells, two crafted-macros and five custom I/O pads, manual clock grid and power distribution design, floorplanning, and global chip P&R.

The addsub is a hybrid CLA/Carry-Select Adder. It is composed of five custom sub-blocks – two 1-bit high cells (PGK and Final) replicated 64 times; two 8-bit high cells (group PGK and local carry) replicated 8 times; and a 64-bit high global carry chain replicated one or two times depending on exact usage of the adder. Carry chain replication and selection were used extensively in both the group PGK blocks and the global carry chain. In addition, the local carry-chains were also replicated. The XOR gates in the local PGK and final cells employ pass gate logic. Twelve instances of the 64-bit addsub crafted-macro are used in the design. Figure 8 shows a die-image and highlights the 12 usages. One 64-bit addsub is required for each of the 6 integer execution units. The floating-point execution unit uses 1 addsub in its divide-square-root unit and 5 in the floating-point multiplier. The execution-unit datapaths are manually tiled and automatically routed.

Figure 9 and Table 4 compare the crafted-macro addsub to other adders in the literature [14][19][24]. The HP (Naffziger) adder employs dynamic circuit techniques and approaches the practical limits of 64-bit addition. It is both faster and smaller than the addsub. Typically, the difference between domino and static logic can range between 25% and 50% in performance. For the remaining examples, the crafted-Macro cell is comparable in area and performance. Note, the Mitsubishi (Ueda) adder employs BiCMOS pass gate logic.

The custom addsub crafted-macro was re-implemented four ways to explore a range of effort and quality choices. Each of these versions employed automatic routing. The first version is a crafted-cell design point, for which, each of the five constituent custom blocks of the original addsub was converted into P&R grid-compatible blocks. The second and third comparison points were synthesized from the addsub netlist. The first of these (ASIC – 1998) targeted a limited cell library. The second of these (ASIC – DW) targeted a more extensive commercial standard cell library and employs the basic Synopsys DesignWare [22] components for adders. The final comparison point was an optimized Synopsys DesignWare Brent-Kung adder. Table 4 shows that the original custom addsub is a factor-of-1.2 faster than the optimized Brent-Kung adder, a factor-of-1.7 faster than the crafted-cell version, and between a factor-of-1.8 and a factor-of-2.7 faster than either synthesized version. The crafted-cell version is slightly faster than the synthesized version mapped to the commercial cell library using basic DesignWare components and is 40% slower than the optimized DesignWare Brent-Kung adder targeting the commercial library. Finally, the performance of the basic ASIC design is more than doubled by the combination of the Brent-Kung DesignWare component and a more extensive set of library cells.

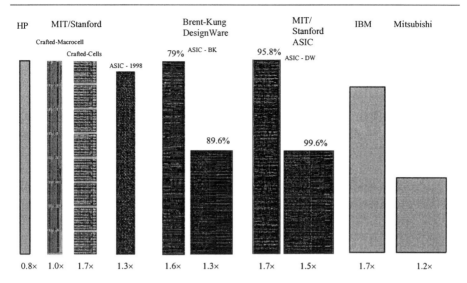

Figure 9. Comparison of layouts - ten 64-bit adders. The percentages above the ASICs are the cell-occupancy.

One additional result highlights the mismatch of standard cells for datapath functions. The aspect ratio of the placement box impacts the utilization of both the Brent-Kung adder and the ASIC- DesignWare adder. The minimum width of these designs was constrained by the requirement to fit large buffers causing between 10-20% area growth.

Adder	Process (L_{drawn})	FO4 (ps)	Delay (ns)	Delay (FO4)	Area (χ^2)	Area (relative)	Number of Instances	Transistors (T) or Gates (G)	χ^2/T
HP (Naffziger)	0.50	140	0.9	6.6	61500	0.8	7000	T	8.8
IBM	0.50	125	1.5	12.0	135802	1.7			
Mitsubishi	0.50	250	4.7	18.8	98400	1.2	4280	T	23.0
MIT/Stanford									
Custom	0.50	250	3.7	14.8	81847	1.0	4666	T	17.5
Crafted	0.50	250	6.3	25.2	136684	1.7	162	Gate	
ASIC (1998)	0.50	250	10.0	40.0	105600	1.3	1041	Gate	
ASIC - DW	0.15	60	1.6	26.8	124848	1.5	1756	Gate	
ASIC - BK	0.15	60	1.1	18.3	105600	1.3	1097	Gate	

Table 4. Comparison of adder implementations.

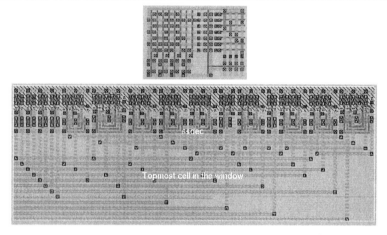

Figure 10. Comparison: PLA and standard cell implementations of a simple 4-input/4-output logic function.

The design and re-use of the addsub crafted-macro provided the combination of custom performance and modest overall design effort as its wide applicability amortized the creation effort. The implementation effort of the addsub could have been further reduced by employing crafted-cells or by using various synthesis approaches at the cost of performance.

2.2.3 PLAs for Finite State Machines

Significant structure exists in circuits that are not datapaths. ROMs, RAMs and PLAs organize wires as a regular grid. The result is a considerably more compact layout. Also, because the characteristics of these wires are well controlled, these types of structured circuits typically employ aggressive logic styles and low swing signaling techniques to further improve speed and reduce energy dissipation. A finite state machine implemented as a well-optimized PLA (or set of partitioned PLAs) requires approximately $4\chi^2$ per logic input. The circuit is laid out on a regular grid; both the product lines and output use reduced voltage swings; the flip-flops, with scan, are integrated into the output sense amplifier. As a result, the power and area are an order of magnitude less than the corresponding CMOS gate circuit and the cycle time is much faster. The same FSM realized in gates takes about $100\chi^2$ per gate input not counting the flip-flops. Figure 10 illustrates the density improvements from this style. The same 4-input, 4-output logic function is implemented with standard cells and with a PLA. Even with the peripheral overhead for such a small design, the PLA version is still a factor-of-8 smaller in the area than the standard cell design. The use of PLA's for control logic can also improve performance. A set of customizable dynamic PLA's was employed by Allen *et al.* to implement the control logic of a 1GHz 64b microprocessor in a 0.22um (0.12 L_{eff}) CMOS process [1].

3. FUTURE DIRECTIONS

Trends in technology scaling will continue to impact both custom and ASIC design choices. ASIC designers will adopt an increasing number of full-custom techniques to counteract the growing impact of wiring on design quality. The increased effect of parasitic resistance, capacitance and inductance for long wires will force ASIC designers to manage structure and explicitly specify key routing. Signal integrity concerns compound this issue as an increasing number of nets will require additional spacing, shielding and/or active anti-miller devices.

Similarly, custom designers will adopt the reuse and automation concepts from ASIC flows to counteract the spiraling expansion in the effort required for full-custom designs. The effort required for full customization scales greater than linearly with the number of unique circuits in a design, while, the number of transistors available on a chip continues to double every 2.5 years [7]. Hence, a "simple extrapolation of a 100-person 'P6' custom microprocessor design team in 1996, results in a 1,600-person 'P10 team' in 2007. Carrying this to the extreme, in less than 70 years every inhabitant of the planet would be an employee of a large microprocessor design company working on the 'P38' chip" [4]. As many of us have other aspirations, it is useful to reduce this effort.

In both cases, designers will need to carefully balance effort and time constraints with the performance, density and energy goals. Successful project teams will consistently identify the key design structures and routing in their designs and clearly focus resources on these critical areas in order to maximize the return on their design effort. Towards this end, the described crafted-cell and crafted-macro techniques can be viewed as either enhancements of ASIC approaches or simplifications of custom methods.

4. SUMMARY

Designers have implementation choices to realize their specific VLSI systems. Both the increasing complexity of designs and the increasing constraints on quality imposed by wires will motivate ASIC designers to adopt a growing number of custom techniques. Unlike traditional ASIC design flows, custom methodologies exploit the physical structure inherent in designs and explicitly manage routing. However, while custom designs can achieve up to a factor-of-3 greater performance and density, they can also require an order of magnitude more design effort. We have presented a sequence of successive custom remedies to improve the quality of ASIC designs: floorplanning, tiling, crafted-cells and, finally, custom crafted-macros. In example applications, these approaches taken together close the gap between ASICs and custom design to within a factor-of-1.1 for performance and to within a factor-of-1.6 for density with only modest additional effort. The underlying theme of our approach is to identify critical components of a design and judiciously employ custom techniques only on these few pieces to maximize quality while minimizing effort.

5. REFERENCES

[1] Allen, D.H. et al., "Custom circuit design as a driver of microprocessor performance," *IBM Journal of Research and Development*, vol. 44, no. 6, November 2000, pp. 799-822.
[2] Bernstein, K. et al., *High Speed CMOS Design Styles*, Kluwer Academic Publishers. 1999.
[3] Cadence, "PKS Concurrent Optimization: Enabling the Cadence SP&R Flow." White Paper.
[4] Chang, A. "VLSI Datapath Choices: Cell-Based Versus Full-Custom." Masters Thesis, Massachusetts Institute of Technology, February 1998.
[5] Circuit Semantics, DynaCore Datasheet. February 2002.
[6] Dally, W. and Poulton, J. *Digital Systems Engineering*, Cambridge Univ. Press, 1998.
[7] Gardini P., "Intel Process Technology Trends," *Intel Developers Forum*, Talk Slides. February 2001.
[8] Gotto, G. et al., "A 54 x 54-b Regularly Structured Tree Multiplier," *IEEE Journal of Solid-State Circuits*, vol. 27, no. 9, September 1992, pp. 1229-1236.
[9] Gotto, G. et al., "A 4.1-ns Compact 54 x 54-b Multiplier Utilizing Sign-Select Booth Encoders," *IEEE Journal of Solid-State Circuits*, vol. 32, no. 11, November 1997, pp 1676-1682.
[10] Gupta, P., "Design and Implementation of the Integer Unit Datapath of the MAP Cluster of the M-Machine." Masters Thesis, Massachusetts Institute of Technology, May 1996.
[11] Hagihara, Y., "A 2.7ns 0.25um CMOS 54x54b Multiplier," *IEEE International Solid-State Circuits Conference*, 1998, pp. 296-297, 449.
[12] Harris, D. and Horowitz, M., "Skew Tolerant Domino Circuits," *IEEE Journal of Solid-State Circuits*, vol. 32, no. 11, November 1997, pp. 1702-1711.
[13] Hartman, D., "Floating Point Multiply/Add Unit for the M-Machine Node Processor." Masters Thesis, Massachusetts Institute of Technology. May 1996, pp. 47-54.
[14] Hwang, W. et al., "Implementation of a Self-Resetting CMOS 64-bit Parallel Adder with Enhanced Testability," *IEEE Journal of Solid-State Circuits*, vol. 34, no 8. August 1999, pp. 1108-1117.
[15] Itoh, N. et al., "A 600-MHz 54x54-bit Multiplier with Rectangular-Styled Wallace Tree," *IEEE Journal of Solid-State Circuits*, vol. 36, no. 2, February 2001, pp. 249-257.
[16] Keckler, S. et al., "The MIT Multi-ALU Processor," *Hot Chips IX*, August 1997.
[17] Khailany, B. et al., "Imagine: Media Processing with Streams," *IEEE Micro*, March/April 2001, pp. 35-46.
[18] Makino, H. et al., "An 8.8-ns 54 x 54-Bit Multiplier with High Speed Redundant Binary Architecture," *IEEE Journal of Solid-State Circuits*, vol. 31, no. 6, June 1996, pp 773-783.
[19] Naffziger, S., "A Sub-Nanosecond 0.5um 64b Adder Design," *IEEE International Solid-State Circuits Conference*, 1996, pp. 362-363.

[20] Rich, S. et al., "Reducing the Frequency Gap Between ASIC and Custom Designs: A Custom Perspective," *Proceedings of the 38th Design Automation Conference*, Las Vegas, NV, June 2001, pp. 432-437.
[21] Rohrer, N. et al., "A 480 MHz RISC Microprocessor in 0.12um Leff CMOS Technology with Copper Interconnects," *IEEE International Solid-State Circuits Conference*, 1998.
[22] Synopsys, *DesignWare Foundation Library Databook Volume 1*. June 2001.
[23] Synopsys, Physical Compiler Datasheet. November 2000.
[24] Ueda, K., "A 64-bit Carry Look Ahead Adder Using Pass Transistor BiCMOS Gates," *IEEE Journal of Solid-State Circuits*, vol. 31, no. 6. June 1996, pp. 810-818.

Chapter 12

Semi-Custom Methods in a High-Performance Microprocessor Design

Gregory A. Northrop
IBM TJ Watson Research Center,
Yorktown Heights, NY 10598, USA

Reprinted with permission of the Association for Computing Machinery. The original work is: Gregory A. Northrop, and Pong-Fei Lu. "A Semi-Custom Design Flow in High-Performance Microprocessor Design," *Proceedings of the 38th Design Automation Conference*, 2001, pp. 426-431.

Permission to make digital or hard copies of all or part of this Chapter for personal or classroom use is granted without fee provided that copies are not made or distributed for profit or commercial advantage and that copies bear this notice and the full citation on the first page. To copy otherwise, or republish, to post on servers or to redistribute to lists, requires prior specific permission and/or a fee.

1. INTRODUCTION

The development of high performance microprocessors requires concurrent design at many levels (logical, circuit, physical) with large teams and tightly interlocked schedules. Often the best design flow is one that most effectively addresses the natural conflicts within this flow (e.g. logic stability versus timing closure), in contrast to one that simply applies the most modern or aggressive approach in each domain. This paper describes such a case, in the development and use of a semi-custom design methodology which has significantly enhanced several generations of IBM zSeries (S/390) processors [2][3][4], as well as the IBM POWER4 processor [1]. The coordinated use of a common parameterized gate representation, standard cell generation capabilities, place and route merged with custom physical design, static transistor level timing and formal circuit tuning, and gain-based synthesis have all led to significant improvements in both quality-of-result and time-to-market in the conventional static CMOS design domain.

2. CUSTOM PROCESSOR DESIGN

The circuit design methodology described in this paper was developed and applied over three generations of the microprocessor family used in the IBM eServer zSeries (S/390 mainframe) [2][3][4]. These processors have achieved frequencies in excess of 1GHz in a 0.18um CMOS technology, combining high frequency with a relatively shallow pipeline (~7 stages) and extensive use of millicode [3] to implement the complex S/390 architecture with a RISC-like micro-architecture.

The physical design of these processors makes extensive use of hierarchy, partitioning the chip into functional units (instruction, FXU, FPU...) and units into macros. There are typically about 6 units and 200 distinct macros (about 600 total instances), and a macro can have anywhere from 1K to 100K or more transistors. The macro serves as the primary partitioning unit for logic entry (HDL), and a common macro connectivity description is used for both functional models for verification and for circuit design. Boolean verification is used to ensure functional equivalence between macro HDL and a schematic representation, which is verified against the physical design. All macros and units are fully floorplanned objects, and the global wiring (chip and unit) is done hierarchically, using a wiring contract methodology. Macros are all characterized for timing, noise, etc., and represented by models at the global level. Timing rules are generated for all macros using static transistor-level simulation. Global timing is run at both the unit and the chip level, using these macro rules and global wire extractions. The resulting timing is routinely used to generate assertions for all macros, which are used with the static transistor level timing to drive timing closure.

Concurrent design with carefully controlled feedback and iteration are the keys to bringing such a design to closure. Circuit and physical design start as soon as sufficient logic is defined, while the early emphasis is on simulation for functional verification. Early floorplanning and initial circuit design are used to check the cycle time feasibility, and as the design matures, the emphasis shifts from strictly functional verification to logic modification and repartitioning as the primary mechanism for achieving timing closure. This means that the efficiency, turn-around-time, and flexibility of the circuit design methods are as important to the ultimate chip cycle-time performance as is the intrinsic circuit performance.

Circuit implementations of macros fall into three classes, with each type occupying about 1/3 of the area: arrays, (cache, table logic) synthesized random logic macros (RLMs) for controls, and full custom dataflow. The dataflow is done predominantly in static logic, with dynamic circuitry reserved for only extremely critical functions, in order to meet power requirements.

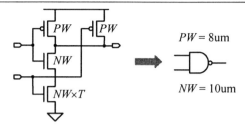

Figure 1. A parameterized gate representation used in the basis set for the primitive standard cell library. The n-channel MOSFET and p-channel MOSFET trees are scaled by two continuous parameters, *NW* and *PW*. There is also an optional tapering factor *T*.

Traditionally, full custom design is used for arrays, register files, and the dataflow stacks that are typical of the instruction and execution units. Custom design is very effective at optimizing performance and achieving high area efficiency, particularly where elements are identical across the bit range of the data stack, as hierarchy and careful tiling lead to highly optimal designs. This clearly applies to register files, working registers, MUXs, and the like. However, many of the most labor intensive and critical functions, particularly those that implement more complex numerical functions (adders, incrementers, and comparators) do not make such a compelling case for full custom design. They are far less regular across the stack, more complex, and often do not tile very easily. They are often timing critical, and although the logical function usually has a stable definition early in the design process, the most appropriate circuit architecture may evolve. It is this class of function which is the primary application for the semi-custom design flow outlined below.

3. SEMI-CUSTOM DEISGN FLOW

3.1 Primitive Parameterized Bookset

The basic building block used in this methodology is a set of parameterized gates, called the primitive bookset, an example of which is shown in Figure 1. These gates are generally a single level of conventional (inverting) static CMOS, with complementary pull-up and pull-down n-channel MOSFET and p-channel MOSFET trees, wherein the parameterization simply scales the p-channel MOSFETs with a parameter *PW* and the n-channel MOSFETs with a parameter *NW*. In general, each transistor can have an additional fixed multiplier, *T*, which is used to define multiple flavors of each logic type providing tradeoffs between the delays from each input pin. With the exception of the XOR and XNOR functions, all these primitive gates are a single level of inverting logic. A complete set of represented topologies can be found in Table 1.

Together, these parameterized gates form a basis set capable of covering most of the design space for combinational static circuitry found in a conventional ASIC library, since more complex functions, such as wide ANDs and Ors are typically composed of multiple levels of these gates. Note that this bookset is only a schematic representation; there is no directly associated layout.

A cell generation tool, described in more detail in Section 3.5, is designed specifically to produce layout in a row-based standard cell image for arbitrary values of *NW* and *PW* for each primitive cell. In addition to its use in semi-custom design, this tool was also used to create a conventional library of discrete sizes for use with synthesis to build the RLMs. This standard cell library had non-parameterized cells with all the conventional views, including timing rules required for synthesis. The sizes were selected by generating a reasonably dense matrix of *NW* and *PW* values for each primitive cell and running it through the cell generator. The *NW* and *PW* values were cast as a power level (*NW+PW*) and a rise/fall ratio (*PW/NW*), also called the beta ratio. The ratio of adjacent power levels was around 1.25, and the range of beta ratios (*PW:NW*) was adjusted to specific rise/fall times. Depending on the complexity of the function, there were from 10 to 25 power levels and from 1 to 4 beta ratios for each primitive type. Including a physically compatible set of cells using low threshold voltage (V_{TH}) transistors, there are about 1200 cells in this library. This type of library, commonly referred to as a "tall thin" library, provides the flexible sizing required for synthesis to achieve maximum cycle time performance in the RLM control logic.

Logic type	Comments
INVERTER	
NAND2, NAND3, NAND4	Multiple Tapers
NOR2, NOR3	Multiple Tapers
AOI21 and OAI21	Multiple Tapers
AOI22 and OAI22	
XOR2 and XNOR2	Pass-gate style
MUX2	2-Way transmission gate multiplexer
PGMERGE	G+P*C for adders (restricted AOI21)
DRXOR2	Dual rail XOR (true and complementary inputs)

Table 1. A list of parametrized logic types.

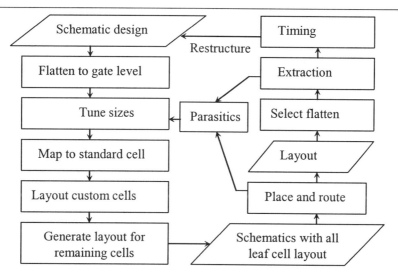

Figure 2. Complete design flow based upon the primitive cell library and cell generation and standard cell place and route.

3.2 Overall Design Flow

The design flow used to implement custom designs built from the primitive bookset is summarized in Figure 2. This flow is largely automated, with most of the work contained in the initial design and the place and route floorplan. After an initial pass, iteration of the design is a relatively rapid process.

3.3 Design Flow Details

This section catalogues some of the details of each step in the flow from Figure 2. Note that not all steps are needed, and a wide variety of combinations have been used, depending upon the details and needs of each design. The primary goal of this flow is to retain as much detailed control of the design at the schematic level while using layout automation, particularly place and route, as much as possible.

3.3.1 Schematic Design (Contents)

The initial schematic design can be all or part of a macro, but generally should encompass functionality that can be floorplanned in a simple block, and whose components can be routed automatically. It can contain hierarchy, with the assumption that it will be flattened to a set of routable leaf cells going into the final physical design. It can contain a mixture of parameterized primitive gates, cells from the standard cell library, and custom cells.

3.3.2 Gate Level Flattening

Even though the design will be flattened to the leaf cells going into physical design, there can be significant advantage to flattening a hierarchical design in schematic form, particularly when tuning a design with low symmetry in its structure or timing constraints.

3.3.3 Static Circuit Tuning

Automated tuning of transistor widths to optimize the slack of a macro is one of the key elements of this methodology. It helps to take full advantage of the sizing flexibility of the primitive bookset to optimize timing and area. An expanded discussion of the tuning method can be found in Section 3.4.

3.3.4 Mapping to the Standard Cell Library

After sizing, often many parameterized gates can be mapped to cells from the standard cell library, allowing only a modest change in transistor sizes and a minimal impact on timing. This helps to control data volume since all remaining primitive gates must be generated specifically for that macro.

3.3.5 Inclusion of Custom Leaf Cell Physical Design

Any required custom leaf cells for the design must be abstracted for place and route. These can be either a standard cell image, or other blocks that are pre-placed.

3.3.6 Automated Cell Generation

Any parameterized cells that remain after step 3.3.4 are generated. Details of the generation process can be found in Section 3.5.

3.3.7 Floorplanning, Place and Route

An interactive floorplanning and place and route environment is used to complete the layout. This environment features:
- A hand constructed floorplan with detailed wire contracts.
- Customizable row configurations for placeable cells, capable of multiple row heights.
- Pre-placement of non-placeable and placeable cells.
- Placement constraints (regions, groups, and net weights).
- Code-based or manual pre-routing.

- Grid based router.
- Incremental (ECO) placement to retain stable timing when iterating a design.

3.3.8 Cell Count and Data Volume Reduction

After completion of place and route, designs that have made heavy use of tuning and cell generation will have a large number of unique cells, many used only once or a small number of times. Selective flattening of these cells in the layout, with a corresponding change back to the parameterized schematic representation, helps strike a balance between a high cell count and the large data volume of a completely flat layout.

3.3.9 Parasitic Feedback into Tuning

The effects of wiring parasitic capacitances have become quite important, accounting for as much as 30% of the delay in even moderate size macros. Lumped capacitances, from either wire length (from place and route) or from a subsequent full extraction of the layout, can be merged into the schematic netlist and the circuit re-tuned to compensate for their effect.

3.3.10 Design Restructuring and Alternate Circuits

The combination of automated sizing and timing-based real physical design allows the designer to try multiple restructurings and circuit architectures, and make a confident comparison of the relative quality of each approach. *This part of the flow, and the associated change in the approach to design, are the most important part of this methodology, and the place where the most benefit will be found when it is fully applied.*

3.4 Circuit Tuning

A pivotal driver of this methodology is the use of circuit tuning to automate the sizing of transistors. While circuit tuning tools have been applied in CMOS design for a number of years, we believe that EinsTuner [5][8], the tool used here, delivers a quality of result that is very important to the overall improvements in design presented in this paper.

Tuning tools can be divided into 2 broad classes, static and dynamic. Dynamic tuning involves simulation with explicit waveforms and measures (delays and slews), while static tuning formulates the optimization problem through static timing, optimizing slack in the presence of timing assertions. The large, non-bitslice circuits for which semi-custom design is best suited present an impractical problem for dynamic tuning, but are an ideal candidate for static tuning, which can keep track of a large number of critical paths as tuning proceeds.

The EinsTuner static tuner is built on top of a static transistor-level timing tool (EinsTLT), which combines a fast event-driven simulator (SPECS) with a timing tool (Einstimer). The SPECS simulator provides timing information (delay and slew) along with first derivatives with respect to circuit parameters, specifically transistor width. EinsTuner uses this to formulate the optimization problem for solution by a large-scale general-purpose non-linear optimization package LANCELOT [6], generally optimizing a linear combination of slack and area, nominally treating all transistor widths as free parameters. Additional features of this tool that make it effective in a practical design environment include:
- Parasitics (lumped capacitance) from physical design.
- Area modeled as the sum of the transistor widths.
- A transistor-width ratioing mechanism, used to constrain transistor widths to match hierarchy or gate parameterization.
- Input capacitance, node slew, effective pull-up and pull-down beta ratio, and min/max transistor-width constraints.
- Complete interactive environment (GUI), including size constraint generation and back annotation.

In its current state of development, EinsTuner is capable of tuning in excess of 3000 gates with run times of normally less than 24 hours, even for large circuits, such as a 64-bit adder (~2000 gates). Experience has shown that tuning results are largely independent of the starting point, meaning that a designer can have a high degree of confidence that the results from a run are optimal for the conditions and design. While heuristic tuning can be faster, such algorithms often need coaching for particular designs, and they offer little certainty as to how close a run is to the "true optimum". This is an important issue when using this methodology to compare and select circuit structures.

3.5 Cell Generation

A second key component of this methodology is the use of a cell generator to create layout corresponding to the parameterized gates. This home-brewed tool, called C-cell, is not a general-purpose cell compiler, but rather a script-based system designed to produce optimal layout, but only for the defined parameterized bookset. The definition of the parameterized gate set is tightly integrated into this tool, delivering a framework that supports semi-custom design in a number of ways:
- Generate a set of layouts and associated views for use as a standard cell library, based upon a list of cell specifications: (primitive name, NW, PW).
- Parse a schematic, form a list of cells to generate to replace parameterized gates, minimizing the number of required cells for a

maximum allowed deviation in size. Generate cells, and create a modified custom schematic referencing the generated cells.
- Includes a facility to convert between parameterized and standard (RLM library) cells.
- Has an integrated floorplanning aid with an interface to the place and route tool.
- Does layout post-processing, including selective layout flattening and shape trimming.

In the cell generation part of the tool, topology and technology specific code takes as input the gate type, size parameters NW and PW, and a global cell image (row height), and generates layout, after selecting the optimal configuration from a range of finger partitionings and topology options. In practice the measure of optimality is cell area, but factors such as wireability, manufacturability, etc., could also be weighted in the selection. While this system is not capable of implementing an arbitrary gate topology, it has been very successful in the domain of conventional static CMOS, where there are a small number of effective topologies, and optimization of the simplest (NAND/NOR) types is vital. To date, the effort required to migrate and modify this approach from technology to technology has been easily justified in its use in both the synthesis environment and in semi-custom design.

4. DESIGN EXAMPLE – 24 BIT ADDER

4.1 Adder Function, Timing, and Floorplan

This 24 bit adder is used in the branch target address prediction in the POWER4 microprocessor [1]. The adder performs an addition between the Current Instruction Address (CIA) and the sign-extended immediate operand from the instruction cache (I-cache). The adder includes a 4-way multiplexer (MUX) to select among four possible target addresses: the link register, the sign-extended immediate field, the adder result, and the counter register. The MUX controls are from an RLM outside the dataflow stack. The adder and MUX, wrapped by latches that were manually designed separately, form a module. There are eight such 24-bit adder modules to handle 8 instructions delivered by the I-cache in parallel. A 10-way MUX (in another macro) selects one of the 8 resulting branch addresses and 2 other sequential addresses, whose lower 12 bits are then sent back to I-cache to fetch the next 8 instructions. This loop path: I-cache→Adder→MUXs→I-cache, takes three cycles to complete. It is one of the cycle-limiting critical paths in the POWER4 design. The allotted timing budget for the adder is about half a cycle.

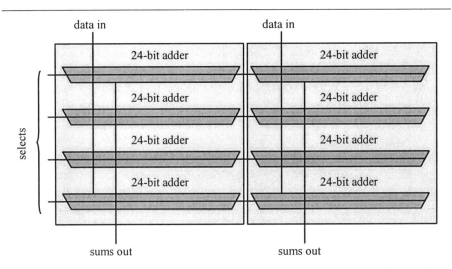

Figure 3. The macro floorplan containing a group of eight 24 bit adders. The multiplexers discussed in Section 4.2 are shown in gray.

The floorplan of the macro containing eight adder modules is shown in Figure 3. The eight adders are arranged as two stacks of four to meet constraints in the chip floorplan and to reduce wiring congestion. The height is dictated by the unit floorplan and needs to be minimized. The width dimension it is fixed at 14 tracks per bit, which is the same width as the rest of the dataflow. The address inputs and outputs flow vertically, while the control signals are from the sides. The MUX inputs for the predicted values of the link register and the count register are common to all eight adders; the CIA is formed by concatenating the word-offset field (0:2) with the I-cache fetch address (IFAR) <38:58> which is also common to all eight adders.

4.2 Circuit architecture

The 24-bit adder is a static, carry-look-ahead (CLA) Ling adder [7], which is a stage faster than a CLA adder. The 24 bits are grouped into six 4-bit groups for carry propagation. The local sum is implemented in parallel with the carry, and the result of the carry bit is used to select the final value.

One key circuit decision made early in the design was to hide the MUX delay behind the adder. The assumption was that the I-cache data through the adder would be the critical path, while the MUX selects and register file data will arrive early. Since the carry bit is the longest path, the partial sums can be pre-multiplexed with the other three inputs and the final result selected by the carry. This led to a skewed 'late-MUX' design as shown in Figure 4(a). Frequently it is assumed early in design that controls are non-critical, however that often turns out not to be true, as will be discussed later.

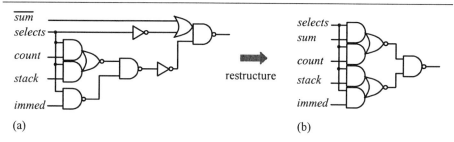

Figure 4. Multiplexer implementations at the output of the adder. The initial implementation (a) assumed *selects* (from controls) were not critical. The final implementation (b) optimizes the *selects* and *sum* paths.

4.3 Circuit tuning and layout

In its final form, the design consisted of 485 gates of the following types and number: INV (164), NAND2 (63), NOR2 (31), AOI21 (15), AOI22 (176), OAI21 (25), and OAI22 (11). Once the schematic design was complete and verified, the physical design followed much of the flow shown in Figure 2. These steps were used: gate level flatten, tune, cell generation, place and route, full extraction, re-tune with actual parasitics, and incremental placement and re-route. Since the schematic was flattened before tuning, all gates could be sized independently by EinsTuner. Wiring use was limited to all of M1, and a portion of the M2 and M3 tracks, as some tracks are reserved for the macro level routing through the stack. Cell occupancy was maintained through EinsTuner's area constraint to about 70% of the floorplan area. No mapping to the standard cell library was done, and 297 cells were generated, since only a ±5% size variation was allowed. If a ±25% size variation had been allowed, this would have still required 143 cells. After place and route, the layout was flattened into shapes, since the maximum reuse of any cell was ten instances, and most cells had only one or two instances. In addition flattening allowed the use of the trimming process to cut excess metal and poly shapes to reduce parasitics, which was found to improve the delay by an additional 2-3%. The turnaround time was less than a day; usually two to three iterations were sufficient to bring the design to convergence after a change in design and/or timing assertions.

4.4 Design iterations and timing convergence

The key issue of the semi-custom design is the timing assertions fed to the tuner. As the control signal timing is unknown in the beginning, the timing assertions are largely estimates that may not be substantiated as the chip timing stabilizes. In the 24-bit adder example, the assumption of the MUX select timing turned out to be wrong after the first tape-out. The gating path was from the selects, thus the circuit construct in Figure 4(a) was suboptimal. The MUX was re-designed using a balanced AOI-NAND scheme in Figure 4(b). The new schematic was then re-tuned and iterated based on the correct timing assertions. The new design was completed in a week, and the negative slack was reduced by more than 80ps, about 20% of the timing budget for the adder. Thanks to the contract based place and route, the adder re-design did not cause any global wiring change.

Another issue is that a new microprocessor design project like POWER4 often straddles across several technology generations. Each technology migration induces device model changes (e.g. different P/N strength ratio), which full-custom designs are difficult to adjust to. With the semi-custom design approach, timing shift can be readily accommodated by the tuner, and with the flexible physical design, adjustments can be made in a timely fashion. We estimate that the semi-custom design at least reduces the total design time by 50%, even for designers unfamiliar with the place and route environment, who need to make an initial investment in learning the tools.

5. OVERALL IMPACT ON CHIP DESIGN

Application of this semi-custom methodology to custom macro design has shown that benefits come in two distinct waves. The first is an improvement in cycle time performance, primarily associated with the circuit tuning process. Generally designers apply it directly to their existing designs, looking for rapid turn-around, combined with some performance or area improvement. The second wave comes when a designer makes a more basic change in approach to design, concentrating more effort on circuit architecture, then using the rapid turnaround to quantify the performance of multiple designs. Each design approach can be tried in a real design context, which is optimally sized, and includes real parasitics and timing assertions. This leads to greater investment in optimizing the circuit architecture, rather than selecting one design early, and optimizing the sizing and physical design manually.

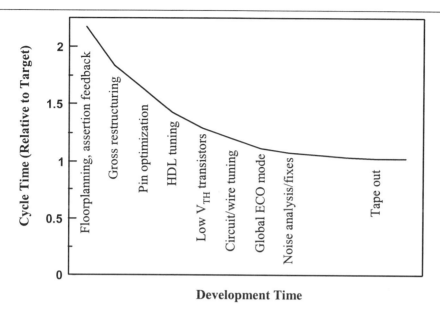

Figure 5. Chip timing closure progress as a function of time. The labels indicate the dominant activities used to improve timing at each stage.

The difference this change in philosophy makes can be put in the context of the overall design closure shown in Figure 5. This shows the convergence of chip cycle time as a function of time for the latest S/390 processor [4], BlueFlame, along with some of the typical design activities driving the improvement at each point, up through tape-out. The custom design phase extends up until the point where the cycle time is about 1.25× the target, by which time full physical design is required for reliable extraction-based timing. With the conventional approach to custom design, the time required to implement physical design often requires commitment to a particular circuit architecture when the chip timing still exceeds the 1.8×-target range. As with the adder in Section 4, this can lead to selecting a less than optimal circuit architecture. Being able to adapt quickly to changes associated with global timing convergence is a major advantage to semi-custom design. Experience suggests that these advantages are more valuable than any loss associated with the use of place and route in the physical design.

Net improvements in designer productivity, above and beyond achieving a superior design, are also of interest. Since this methodology was most heavily applied in the instruction unit of the BlueFlame processor, we have done some analysis there to try to understand its overall impact. This involved a code-based and manual survey of the design to gather specific statistics, followed by some interviews with designers to make work time estimates. Table 2 breaks down the circuits in this unit by their construction method. While the total semi-custom area was fairly small, the impact on the design work was significant, as it was applied to the most difficult functions. The survey indicated that most of the remaining full custom design was bit-slice in nature (working registers, MUXs, and three register files), having relatively large areas implemented with a small number of cells. Table 3 lists the 23 semi-custom blocks by function type, giving the number of unique designs for each.

No truly valid quantification of the productivity improvements associated with semi-custom design could be made, due to a lack of both a detailed recording of design effort and any form of a control, since previous design points were different and the design teams and overall methodology have changed too. Instead, we made an effort to estimate the improvement, giving quantities that were derived from discussions with several designers. A proto-typical set of times are given in Table 4 for seven stages of design, for the full custom approach and the equivalent design in semi-custom. Once a design has been through the complete semi-custom flow once, an iteration typically only requires a couple of days, including structural changes to the design. When one includes all stages of circuit and physical design and analysis, the semicustom approach requires roughly one half the time, yields as good or better results, and provides the ability to change the design quickly and reliably late in the design process.

Circuit Construction	Area %
Custom	55.3
Semi-custom	15.2
RLM (synthesized)	29.5

Table 2. Macro circuit area by construction method in BlueFlame instruction unit.

Function Type	Number of Blocks
Adder	3
Increment	4
Compare	8
Error check	2
Other	6

Table 3. Number of semi-custom functions by type in BlueFlame instruction unit.

Design Steps	Full-Custom	Semi-Custom
Circuit arch, initial schematic	5	5
Floorplan, area estimate, wire util.	3	3
Parasitic estimation, circuit sizing	3	-
Post-initial timing circuit struct/sizing	6	-
Physical design – leaf cells	12	2
Automated tuning	-	2
Physical design – assembly	6	2
TOTAL	35	14
Time for each additional iteration	?	2

Table 4. Estimated time (arbitrary units) by design phase for full- and semi-custom design flows.

6. ACKNOWLEDGMENTS

Our thanks to Chandu Visweswariah, Phil Strenski, and Ee Cho (circuit tuning), Joe Nocerra and Ching Zhou (cell generation), Keith Barkley (place and route), Brian Curran, Tom McPherson (timing convergence), and many designers for their patience, experiences, and suggestions.

7. REFERENCES

[1] Anderson, C.J., et al., "Physical Design of a Fourth-Generation POWER GHz Microprocessor," *International Solid State Circuits Conference*, Digest of Technical Papers, pp. 232-233, Feb 2001.
[2] Averill, R.M. et. al., "Chip integration methodology for the IBM S/390 G5 and G6 custom microprocessors." *IBM Journal of Research and Development*, vol. 43, 1999, pp. 681-706.
[3] Check, M.A., Slegel, T.J., "Custom S/390 G5 and G6 microprocessors," *IBM Journal of Research and Development*, vol. 43, 1999, pp. 671-680.
[4] Curran, B., et al., "A 1.1 GHz First 64b Generation Z900 Microprocessor," *International Solid State Circuits Conference*, Digest of Technical Papers, pp. 238-239, Feb 2001.
[5] Conn, A.R., et al., "Gradient-based optimization of custom circuits using a static-timing formulation," *Proceedings of the Design Automation Conference*, pp. 452-459, June 1999.
[6] Conn, A.R. Gould, N.I.M., and Toint, P.L., *LANCELOT: A Fortran Package for Large-Scale Nonlinear Optimization*. (Release A). Springer Verlag, 1992.
[7] Ling, H., "High-Speed Binary Adder," *IBM Journal of Research and Developopment*, vol. 25, no. 3, May 1981, pp. 156-166.
[8] Visweswariah, C., and Conn, A.R., "Formulation of static circuit optimization with reduced size, degeneracy and redundancy by timing graph manipulation," *IEEE International Conference on Computer-Aided Design*, pp. 244-251, November 1999.

Chapter 13

Controlling Uncertainty in High Frequency Designs

Stephen E. Rich, Matthew J. Parker, Jim Schwartz
Intel Corporation,
RA2-401,
2501 NW 229th Avenue,
Hillsboro, OR 97124, USA

1. INTRODUCTION

This chapter introduces the concept of uncertainty and examines its negative impact on frequency as well as time-to-market. The reader will be introduced to three distinct types of uncertainty with real world examples given in the context of high frequency microprocessor design. The chapter will provide tools and processes to create an uncertainty plan and demonstrate how it can be used to allow design teams to take on more aggressive frequency goals without additional time or effort.

2. FREQUENCY TERMINOLOGY

The longer the design project the harder it is to predict the final frequency of the part. On projects as complex as microprocessors, which have traditionally taken up to four years, four frequency terms are used to manage the performance of the design.
- **Actual Frequency:** This is the final frequency that is realized in silicon within the targeted system.
- **Market Frequency:** This is the minimum acceptable frequency committed to the end customer. Depending on the type of design, having an *Actual Frequency* higher than *Market Frequency* may enable a higher price, as is the case with microprocessors. In other types of design having the *Actual Frequency* higher than the *Market Frequency* is of no value.

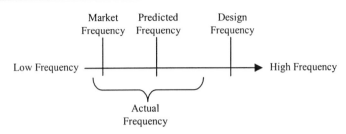

Figure 1. Use of multiple frequency targets.

- **Design Frequency**: This is the frequency given to the designers and the design tools. This frequency is used to make all design tradeoffs and is set to a higher value than the *Market Frequency* to ensure the minimum *Market Frequency* is met.
- **Predicted Frequency**: This is the frequency of the current state of the design as determined by transistor level timing simulations. Initially, the *Predicted Frequency* will fluctuate widely during the design process as features are added into the product. As the design matures, the *Predicted Frequency* should stabilize near the *Design Frequency*.

Due to uncertainties that will be described in this chapter, the *Actual Frequency* of the design may range from slower than the *Market Frequency*, resulting in an unmarketable design, to faster than the *Predicted Frequency*. This is shown in Figure 1.

Reducing the gaps between each of these frequencies will increase the *Actual Frequency* of the design, but will also increase the design risk. Managing these gaps effectively is critical to the success of high frequency design.

3. UNCERTAINTY DEFINED

There are three major categories of uncertainty, or variations, in the design/development process: process uncertainty, tool uncertainty, and design uncertainty.

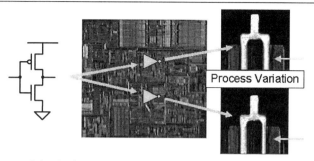

Identical Inverter used in two parts of the chip.

Figure 2. Process variation causes design uncertainty in the inverter.

3.1 Process Uncertainty

Process uncertainty is caused by variations in the manufacturing process. One example of process uncertainty is the in-die variation of transistor geometries as shown in Figure 2. In this example, an inverter is duplicated in two different parts of the chip. Although these two inverters have identical drawn dimensions, process variations can cause drive strengths to differ. These variations must be taken into account when designing the chip.

3.2 Tool Uncertainty

Tool uncertainty is caused by inaccuracies in the simulation and extraction tools. For instance, if inductance is not extracted, the *Predicted Frequency* of the design will be optimistic relative to the *Actual Frequency* of the design. Since inductance varies from wire to wire, the tool predicted ordering of the critical paths will differ from the actual ordering of critical paths. This will cause the design team to focus on the wrong paths.

3.3 Design Uncertainty

Design uncertainty is caused by unpredictable variations in the design process between design iterations as well as variations in the execution of design methods across the chip. For example, assume an automated tool is used to route a clock tree. The unpredictable nature of the clock routing between design iterations causes variations in the RC delay as shown in Figure 3.

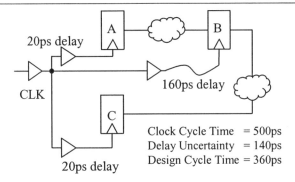

Figure 3. Clock uncertainty caused by scenic routing.

Figure 3 shows the clock cycle time as 500ps. In an ideal world, all 500ps would be available for the logic functionality. However, the variations in the clock routing result in 140ps (160ps – 20ps) of uncertainty in the cycle time. To take into account the worst case clock delay, the design cycle time is limited to only 360ps. In other words, the clock cycle time must be guard banded by 140ps in order to account for the uncertainty of the clock routing (clock skew).

The unpredictable nature of the clock routing between place and route iterations causes significant variations in the clock tree. In turn, this causes the ordering of the critical paths to change between iterations. The most critical path, the path with the worst negative slack (WNS), changes between iterations. This makes it very difficult to focus on, and fix, the worst path as the clock delay changes in every iteration of place and route. Using a more predictable design method, such as pre-routing the global clock tree, reduces the uncertainty of the clock routing between iterations, thus increasing the cycle time available to logic functionality as shown in Figure 4.

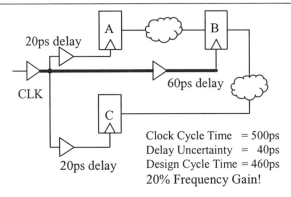

Figure 4. Use of clock pre-routing to reduce clock uncertainty

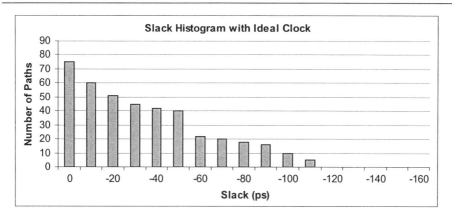

Figure 5. Slack histogram with no clock uncertainty.

As the frequency of the design increases the design uncertainty consumes a larger and larger percentage of the clock cycle time, thus reducing the amount of time available for logic functionality. As the frequency of the design increases second order design uncertainties become first order effects. Thus, the higher the frequency of the design the harder the design task becomes. Learning to efficiently deal with design uncertainty in a uniform manner becomes exceedingly important as the frequency of the design increases. In microprocessors with frequency in the multiple Gigahertz range it becomes essential. *(For additional sources of design uncertainty see Table 2 at the end of this chapter.)*

Key Point: Uncertainty in the manufacturing, tools and design processes cause a gap between the predicted and actual frequencies, thus reducing the cycle time available for logic functionality.

4. WHY UNCERTAINTY REDUCES THE MAXIMUM POSSIBLE FREQUENCY

Uncertainty in the design process causes the design team to waste resources and energy working on non-critical parts of the design. This ultimately results in either a slower part or an unnecessary delay in the time-to-market.

To illustrate this, first look at the typical slack graph shown in Figure 5 (slack = *Design Frequency - Predicted Frequency* on a per paths basis). This graph represents the *Predicted Frequency* as simulated with an ideal clock and no forms of uncertainty.

Now interject 40ps of clock uncertainty as described in Section 3.3. This means that the design could run 40ps faster than expected or 40ps slower than expected. If the clock edge at the sampling sequential is 40ps late for the critical path, the design's *Actual Frequency* may be faster than the *Design Frequency*. In other words, the part would run faster than predicted, such is the case with Path (A→ B) in Figure 4.

However, if the clock at the sampling sequential is 40ps earlier due to the uncertainty in the design process, the *Actual Frequency* would be slower than the *Design Frequency*, which translates into a zero percent raise at the end of the year... (if you're lucky). Such is the case with Path (B→C) in Figure 4.

The simplest way to deal with this uncertainty, both in the design and in your raise, is to add 40ps of guard banding into the design flow to account for the worst case clock skew. The next graph (see Figure 6) illustrates a shift of the slack histogram to the right to account for the guard banding. This will force the tools to work harder on at least the critical paths thus limiting the risk of the *Actual Frequency* being slower than the *Design Frequency*.

Side Note: In real life it should be noted that there are several sources of uncertainty in the design process simultaneously. Simply applying all of the uncertainty factors to every path usually leads to an unrealistic design window. Statistics are generally applied to determine how much guard banding should be added for each source of uncertainty. The net result, even after guard banding for the uncertainty, is that any given path could be hit by compounding uncertainty factors greater than the guard banding. If this unlikely event occurs, it is usually found in silicon and is labelled a "silicon escape".

Now there is a danger of falling into a common design trap. Take a look at two paths from the example in Figure 4. Assume path (A→B) falls into the -160ps slack bucket and path (B→C) falls into the -120ps slack bucket, as shown in Figure 6. It is very tempting to spend effort fixing the worst negative slack (WNS) of path (A→B) regardless of its complexity. However, since both paths (A→B) and (B→C) fall within the uncertainty range, either one could be the worst path in the design.

Assuming that fixing path (A→B) is significantly harder than fixing path (B→C), the designer could focus their effort on path (B→C) ignoring (A→B) and still possibly improve the frequency of the design. Thus, the slack graph can be very misleading when used to direct effort.

In the early stages of a project a modified slack graph should be used when directing resources. All paths that have a slack delta of less than the uncertainty in the design should be grouped into the same bucket. The modified slack graph is show in Figure 7.

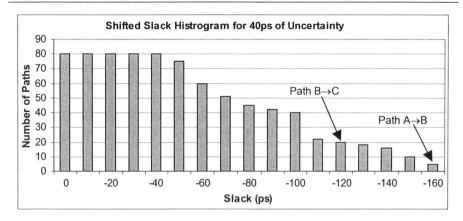

Figure 6. Shifted slack histogram with 40ps of clock uncertainty

When viewed this way, the tail of the slack wall looks more like a wall; the WNS bucket went from having 5 paths to over 30 paths. This more accurately depicts the number of paths that need to be addressed to be certain of frequency improvements. In order to guarantee an increase in the *Actual Frequency,* all paths within the WNS bucket must be fixed. Conversely, any path that does not fall into the WNS bucket cannot possibly affect the maximum frequency of the design and can be safely ignored until all of the paths in the WNS bucket have been fixed. Any time spent working on paths outside of the WNS bucket is a waste of effort in terms of frequency.

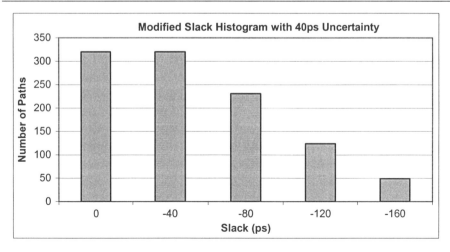

Figure 7. Modified slack bucket histogram for 40ps of clock uncertainty.

Assume there are only enough resources to fix a couple of the paths in the WNS bucket. Which ones should be fixed? The ones with the worst slack? The only way to guarantee an increase in frequency is to fix all of the paths in the WNS bucket. Therefore the width of the WNS bucket, and thus the number of paths that must be fixed, becomes exceedingly important. Decreasing the uncertainty in the design decreases the width of the WNS bucket and the number of paths that need to be addressed in order to increase the *Actual Frequency*.

If the uncertainty is small enough, meaning the design team is very confident it is working on the frequency limiting paths, high effort design techniques can be applied. For instance, the effort required to use dynamic logic on the worst paths can be justified if the probability of an *Actual Frequency* increase is high. If the design uncertainty is larger, a larger percentage of paths in the design would need to be converted to dynamic logic in order to guarantee a higher *Actual Frequency*. As a result, the effort required to use dynamic logic on a large number of paths may become prohibitively expensive.

5. PRACTICAL EXAMPLE OF TOOL UNCERTAINTY

Figure 8 shows a practical example of how tool uncertainty can affect the design process. It is typical for cell based static timing tools to propagate the worst case slope when timing through a path. This injects a large amount of uncertainty into the design process. For example, if a two input OR gate has an early arriving signal with a poor slope and a late arriving signal with a fast slope, many popular timing tools will combine the worst slope with the worst case timing. This is, of course, pessimistic.

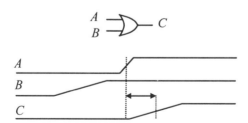

Delay from *A* combined with slope from *B*.

Figure 8. Worst case slope propagation through an OR gate.

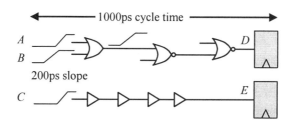

Figure 9. Tool uncertainty leading to wasted design effort.

At first glace this does not appear to be a problem because the *Actual Frequency* will be higher than the *Predicted Frequency*. But, the tool uncertainty generated by the worst case slope propagation increases the amount of uncertainty in the design, which in turn increases the effort required to achieve higher *Actual Frequencies*.

To illustrate this further look at the simplified design shown in Figure 9. The slacks are:

Path C→E		Path A→D	
1000ps	Cycle Time	1000ps	Cycle Time
-950ps	Gate Delay	-850ps	Gate Delay
-100ps	Setup	-200ps	Worst Case Slope
		-100ps	Setup
-50ps	Reported Slack	-150ps	Reported Slack
-50ps	Actual Slack	+50ps	Actual Slack

When timing the path from (A→D), the typical timing tool will combine the worst timing from path (A→D) with the worst slope from (B→D). As a result, the timing tools will show that the design is missing the timing targets by -150ps. The natural course is to spend resources fixing this path. In actuality, the path will run with +50ps of slack and any resources applied to it are simply wasted.

To make matters worse, the real critical path (C→E) largely goes unnoticed since it is not reported as the WNS path. If the designer focuses on path (C→E) any improvement made, up to 50ps, results in a higher frequency design. Another alternative is, of course, not to spend any additional resources and go straight to manufacturing. This is the kind of work that great raises are made out of.

This principle holds true for all sources of uncertainty. The inevitable result of uncertainty is a gap in the *Predicted* and *Actual Frequency* which results in the misdirection of resources that could otherwise be working on improving the time-to-market or *Actual Frequency* of the design.

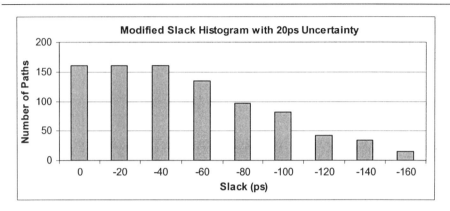

Figure 10. Modified slack bucket histogram with 20ps uncertainty.

Key Point: Any uncertainty in the design process will cause a gap between the *Predicted* and *Actual Frequencies*. This will cause the design team to work on the wrong paths increasing the effort required to obtain higher frequencies.

6. FOCUSED METHODOLOGY DEVELOPMENT

In order to correctly apply resources, the uncertainty in the modified slack graph must be reduced. Figure 10 shows that, as the uncertainty window is decreased, the modified slack graph starts to look more and more like the actual slack graph proposed in Figure 5. The number of paths in the WNS bucket has now been decreased to 5 paths again. Any work applied to these paths will be more likely to increase the frequency of the design. The smaller the uncertainty, the more likely the effort applied will affect the *Actual Frequency* of the part.

Reducing the uncertainty increases the probability of frequency returns for effort spent. Thus, early in the project, the design team should aggressively look for ways to reduce the level of uncertainty in the design in order to reduce the effort required to fix the critical paths during the later phases of the project.

Key Point: Reducing the uncertainty will minimize the number of paths in the WNS bucket and thus the effort required to address them.

It has been shown that the closer the correlation between the *Predicted Frequency* and the *Actual Frequency* for the paths in the WNS bucket the higher the *Actual Frequency* will be.

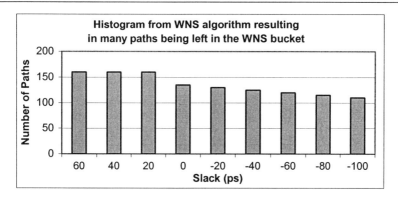

Figure 11. WNS algorithms leave many paths in the WNS bucket.

Any gap between the *Predicted* and *Actual Frequency* implies a source of uncertainty in the development process. Any uncertainty in the development process will limit the maximum potential *Actual Frequency* of the design and thus should be the focus of high speed methodology development. The remaining sections of this chapter will focus on methodologies for reducing design uncertainty.

7. METHODS FOR REMOVING PATHS FROM THE UNCERTAINTY WINDOW

It is customary for CAD algorithms, and designers, to expend all available effort on the path that exhibits the worst negative slack (WNS). In CAD, this practice is referred to as a WNS algorithm, since the algorithms are unlikely to expend resources or effort on paths once their slack is better than the WNS path. As can be seen in Figure 11, these algorithms tend to result in a large number of paths having slack values just slightly better than the WNS path.

Unfortunately, the WNS algorithm has left large numbers of paths in the WNS bucket, many of which would be removed with additional effort. As discussed earlier, any path left in the WNS bucket may limit our *Actual Frequency*, so it is a good practice to remove all easily fixed paths from the WNS bucket before manufacturing the part.

To accomplish this task automatically, use a total negative slack algorithm, rather than a WNS algorithm. Total negative slack (TNS) is defined as the sum of negative slacks over the entire design. TNS algorithms will continue expending effort and resources to improve any negative path until it reaches zero slack, regardless of the value of the worst negative slack.

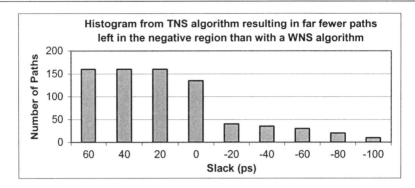

Figure 12. TNS algorithm removes more paths from the negative region.

As Figure 12 depicts, TNS algorithms result in moving many paths left in the negative region by the WNS algorithm into the positive region.

Note that the goal of the TNS algorithm is to remove all paths from the negative region. Whereas, the design goal is to remove all paths from the WNS bucket in order to achieve a higher *Actual Frequency*. In order to direct the TNS algorithm to achieve this design goal, it is necessary to set the *Design Frequency* such that all paths in the WNS bucket are in the negative region. Following this practice can usually be accomplished by simply toggling a few switches in most synthesis compilers. This also sets a lower limit on the *Design Frequency's Cycle Time* with respect to the *Market Frequency's Cycle Time*.

Design Cycle Time = Market Cycle Time – Width of WNS bucket

It is clear that the Design Frequency should never be set lower than this value, or paths will be left in the WNS bucket unnecessarily. However, setting the Design Frequency above this value will cause design problems as well. Once a path is moved out of the WNS bucket, any further slack improvement is over design. Over design leads to unnecessary use of resources and effort, which could otherwise be used to remove more paths from the WNS bucket. A classic example of this is where capacitive loads on a path in the WNS bucket are increased due to oversized transistors in a related path, which is well out of the WNS bucket.

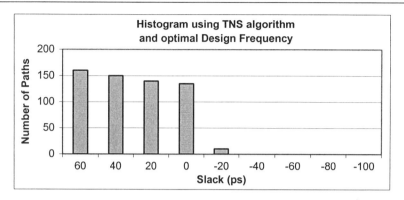

Figure 13. Design frequency set such that only the WNS bucket is negative.

Another method for removing paths out of the uncertainty window is to simply reduce the uncertainty window. This is the subject of the final section.

8. THE UNCERTAINTY LIFECYCLE

When developing a plan to address the sources of uncertainty within a project, it is important to understand the basic lifecycle. Figure 14 shows the uncertainty lifecycle. The two rings depict design and development effort costs. Design effort is defined as the work done by electrical engineers on critical paths. Development effort is defined as work done while defining the methodology and assembling the tools required for the design effort. These are usually two different sets of resources that can be applied to the goal of higher frequencies. Let's examine it in the context of clock skew.

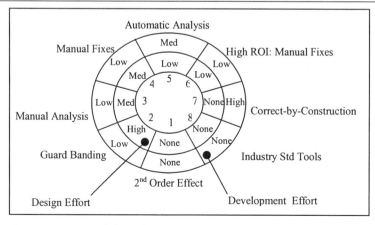

Figure 14. Uncertainty lifecycle.

8.1 Stages of the Uncertainty Lifecycle

Stage-1, Second Order Effect: During Stage-1 of the lifecycle, assume that the clock skew is less than 1% of the total cycle time. At this point in time, the clock skew can be safely ignored. Effort during this stage is zero.

Stage-2, Guard Banding: As the frequencies of the designs increase over time, the team shifts into Stage-2. Here the effects of clock skew can no longer be ignored so guard banding is used to account for the clock skew uncertainty. The development effort is low, as most tools support guard banding in some form, but the design effort is very high. Every path not requiring guard banding results in effort wasted since guard banding is simply safeguarding for uncertainty and does nothing to reduce it.

Stage-3, Manual Analysis: As the frequencies continue to increase, it becomes critical to more accurately analyze the clock tree - at least locally for the critical paths. During this stage, it would be typical to override the clock skew guard banding on the critical paths after careful manual analysis (accurate device level analog simulation). This has the effect of reducing the uncertainty on the critical paths. The development effort is low; however, the design effort of performing the manual analysis on the critical paths is med-high. A common mistake is to assume that manual analysis is simply additional design effort. As previously noted, accurately analyzing the path can remove it from the WNS bucket just as effectively as fixing the path. Breaking through the global guard banding methodologies of stage-2 is difficult and can become a major barrier to many design teams. It is, however, critical to reducing the uncertainty in the design and, therefore, the frequency of the design. This is where many design teams fall behind due to limited resources.

Stage-4, Manual Fixes: Now that the clock skew can be accurately analysed on local portions of the design, the number of paths in the WNS bucket has been reduced. Manual fixes to the clock tree can now applied to the critical paths. For example, for the WNS paths, the clock tree may be routed manually. It should be noted that, since the analysis and fixes are being applied locally, the uncertainty across the entire design has not been reduced. Thus, the design effort may not be spent on the critical paths. To avoid this, it is important to make sure that the local/manual analysis and fixes cover all of the paths in the uncertainty window. The development effort is, again, low and the design effort is med-high.

Stage-5, Automatic Analysis: As the frequencies of the designs increase, the number of paths in the WNS slack bucket will increase to the point where manual analysis is too costly. During this stage automated CAD tools and methodologies are developed to reduce the uncertainty of the analysis of the entire clock tree. Reducing the uncertainty will reduce the number of paths in the WNS slack bucket, allowing local manual optimizations to be applied with high confidence.

For example, the clock skew calculations may now take into account the point of divergence of the sampling and generating clocks instead of one simple guard banding value. This analysis would be done globally to reduce the uncertainty and thus reduce the number of paths in the WNS slack bucket. The development can be high depending on the skill set within the team. The design effort starts to decrease as the automatic global analysis reduces the uncertainty and accordingly the number of paths in the WNS bucket.

Stage-6, High ROI Manual Fixes: Now that the number of paths in the WNS bucket has been reduced local fixes can be resumed. The reduced uncertainty improves the ROI as the confidence of predicting the critical path increases. The development and design effort at this stage are med-low.

Stage-7, Correct-by-construction: At this point the design team is using clock tree CAD tools that can automatically hit their skew targets. These tools are often developed within the company or co-developed with an external EDA firm. The tool development effort is now extremely high but the design effort has fallen dramatically. It is important to note that correct-by-construction can be achieved by tool enhancements like better clock tree algorithms or via design methodology, such as using a latch-based design. Methodology solutions generally have lower development costs.

Stage-8, Industry Standard Tool: The ultimate goal is to have the clock tree algorithms which are capable of hitting the skew targets become part of the industry standard tools. This frees up resources to work on reducing the clock uncertainty even further or work on the next source of uncertainty. As the frequencies increase, the uncertainty introduced by tools, even those which are standard in the industry, will be too much and the uncertainty lifecycle will continue.

8.2 Developing an Uncertainty Plan

With a firm understanding of the uncertainty lifecycle and the effort profiles for each stage, a plan to address each source of uncertainty can be developed. First list all of the sources of uncertainty in the design. Then, based on the availability and skill set of the design team, address each source of uncertainty by assigning it to a stage in the lifecycle. The aim is to minimize the amount of uncertainty in the design with a plan that optimizes the combination of development versus design effort.

Source of Uncertainty	Value (ps)	Course of Action
Clock Skew	40	90% Guard Band
		Manual analysis and fix for WNS bucket
Worst Case Slope Propagation	400	No Guard Banding
		Move to Depth First Search Timing Tool
		(correct-by-construction)
Fringe Effect	10	100% Guard Band
Inductance Extraction	1	Treat as Second Order Effect

Table 1. Uncertainty plan.

Accurate analysis is the key, even if the methodology does not apply fixes. Moving paths out of the guard banding stage will reduce the number of paths in the WNS bucket and the total design effort. Using these guidelines, it is possible for the design team to develop a plan for minimizing uncertainty in the design process based on the available resources. Table 1 shows a simple example of an uncertainty plan.

8.3 UNCERTAINTY AND THE PROJECT LIFECYCLE

Deciding where to spend effort is an art form that is project as well as team specific. In this section, simple guiding principles are proposed to help determine where effort may be best spent throughout the different phases of the project.

Beginning of the Project: In the early project development stage, emphasis should be placed on reducing the sources of uncertainty. Reducing the uncertainty window is equivalent to removing the path from the uncertainty window through design effort. It should also be noted that manual analysis and/or fixing should not be attempted until the design has stabilized between iterations. If the ordering of the paths is not consistent, any effort spent on manual optimizations would be wasted.

Middle of the Project: During this phase of the project, the methodology should be stabilized. Focus should shift to adjusting the *Design Frequency* as described in Section 7. This will force the synthesis tool/design team to remove as many paths out of the uncertainty window as possible while avoiding over-design. Manual optimizations can start on the critical paths within the uncertainty window as soon as the design starts to stabilize between iterations.

End of the Project: During the final stage of the project, the methodology and design frequencies need to be fully stabilized. Any change in the methodology or *Design Frequency* at this point can cause a significant amount of re-work. The design team should instead focus their effort on all of the paths within one sigma of the predicted WNS path. The design team should treat all of the paths within this range as equal. Effort should be spent fixing not just the worst path within this range but all paths within this range. Each path fixed increases the probability of a higher frequency design. Reducing the automation used at this point will also reduce the tool and design uncertainty helping the team to focus on the critical paths.

9. CONCLUSION

The closer the correlation between the *Predicted Frequency* and the *Actual Frequency* for the paths in the WNS bucket, the higher the *Actual Frequency* will be. Any gap between the two frequencies implies uncertainty in the manufacturing, tool, and design processes. This causes design teams to work on the wrong paths, increasing the effort required to obtain higher frequencies.

Reducing the uncertainty in the analysis of the design also opens the way to more costly local optimization techniques such as dynamic circuitry, datapath placement, and so on. The smaller the uncertainty in the design, the higher the return for these local techniques.

To control design and tool uncertainty take the following steps:

1. List all sources of uncertainty in the design.
2. Develop a plan to reduce the uncertainty as much as possible using the modified slack graph and the process described in Section 8.2.
3. Reduce as much guard banding as possible {remember: Having an *Actual Frequency* higher then the *Predicted Frequency* is still bad}.
4. Use a TNS-based cost model.
5. Tune the *Design Frequency* as described in Section 7.
6. Toward the end of the design, treat all paths within one sigma of the design as equal. Reduce the uncertainty by reducing the level of automation.
7. Push CAD tool vendors to develop algorithms that reduce the design uncertainty in order to free up resources.
8. Finally, view all of the methodologies decisions in terms of their effects on the gap between the *Predicted* and *Actual Frequencies*: Any gap either positive or negative is costing the design frequency.

Categories of Uncertainty	Source of Uncertainty	Details	Type of Uncertainty
In-Die Variance	Transistor geometry variations	Transistor geometry variation across the die causes uncertainty in transistor speed. Particularly important for min-delay analysis.	Process
	Temperature Variations	Hot spots cause parts of the die to run slower than expected effecting max-delay. Cold spots cause min-delay issues.	Tool
	Power Delivery	Voltage drop across power distribution leads to slower device switching.	Design/Tool
Timing Analysis	Slope propagation	Worst Case Slope propagated by timing tools causes pessimistic timing. See example in Figure 8.	Tool
	Cell Based Characterization	Timing tool forced to interpolate tables causing error. Not as accurate as transistor level simulation.	Tool
	Timing Arc Characterization	Multiple switching effects can not be accurately modelled with single input timing arcs.	Tool
	Distributed vs. Lumped RC networks	Timing analysis inaccuracy introduced by using reduced distributed RC networks (or worse yet lumped RC).	Tool
	Miller Effect	Cross-coupling capacitance should be increased for neighboring signal switching.	Tool
Clock Analysis	Edge Skew	Differing RC delays between sampling devices.	Design/Tool
	Edge Jitter	PLL locking shifts between cycles.	Design/Tool
	Point of Divergence	Better than worse case guard banding, but still not as accurate as full analysis.	Process and Design
Extraction	Resistance	Reduction of network causing a large number of small resistors to be dropped.	Tool
	Capacitance Fringe effect	Large signal spacing leads to increased orthogonal routing cross-coupled capacitance.	Tool
	Inductance	Not accounted for in most tools/methodologies.	Tool
Noise	Timing Push Out	Noise on sensitive node can cause setup failures.	Tool

Table 2. Sources of uncertainty (subset).

Chapter 14

Increasing Circuit Performance through Statistical Design Techniques

Michael Orshansky
Department of Electrical Engineering and Computer Science,
University of California at Berkeley

The standard ASIC design methodology is pessimistic in its approach to modeling timing. In order to design high-performance circuits, it is necessary to improve the accuracy of delay modeling while remaining conservative. Intra-chip (i.e. within-chip) parameter variation, which increases as a portion of the overall variation budget, makes accurate timing modeling more difficult. Intra-chip parameter variation is caused by a variety of manufacturing phenomena that will be detailed in this chapter, and affects both the devices and interconnect structures.

Inter-chip parameter variation leads to variation of chip clock frequencies around the average value. In contrast, intra-chip parameter variation forces the average clock period to uniformly increase, degrading the overall circuit speed. Also, due to intra-chip variation, the worst-case timing behavior predicted by the standard ASIC design methodology becomes excessively conservative. This leads to over-design and loss of performance. More sophisticated statistical modeling techniques help to recover much of the lost performance.

A variety of corrective techniques and modeling methodologies that help reduce intra-chip variation of design parameters, or uncertainty about their values, is available. These techniques reduce the intra-chip portion of parameter variability and thus improve average circuit performance. Some of these techniques have been successfully used by the designers of high-end custom chips. Making them part of the standard ASIC design methodology would increase the achievable circuit performance levels.

In this chapter we consider the various sources of process variability and describe ways to deal with their effect in order to improve performance. Section 1 quantifies the rise of the intra-chip component of process variability, considers the various sources of circuit performance variation, and explains why intra-chip variability degrades performance. Section 2 discusses how one can assess the true clock-cycle distribution and how much circuit performance improvement is thereby enabled. Section 3 describes some of the ways to alleviate the impact of process variability on performance of high-speed ICs. Finally, Section 4 presents an example of accounting for impact of gate length variation on circuit performance, including the required characterization effort. Admittedly, this chapter outlines more problems than it solves. As such, it is likely to be of less interest to the ASIC designer anxious for immediate performance relief. On the other hand, we do feel that the problems raised in this chapter will be addressed by tools in time. Thus, this chapter gives a number of interesting directions for high-impact research.

1. PROCESS VARIABILITY AND ITS IMPACT ON TIMING

Among the difficulties brought about by the continued scaling of CMOS technologies is the increasing magnitude of variability of the key parameters affecting performance of integrated circuits [4][34]. The growing extent of variation of these parameters can be attributed to several factors. The first is the rise of multiple systematic sources of parameter variability due to the interaction between the manufacturing processes and the properties of the design. For example, optical proximity effects cause the polysilicon feature size to vary depending on the local layout surroundings, and the inter-layer dielectric thickness varies due to the dependence of chemical-mechanical polishing on the local wire density [10][16][36]. Also, while the nominal target values of the key process parameters, such as effective channel length of the CMOS transistors or the interconnect pitch, are being reduced, our ability to improve the manufacturing tolerances, such as mask fabrication and overlay control, is limited [34].

The most profound reason for the future increase in parametric variability is that the technology is approaching the regime of fundamental randomness in the behavior of silicon structures. In a few years, the shrinking volume of silicon under the channel of the MOS transistor will contain a very small number of dopant atoms. Because placement of dopant atoms is random, the final number of atoms that end up in the channel of each transistor is a random variable. Thus, the threshold voltage of the transistor that is determined by the number of dopant atoms is also going to exhibit a significant variation, leading to variation in the transistors' circuit-level properties, such as delay and power [9][41].

All these reasons lead to the greater relative variability of device parameters around the nominal value (Figure 1).

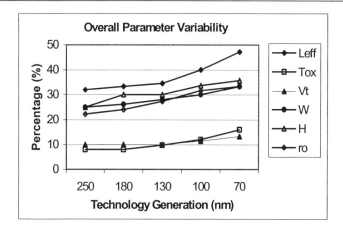

Figure 1. The overall parameter variation increases as a percentage of three standard deviations (3σ) from the nominal. (From S. Nassif [23].)

1.1 The Rise of Intra-Chip Process Variation

The patterns of variability are also changing. It used to be the case that process-caused parameter variability would exclusively lead to differences in the chip-to-chip properties, so that within the chip the variation of the parameters could be safely neglected. This is no longer true [25]. A recent study estimates that for 0.13um CMOS technologies, the percentage of the total variation of the effective MOS channel length that can be attributed to intra-chip variation is 35% [17]. Moreover, the contribution of the within-chip variation component is rapidly growing. By the time of 0.07um CMOS technology, the percentage of total variation of effective channel length L_{eff} that is accounted for by intra-chip variation will grow to almost 60%. Similarly, intra-chip variation of the interconnect geometries is rising in relation to the total variation budget. Variation of wire width W, height H, and thickness T will go from 25% to about 35% (Figure 2).

The increase of intra-chip parameter variation is caused by the emergence of a number of variation-generating mechanisms located on the interface between the design and process. For example, variation of the effective channel length can be largely attributed to the optical proximity effect. As the transistor feature size gets smaller compared to the wavelength of light used to expose the circuit, the light is affected by diffraction from the features located nearby. As a result, the length of the resulting polysilicon line becomes dependent on the local layout surroundings of an individual transistor [45].

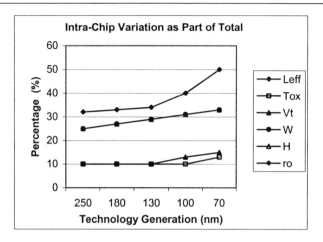

Figure 2. The intra-chip component of variation grows as a percentage of total variability (From S. Nassif [23]).

The other source of large intra-chip parameter variation is the aberrations in the optical system of the stepper used to expose the mask [37]. These aberrations lead to predictable, systematic, spatial variation of the MOS gate length across the chip. For interconnect, an important source of variability is the dependence of the rate of chemical-mechanical polishing (CMP) on the underlying density of interconnect [37]. The most significant problems that may arise when polishing are dishing and erosion, which happen when some areas of the chip are polished faster than others. In dishing, the metal (usually copper) is "dished" out of the lines. Erosion happens when some sections of the inter-level dielectric are polished faster than others.

1.2 Identifying Sources of Circuit Performance Variation

In coming up with a strategy to deal with parameter variation, we need to decide which of the multiple sources and patterns of variation deserve the most attention. This can be done by assessing their impact on a circuit performance metric, for example, path delay. This is necessary because the actual circuit performance variation as a result of a particular underlying process variation component may be either amplified or reduced. The exact variability contribution of a process parameter is determined (a) by the sensitivity of a circuit performance variable to the change in the parameter, and (b) by the magnitude of variation of this parameter. To find a parameter's contribution to variability, we need to consider how it affects speed of some representative circuit: a simple gate or a canonical critical path. We can represent the individual variation contributions to overall delay variability for any given parameter P_i by

Increasing Circuit Performance through Statistical Design Techniques

(1) $$Var(d_{P_i}) = (\frac{\partial d}{\partial P_i})^2 \sigma_{P_i}^2$$

where d is the delay of the delay element (gate or wire segment) or an entire path, and P_i is the process parameter of interest [2][29].

Nassif [23] performs the analysis of delay variation and trends using a circuit consisting of a buffer driving a long minimum-width interconnect line. Using the estimates of process variability from the roadmap and some early experiences with the advanced technologies, he arrives at the following values for the $\pm 3\sigma$ of gate delay variation, expressed in percentages.

Assuming that all these sources of variation are statistically independent, the total variation of gate delay due to each of the above parameters is a sum:

(2) $$V_{tot} = \sum_{i=1}^{n} V_i$$

where V_i is the variance of delay due to the ith variation source. Based on Equation (2) and Table 1, gate delay has a 3σ variation of about 40%. Table 1 suggests that delay performance of the gate is most sensitive to, and affected by, variation of effective channel length of the MOS transistor. However, delay variation due to effective channel length L_{eff} is projected to decrease with technology scaling from $\pm 16\%$ for 0.25um generation to $\pm 12\%$ for 70nm generation [23]. Even though the magnitude of L_{eff} variation is increasing, with scaling the transistor current drive becomes less sensitive to L_{eff} [12]. The other entries in Table 1 are transistor oxide thickness (t_{ox}) and threshold voltage (V_{TH}), and interconnect width (W), spacing (S), thickness (T), height (H), and resistivity (ρ_{wire}).

	Delay Variation (%)								
	Device			Wire					
Generation (um)	L_{eff}	t_{ox}	V_{TH}	W	S	T	H	ρ_{wire}	Total
0.25	32.4	1.4	3.8	13.4	9.4	6.8	7.8	16.0	41.2
0.18	28.4	2.6	5.4	12.0	9.4	7.0	8.0	16.6	38.2
0.13	25.6	3.2	5.6	11.8	10.0	8.0	8.2	18.0	37.0
0.10	24.6	4.0	6.6	11.4	9.4	8.2	8.4	18.4	36.8
0.07	23.8	5.0	7.2	10.6	9.4	8.2	7.2	20.2	37.0

Table 1. Delay variation due to intra-chip parameter variations (From S. Nassif [23]).

Generation	Local Interconnect (um)			Global Interconnect (um)		
(um)	W, S	T	H	W, S	T	H
0.25	0.30	0.50	0.65	2.0	2.5	1.4
0.18	0.23	0.46	0.50	2.0	2.5	1.4
0.13	0.17	0.34	0.36	2.0	2.5	1.4

Table 2. Difference in the values of local and global interconnect lines that has to be taken into account when considering the relative impact of variations due to local and global wires [33].

The relative importance of gate and interconnect variability depends on a number of design choices and assumptions. If we assume that the long wires are laid out in the minimum pitch interconnect layers, the interconnect variability is projected to increase and contribute almost an equal proportion of the overall variability [23]. Different variability decomposition results if global interconnect is routed in the top metal layers, while local interconnect is routed in the lower-level metals [29][32][35]. This scenario can be modeled by a canonical critical path consisting of a gate chain, connected by local interconnect, with length determined by the technological choices, and a single global interconnect line buffered by repeaters. Note the difference in the characteristics of global and local interconnects, Table 2.

The resulting variability decomposition depends on a number of design and technology choices. For example, adopting the "fat" wire approach [32] helps to reduce total delay as well as dramatically reduce the contribution of global interconnect to overall path delay variability. In the evaluation performed for a generic 0.18um CMOS technology, the design that uses "fat" wires for global interconnects is heavily dominated by device-caused delay variability that contributes 88% of variability of the canonical path delay. For the design in which all the metal layers are laid out using the minimum size closely-pitched metal layer (to achieve the highest density), the contribution of device-caused delay variation is still dominant – 65%, with interconnect variability accounting for 35%, Table 3. The results of this analysis indicate that device caused variability is likely to remain the dominant source of path delay variation, because circuit design practices universally used to reduce the delay of long interconnect lines also help in reducing delay variability due to global interconnect [29].

Design	Delay (%)		Variability (%)	
	Gates	Wires	Gates	Wires
Fat wires	59	41	88	12
Min-pitch wires	48	52	65	35

Table 3. Delay and variability decomposition for different designs [29].

1.3 How Intra-Chip Variability Affects Circuit Timing

Process variability and ways to account for it in the course of circuit design have always been a concern for IC designers. In order to predict the spreads of circuit performances (e.g. microprocessor speeds) after manufacturing, or in order to guarantee the robustness of a design under variations, a range of worst-case or Monte-Carlo modeling approaches have been proposed [3][24][31][44]. All these methods made two key assumptions: (1) intra-chip parameter variability within the chip is negligible compared to inter-chip variability, and (2) all types of digital circuits, e.g. all cells and blocks, behave statistically similarly (i.e. in a correlated manner) in response to parameter variations. For a long time these assumptions were valid.

However, as we found in the sections above, the intra-chip component of variation is steadily growing. The second assumption is also not holding up well, as many practicing engineers know. Specifically, different cells have different sensitivities to process variations [24]. This is especially true of cells with different aspect ratios, tapering, and cells designed within multi-threshold and multi-V_{DD} design methodologies. It is also clear that interconnect and gate delays are not sensitive to the same parameters, and with the increasing portion of delay being attributed to interconnect, this will have to be taken into account.

The breakdown of these two assumptions leads to significant intra-chip variation of gate and wire delays. In the presence of large intra-chip delay variation, the standard timing analysis is bound to result in an unreasonable level of conservatism. Indeed, the worst-case static timing analysis proceeds by setting each gate to its worst-case timing value, and performing a longest path computation to arrive at the worst-case critical path delay. The assumption that is implicit in this approach is that delay elements (gates and wires) are perfectly correlated with each other. The failure to consider the validity of the assumptions makes the probability of finding a chip, with characteristics assigned to it by the worst-case timing analysis, very small, leading to lost performance and expensive over-design.

2. INCREASING PERFORMANCE THROUGH PROBABILISTIC TIMING MODELING

It is important to realize that intra-chip variation is fundamentally different in its impact on circuit design and modeling from the previously dominant inter-chip parameter variation [5][6][15][27]. The analysis and design techniques increasingly have to take this fact into account.

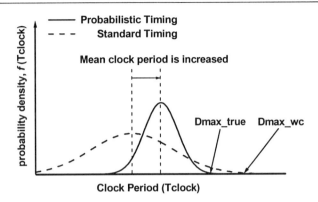

Figure 3. If we perform a more accurate analysis of path delay distribution, we find that clock period (binning) distribution is modified compared to the standard timing approaches. A sizable reduction of conservatism is possible.

Why is intra-chip variability different from inter-chip variation? Consider a high-speed digital chip being manufactured in volume production. Delay optimization techniques aim to move delay off critical paths onto paths with timing slack. As a result, high-performance chips are designed in such a way that there are a large number of paths with delays close to the maximum delay D_{max}. The value of the maximum delay is dependent on the assumed process conditions. Inter-chip parameter variation affects each path in a similar way, slowing down or speeding up each path. In contrast to that, intra-chip variation will make some paths slower, and others faster, depending on their spatial location and composition. Inter-chip variation results in a traditional binning distribution, with the average clock speed given by D_{max}^{typ}, the maximum delay evaluated at the nominal (typical) process conditions. On the other hand, because clock period is always defined by the maximum path delay, intra-chip parameter variation forces the average maximum path delay to increase, degrading the overall circuit speed. At the same time, the variance of the distribution is reduced [29]. As a result, the true (probabilistically assessed) maximum path delay is smaller than the worst-case timing number of the standard timing approach, $D_{max}^{true} \leq D_{max}^{wc}$ (Figure 3).

In the standard ASIC methodology, timing analysis and optimization are done for the worst-case process corner, and the timing number (D_{max}^{wc}) represents the worst possible timing behavior of a circuit: all chips are expected to achieve the delay D_{max}^{wc}, thus maximizing the yield. Being able to exactly predict the shape of the clock speed distribution would allow significant reduction of the timing conservatism of the standard timing methodology, improving circuit performance and reducing over-design.

2.1 How to Approximate the True Clock-Cycle Distribution

In this section, we present a simple framework for bounding the true distribution of the maximum achievable clock speed through probabilistic delay analysis [29]. We then use the resulting bounds to estimate the possible reduction of conservatism of the standard timing methodologies.

The clock cycle of a chip is constrained by the maximum path delay (plus set-up time, which we will leave out of the probabilistic analysis for the sake of simplicity):

(3) $\max\{D_1...D_N\} \leq T_{clock}$

where D_i is the delay of the ith path in the circuit. In our probabilistic framework, path delays are random variables. The statistical properties of chip's timing are specified by the distribution of $\max\{D_1...D_N\}$, whose cumulative probability function is given by the integral:

(4) $F(T_{clk}) = \int_{-\infty}^{T_{clk}} (N-1) \int_{-\infty}^{T_{clk}} f(D_1, D_2,...D_N) dD_1 dD_2 .. dD_N$.

where $f(D_1, D_2,...D_N)$ is the joint probability density of $\{D_1...D_N\}$. In general, the exact analytical solution of this equation is not available. However, under certain assumptions we can establish some useful bounds on the shape of $\max\{D_1...D_N\}$ distribution, without solving Equation (4).

Let D_o be the maximum designed-for delay in the absence of variation. We assume that there are N paths with delays close to D_o, and that path delays are Gaussian random variables whose joint distribution is fully characterized by the pair-wise covariance terms between path delays. The pair-wise covariance for paths D_i and D_j is given by

(5) $Cov\{D_i, D_j\} = \sum_{k_i=1}^{m_i} \sum_{k_j=1}^{m_j} Cov\{d_g(i,k_i), d_g(j,k_j)\}$ (3)

where m_i is the number of gates along the path i, and $d_g(i,k_i)$ is the delay of gate k_i of path i.

Let the delay of a gate be given by an arbitrary function of a vector of process parameters P: $d_g = f(P)$. In order to establish an expression for the pair-wise covariance of gate delays, we assume the linearity of delay response to the localized variation of process parameters. Then,

(6) $d_g = d_g(P_o) + \varphi(P_o)^T P$

Here $\varphi(\cdot)$ is Jacobian of the delay function: for example, $\varphi(\cdot) = (\partial d_g/\partial L, \partial d_g/\partial V_{th})$. If $l = \dim\{P\}$, the covariance of two gate delays i and j is given by:

(7) $$Cov\{d_g(i), d_g(j)\} = \sum_{t_i=1}^{l} \sum_{t_j=1}^{l} \varphi_{t_i} \varphi_{t_j} Cov\{P_{t_i}, P_{t_j}\}$$

Equations (3) to (7) are completely general with respect to the particular nature of process variability and the types of delay elements we want to include in the analysis. These equations allow finding the joint probability density function of the random delay vector reflecting any design or layout dependence, for example, gate's spatial location, distance to other gates, orientation, inter-gate spacing, and so on. Wire delays as well as gate delays can be naturally accommodated into this analytical framework.

Simple bounds can be established on the distribution of $\max\{D_1...D_N\}$ using the path variances of Equation (3), and assuming constant mean path delays of D_o. The bounds are most accurate in the range of $10 \le N < 1000$. The bounds are conservative and therefore may be used in place of timing numbers provided by the standard timing analysis tools. It is possible to show that the value of $T_{clk} = \max\{D_1...D_N\}$ at the kth percentile of the distribution is bounded by [29]:

(8) $$T_{clk}^k \le E\max\{D_1...D_N\} + z_k \sigma_D$$

where $E\max\{D_1...D_N\}$ is the expected value of $\max\{D_1...D_N\}$ and z_k is the value of the standard normal at the kth percentile of the normal distribution. For path delays with constant variance, the expected value of $E\max\{D_1...D_N\}$ can be given by an especially simple expression:

(9) $$E\max\{D_1...D_N\} = D_o + \eta \sigma_D$$

In this equation, η is the deviation factor that can be calculated from $\Phi(\eta) = (2N-1)/2N$, where $\Phi(\cdot)$ is the Laplace function [28]. In the next section, we use Equations (8) and (9) to assess the reduction of conservatism permitted by the probabilistic analysis when compared to the standard timing methods.

2.2 Increasing Chip Performance through Reduction of Timing Conservatism

The traditional worst-case timing analysis is inherently non-probabilistic. This fact makes it impossible to quantify the likelihood that its timing predictions will appear in the actual silicon implementation of a circuit. In contrast to that, the approach outlined above is entirely probabilistic, seeking to construct the probability distribution of an achievable clock period for a given circuit. The likelihood of a 'worst-case' scenario is thus quantified, enabling the avoidance of expensive redesigns.

Note that conservatism of the traditional timing tools is more disadvantageous for ASICs than for custom circuits. Because ASICs are not typically tested at full speed, they cannot afford the risk of having some chips suffer from parametric yield problems, which can only be identified with a full speed test. As a result, they are forced to live with the existing pessimism of worst-case timing analysis. It is certain that much of the time and effort associated with achieving timing closure could be obviated if better process-related timing information were migrated up-stream in the design process.

By implementing a probabilistic timing analysis methodology, the conservatism built into the standard ASIC design methodology can be reduced. Once the true distribution of the maximum path delay is found (Figure 3), we can use a less conservative D_{max}^{true} in place of an overly conservative D_{max}^{wc} provided by the standard timing tools. Contrary to the worst-case approach, however, the probabilistic estimate of D_{max}^{true} is determined by the level of parametric yield one wishes to achieve. Requiring 99.99% yield forces the chip to be significantly slower than possible at the acceptable 98% yield.

We can use Equations (8) and (9) to evaluate the reduction of conservatism permitted by the probabilistic timing analysis. To assess the worst-case conservatism, let us assume that parameter variability is entirely within the chip, e.g. that the inter-chip variation component is negligible. Also, to simplify the analysis, we consider a set of paths that have no shared gates. Then, joint probability density of path delays is described by a multivariate normal distribution with the diagonal covariance matrix. Let us assume that all paths have the same typical delay and the same variance, σ_D^2. Let $m = 25$, $N = 100$, and $3\sigma_d / d_o = 30\%$. Then, at 98% parametric yield, the analysis above shows that we can reduce conservatism by at least 17%. More accurate probabilistic timing algorithms that are currently under development will allow bigger reductions in conservatism, compared to the simple loose bounds described in this section. Also, as the relative magnitude of variation of delay grows (see Section 1.2), probabilistic timing analysis will become even more valuable as a tool to reduce timing conservatism and increase chip performance.

2.3 Trading Parametric Yield for Performance

Having an accurate prediction of the speed binning distribution is also critical for a different reason. Apart from reducing conservatism of the standard timing methodology, another way to improve circuit performance of ASICs in comparison to custom chips is to allow them to trade parametric yield for performance.

It has been noted that an ASIC chip produced in the foundry may run up to 40% faster than predicted by standard timing analysis [13]. Figure 3 shows that there is nothing surprising about this fact. Assuming that gate delay variation due to process parameter variations is $\pm 20\%$, half of all the chips will run at least 20% faster than predicted. And, about 2% of all the chips would run 35% faster.

This performance difference is quite significant. One way to recover it is to accurately predict the binning distribution, rather than a single worst-case timing number. Then, one could implement a full speed testing procedure and enable testing ASICs at a specifiable speed, which would be chosen to satisfy some set of yield performance constraints. The complexity and overhead of doing full speed testing hampers the adoption of this procedure by the ASIC suppliers. ASIC vendors will trade yield for performance if revenue from faster chips will justify the additional expense in lost yield and testing overhead. The adoption of this strategy is further hindered by the lack of tools for predicting yield-performance trade off, needed for a better understanding of the results of mass chip production [C. Hamlin, personal communication].

3. INCREASING PERFORMANCE THROUGH DESIGN FOR MANUFACTURABILITY TECHNIQUES

The sections above indicate that the increase in intra-chip parameter variability will degrade the achievable maximum clock frequency. In this section we consider the strategies for reducing the impact of intra-chip parameter variation. These strategies fall into two major categories. The first is to improve the process quality to eliminate intra-chip variation. The second is to change the circuit analysis and design methodologies to account and, possibly, compensate for intra-chip process variability at the circuit level.

Improving intra-chip uniformity or reducing uncertainty about parameter values can mitigate the detrimental impact of variability on circuit performance. Most of the techniques discussed in this section have been applied in design of high-end custom chips. By making them a more accepted part of the standard ASIC flow, a sizable performance boost can be made available to ASIC designers.

3.1 Improving Uniformity through Process Measures

Obviating a problem is better than solving the problem after it arises. A number of measures have recently been offered to prevent or minimize intra-chip variation of certain parameters. These measures work by identifying the physical mechanism responsible for a particular pattern of variation, and then compensating for it.

Among the most successful and widely adopted measures is the use of optical proximity correction (OPC) schemes [8][42]. OPC may cover a wide range of reticle enhancement techniques, in which additional geometrical structures are added to the mask to correct for critical dimension (L) and resolution variation. For example, the dependence of the width of the printed polysilicon line on the local layout surrounding is corrected by increasing (or decreasing) the ideal-case dimension of a feature on the mask, so that it is properly printed on the wafer. Similarly, additional structures are added onto the mask to compensate for the corner-rounding and line pull-back. OPC has proven itself as an efficient way of reducing intra-chip L variation and increasing yield, and has been successfully used in many industrial designs. It is now routinely offered through the foundries, which perform the corrections in the mask-making step.

The other powerful method of ensuring better process uniformity in the era of sub-wavelength photolithography is the use of phase-shift masking (PSM) technologies [19][20]. Phase shifting is based on optical interference to improve depth-of-field and resolution in lithography. The problem is that it is extremely difficult to reliably produce on the wafer geometries, whose feature sizes are comparable or smaller than the wavelength of light used to print them. This leads to lower yield and higher variation of the final geometries. PSM technology utilizes phase-shifters to improve the sharpness of the exposed image by canceling out light interference caused by the nearby structures (Figure 4). Phase shifting makes it possible to consistently resolve features one half the size of the exposing wavelength. It also improves the uniformity and, by enabling a reduction of the feature size, gives a significant performance benefit: By shrinking gate-length by 0.10um, users can expect a 100MHz speedup, on average [26].

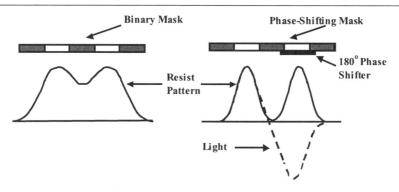

Figure 4. Phase shifting masks are a powerful way to improve manufacturability of nanometer silicon circuits [26].

Currently, a significant effort is under way to provide cell libraries that are OPC- and PSM-compliant. Having access to such libraries will allow ASIC designers to exploit the benefits offered by OPC and PSM, sometimes combined under the term resolution enhancement techniques (RET). Implementing RET within the standard cell flow is challenging because interactions between adjacent cells have to be taken into account. Therefore, the optimal point of inserting RET is at the physical verification stage, where standard cell placements are finalized. Cell library vendors are now coming up with RET-compliant cells, so that designers have an opportunity to make optimal decisions without committing to RET too early [22].

In Section 1.2, we observed that the rate of CMP is dependent on the underlying density of interconnect structures. We can increase the uniformity of wires by inserting dummy features in the regions of lesser density. Intentional filing of dummy features improves process uniformity by increasing the uniformity of the chemical mechanical polishing (CMP) [14][18]. The downside is the increased coupling capacitance, and, the corresponding, delay and signal integrity dangers.

Spatially correlated variation is especially troublesome, because it increases path delay variance more than un-correlated intra-chip variation [27]. Lens aberrations, which become more pronounced as the optical photolithography is pushed to its limits, lead to systematic spatially correlated variation of L across the chip. One approach to dealing with this pattern of variation is through a mask-level spatial correction algorithm performed in conjunction with OPC. After a comprehensive characterization, a 2-D correction profile would be applied to the mask, compensating for the observed spatial L dependency to achieve superior uniformity. The existing commercial OPC software can be enhanced to enable spatially dependent correction [27].

3.2 Systematic Circuit Modeling Approaches for Intra-Chip Variability

The existence of substantial intra-chip parameter variability requires new circuit analysis and modeling approaches to ensure that the circuit will operate properly when manufactured. The choice of strategy depends on whether the specific variation component is systematic or random. A particular pattern variation is considered to be systematic if we know its generating mechanism and can systematically predict its properties. For example, the dependence of ILD thickness on the material and layout properties can be described by an analytical expression [40]. Similarly, the spatial variation of L due to the lens aberration can be systematically described by modeling the mean component of variation through an empirical equation [27]. The patterns of variation, which are too complex for deterministic modeling, or which are truly random, can at best be described by statistical means, e.g. through probability density functions [29]. However, using a probabilistic description, instead of a systematic one, leads to a sub-optimal, more conservative, result. In order to reduce the large conservatism built into our modeling approaches, we should strive for a systematic description of intra-chip parameter variation whenever possible.

Implementing a systematic modeling methodology is possible for some patterns of variation, and appears to be difficult for others. For example, Stine [38] establishes a simulation framework for modeling the impact of systematic pattern-dependent variation of L to assess the impact of this variation on the performance of an SRAM circuit. The methodology involves performing an optical aerial simulation to determine the impact of proximity on the final silicon, and then using layout extraction and circuit simulation to assess the impact on circuit performance. For a particular SRAM example, the authors find that including the modeling of pattern-dependent L variation results in an approximately 10% skew in the write access time.

Systematic interconnect variation due to CMP is studied by Mehrotra [21] and Stine [39]. The concern of both authors is the impact of systematic ILD thickness variation on the skew of the global clock network. Specifically, Mehrotra describes a methodology to incorporate modeling of the pattern-dependent interconnect variation into a standard CAD flow through the sensitivity-based analysis. The authors find that the performance benefit obtained by improving process uniformity (adding metal fill) is less than that obtained by better modeling. The systematic modeling allows reducing conservatism by the amount sufficient to insert an additional level of logic on some paths [21].

So far we have only been concerned with systematic intra-chip variation of physical process parameters. We also need to consider the impact of systematic variation of environmental factors, e.g. V_{DD} and temperature. The impact of variation of supply voltage across the chip is now routinely assessed through tools performing IR-drop analysis. In deep sub-micron designs, temperature variation becomes an increasingly important issue. The increase in power dissipation and the current densities leads to an increase in chip temperature. Thermal gradients are due to different activities and sleep modes of the functional blocks; gradients of up to 40°C have been reported in a high-performance design [43]. Certain low power design techniques, e.g. dynamic power management and clock gating, also may lead thermal gradients. One study finds that large circuit performance improvements are possible by accounting for the systematic across-the-chip thermal gradient. Specifically, thermally aware buffer insertion may result in 5-10% speed improvement [1].

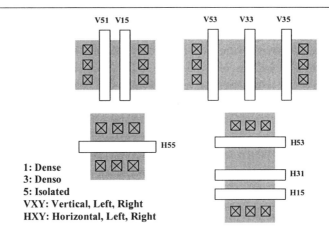

Figure 5. Multiple geometric patterns must be considered separately: vertical vs. horizontal gates, left vs. right nearest neighbor, different spacing: dense, denso, and isolated.

It is likely that tools will soon address many of the effects considered in this section. That will allow ASIC designers to fully exploit the potential of nanoscale silicon technologies.

4. ACCOUNTING FOR IMPACT OF GATE LENGTH VARIATION ON CIRCUIT PERFORMANCE: A CASE STUDY

As we observed in Sections 1.1 and 1.2, the parameter that leads to the greatest variation in circuit properties is the effective channel length of the MOSFET transistor. Intra-chip L variation is spatially correlated, which results in a particularly strong degradation of clock speed. The patterns of variation are complex and require careful characterization methodology. Below we consider a case study of the characterization and simulation experiment to completely analyze the patterns of L variation in an advanced 0.18um logic process technology and predict the impact of this variation on circuit performance [27].

This section is intended to give some practical examples of the methods and models discussed in the chapter. We consider characterization needs and the impact of L variation on circuit behavior, and estimate performance improvements possible through systematic modeling. We then use the probabilistic timing framework of Section 2.1 to predict the magnitude of average clock speed degradation due to L variation if it is not addressed through systematic means.

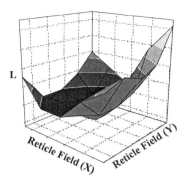

Figure 6. Aberrations of the stepper machine lenses lead to the systematic spatial variation of L across the reticle field and the chip.

4.1 Characterization and Analysis of Intra-Chip Parameter Variability

The previously used statistical modeling approaches to circuit analysis and design assumed that inter-chip variation component is dominant [3][24][31][44]. Therefore, the characterization effort required collecting a statistically significant population of process parameter measurements, with a single measurement of a parameter per chip. Over the years, multiple techniques to speed up such data collection have been developed, including a fully automated device parameter extraction from the electrical measurements [11]. If we want to account for intra-chip parameter variation as much as possible, a much more elaborate characterization procedure is required.

Characterization has to address the possible interaction between the global lens aberration and the local layout pattern-dependent non-uniformities due to the optical proximity effect. Toward this end, all the gates were classified into 18 categories depending on their orientation in the layout (vertical or horizontal) and the spacing to the nearest neighboring gate (Figure 5). To capture a particular lens aberration, the coma effect, the relative position of the surrounding gates, i.e. the neighbor being on the left vs. right, was also distinguished. In order to characterize the spatial L profile, a test-chip was used which contained a 5×5 grid of test-modules located across the area of the reticle field. Each module contained long and narrow polysilicon resistors, with a variety of distances to adjacent polysilicon lines.

The spatial L maps were measured separately for each gate category (Figure 6). The range of systematic spatial variation of L is 8-12% depending on the category. It was also found that the L maps for different categories have quite distinct spatial behaviors, due to the interaction between the global lens aberrations and the pattern-dependent optical proximity effect.

The most direct impact of spatial L variation is the resulting variation of CMOS gate delay: ring oscillator speed variation across the chip is 14.5%. Also, circuit paths with identical designed-for delays will, in reality, have considerably different delays, depending on the physical location of the path within the chip. Timing analysis of a benchmark combinational circuit from ISCAS'85 [7] was performed for 9 spatial locations on the reticle field in a uniform 3x3 grid. This showed that for the chip located in the lower-right quadrant, the delay of the same path placed at different corners (fast and slow) of the chip varied by 16%.

The path delay distribution has also broadened by almost 50%, with some slower and some faster paths (Figure 7). Importantly, the order of critical paths also changes depending on the location of the combinational block within the chip: only 6 out of top 20 paths found in the "fast" corner of the chip are also found in the top 20 paths of the "slow" corner. This regrouping significantly complicates the use of pre-designed and pre-characterized circuit blocks physically localized within the chip, such as hard IP blocks, since their pre-characterized behavior will not adequately correspond to the location-dependent L.

Systematic across-chip L variation also affects the global circuit properties such as clock skew in clock distribution networks containing buffers for driving and restoring the signal. Control of clock skew is critical, since in determining a conservative clock cycle time, a percentage delay due to clock skew is additive to the set-up times and hold times of the circuitry. The clock skew due to the systematic L variation only is 8% of the total clock cycle (Figure 7), which is significant.

Because spatial L variation is largely systematic, it is most effectively dealt with through design for manufacturability techniques, such as discussed in Section 3. These techniques will help improve performance by accounting for the spatial dependence of gate and interconnect delay that we outlined above.

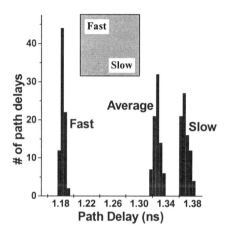
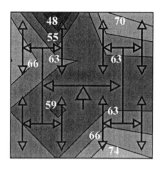

Figure 7. Distribution of critical paths of a combinational logic cell changes significantly if placed at different locations within chip. Global skew map for H-tree clock: max skew is 8% of clock cycle.

4.2 Impact of Gate Length Variation on Clock Speed

We now utilize the detailed process characterization data to evaluate the impact of channel length variation on the clock speed distribution. For that we use the probabilistic timing framework of Section 2.1. The results then indicate the amount of degradation we can expect from L variation if it is not addressed through systematic corrective means.

After relating gate delay to the value of channel length through a compact gate delay model, we propagate the statistical information about L variation to the level of gate and path delays. Following [27], we use a statistical model of intra-chip L variation that decomposes overall variation into three distinct components: proximity-dependent, spatial, and random residual. We simplify the analysis by making the following assumptions: (i) the spatial correlation is significant in the short range; and, (ii) proximity-dependent and residual variations are spatially uncorrelated. Then, the variance of the full path delay is [27]:

$$(10) \quad \sigma_D = \frac{1.5 D_o}{L_o} \left(\frac{\sigma_{Lprox}^2 + 0.28\sigma^2}{m} + \sigma_{Lspat}^2 \right)^{1/2}$$

where m is the number of gates in the path, D_o is the nominal path delay, L is the nominal L value, and σ_{Lprox}, σ_{Lspat}, and σ are the standard deviations of the proximity-dependent, spatial, and residual components of variation correspondingly. We use the empirically determined parameter values to evaluate Equation (10). Then Equation (9) is used to predict the impact of L variation on clock speed.

The model predicts an up to 21% degradation of the average circuit speed as a result of intra-chip L variation. Speed degradation is worse for more complex chips, since they contain more critical paths (larger N), and for shorter paths (smaller m). Thus, the degradation is likely to be worse in the future. The model also suggests that spatially correlated intra-chip variation has a much stronger effect on degradation of circuit speed than the uncorrelated proximity-dependent and random variation. This is because the averaging of the lengths, L, of the gate stages within the path reduces the delay variation due to the proximity effect.

The result implies that from the perspective of improving circuit speed, much more attention should be paid to improving the spatial intra-chip uniformity. As we discussed in Section 3.1, spatial L non-uniformity is most effectively addressed through mask-level corrective techniques, such as spatially aware optical proximity correction [27].

5. CONCLUSION

In this chapter we considered the impact of process variability on circuit performance, and reviewed in detail a number of effects and methods that ASIC designers can use to better understand and design chips. Intra-chip variation of process parameters is increasing as a portion of the overall variation, as well as in absolute terms. Intra-chip variation significantly degrades the distribution of the achievable clock speed. At the same time, it makes the timing estimates provided by the standard design methodology overly conservative. ASIC circuits suffer more from this effect than do custom circuits. This is because the lack of at-speed testing for ASICs does not permit even the least parametric yield loss. As a result, ASIC designers are forced to use extremely conservative lumped worst-case models that assume a simultaneous set of all the worst-case parameters, both physical (e.g. process-related) and environmental (temperature, V_{DD}). Fundamentally, as long as ASIC designers cannot tolerate any parametric yield loss, they will have to live with the fact that the same design may have run, on average, 20-30% faster, and sometimes, 40-50% faster.

Still, ASIC designers may win back some performance through improved delay modeling using techniques indicated in this chapter. While these techniques are not currently commercially available, it seems likely that statistical information will eventually be integrated into commercial timing tools and made available to ASIC designers.

6. REFERENCES

[1] Ajami, A., Banerjee, K., and Pedram, M., "Analysis of Substrate Thermal Gradient Effects on Optimal Buffer Insertion," *Proc. of IEEE/ACM International Conference on Computer Aided Design*, p. 44, 2001.
[2] Bernstein, K., et al., *High Speed CMOS Design Style*, Kluwer Academic Publishing, 1998.
[3] Bolt, M., Rocchi, M., and Angel, J., "Realistic Statistical Worst-Case Simulations of VLSI Circuits," *Trans. on Semicon. Manufacturing*, n.3, pp.193-198, 1991.
[4] Boning, D., and Nassif, S., "Models of Process Variations in Device and Interconnect," in *Design of High-Performance Microprocessor Circuits*, A. Chandrakasan (ed.), 2000.
[5] Bowman, K., and Meindl, J., "Impact of within-die parameter fluctuations on the future maximum clock frequency distribution," Proc. of CICC, 2001.
[6] Bowman, K., Duvall, S., and Meindl, J., "Impact of within-die and die-to-die parameter fluctuations on maximum clock frequency," *Proc. of ISSCC*, 2001.
[7] Brglez, F., and Fujiwara, H., "A Neutral Netlist of 10 Combinational Benchmark Circuits and a Target Translator in Fortran," distributed on a tape to participants of the Special Session on ATPG and Fault Simulation, International Symposium on Circuits and Systems, June 1985.
[8] Burggraaf, P., "Optical lithography to 2000 and beyond," *Solid State Technology*, Feb. 1999
[9] Burnett, D., et al., "Implications of Fundamental Threshold Voltage Variations for High-Density SRAM and Logic circuits," *Symp. VLSI Tech.*, pp. 15-16, Jun. 1994
[10] Chang, E., et al., "Using a Statistical Metrology Framework to Identify Systematic and Random Sources of Die- and Wafer-level ILD Thickness Variation in CMP Processes," *Proc. of IEDM*, 1995.
[11] Chen, J., et al., "E-T Based Statistical Modeling and Compact Statistical Circuit Simulation Methodologies," *Proc. of IEDM*, pp. 635-638, 1996.
[12] Chen, K., et al., "Accurate Models for CMOS Scaling in Deep Submicron Regime," *1997 International Conference on Simulation of Semiconductor Processes and Devices (SISPAD '97)*, Boston, September 1997.
[13] Chinnery, D., and Keutzer, K., "Closing the Gap Between ASICs and Custom," *Proc. of DAC*, 2000.
[14] Divecha, R. R., et al., "Effect of Fine-line Density and Pitch on Interconnect ILD Thickness Variation in Oxide CMP Process," *Proc. CMP-MIC*, 1998.
[15] Eisele, M., et al. "The impact of intra-die device parameter variations on path delays and on the design for yield of low voltage digital circuits," Proc. of ISLPED, pp. 237-242, 1996.
[16] Fitzgerald, D., "Analysis of polysilicon critical dimension variation for submicron CMOS processes," M.S. thesis, Dept. Elect. Eng. Comp. Sci., Mass. Inst. Technol., Cambridge, June 1994.
[17] Kahng, A., and Pati, Y., "Subwavelength optical lithography: challenges and impact on physical design," *Proceedings of ISPD*, p.112, 1999.
[18] Kahng, A. B., et al., "Filling Algorithms and Analyses for Layout Density Control," *IEEE Trans. Computer-Aided Design* 18(4) (1999), pp. 445-462.
[19] Lenevson, M.D., Viswanathan, N. S., and Simpson, R. A. "Improving resolution in photolithography with a phase-shifting mask," *IEEE Trans. On Electron Devices,* 29 (11) :1828-1836, 1982.
[20] Liu, H., et al., "The application of alternating phase-shifting masks to 140nm gate patterning: line width control improverments and design optimization," *Proc. of SPIE 17th annual BACUS Symposium on Photomask Technologies*, volume 3236 of SPIE, 1998.
[21] Mehrotra, V., et al. "Modeling the effects of manufacturing variation on high-speed microprocessor interconnect performance," *IEDM Technical Digest*, p.767, 1998.

[22] Mentor Graphics, Subwavelength-Compliant Cell Generation Flow, 2001.
http://www.mentor.com/press_releases/feb01/prolific_calibre_pr.html
[23] Nassif, S., "Delay Variability: Sources, Impact and Trends," Proc. of International Solid-State Circuits Conference, 2000.
[24] Nassif, S., "Statistical worst-case analysis for integrated circuits," Statistical Approaches to VLSI, Elsevier Science, 1994.
[25] Nassif, S., "Within-chip variability analysis," *IEDM Technical Digest*, p.283, 1998.
[26] Numerical Technologies, *"Phase shifting technology,"* 2002.
http://www.numeritech.com/nttechnology
[27] Orshansky, M., et al., "Characterization of spatial CD variability, spatial mask-level correction, and improvement of circuit performance," *Proceedings of the International Society for Optical Engineering*, vol.4000, pt.1-2, 2000, 602-611.
[28] Orshansky, M., et al., "Impact of Systematic Spatial Intra-Chip Gate Length Variability on Performance of High-Speed Digital Circuits," *Proc. of IEEE/ACM International Conference on Computer Aided Design*, pp. 62-27, San Jose, CA, 2000.
[29] Orshansky, M., Spanos, C., and Hu, C., "Circuit Performance Variability Decomposition," Proc. of 4th IEEE International Workshop on Statistical Metrology for VLSI Design and Fabrication, p.10-13, Kyoto, Japan, 1999.
[30] Orshansky, M., and Keutzer, K., "A Probabilistic Framework for Worst Case Timing Analysis," *Proc. of DAC,* 2002.
[31] Power, J. et al., "An Approach for Relating Model Parameter Variabilities to Process Fluctuations," *Proc. ICTMS*, 1993, p. 63.
[32] Sai-Halasz, G., "Performance trends in high-end processors," *Proc. IEEE*, 1995.
[33] Semiconductor Industry Association, *International Roadmap for Semiconductors*, 1999.
[34] Semiconductor Industry Association, *International Technology Roadmap for Semiconductors*, 2001.
[35] Sylvester, D., Keutzer, K., "Getting to the bottom of deep-submicron," *Proceedings of the International Conference on Computer-Aided Design*, 1998.
[36] Stine, B., et al., "Inter- and intra-die polysilicon critical dimension variation," *Microelectronic Manufacturing Yield, Reliability, and Failure Analysis II, SPIE 1996 Symp. Microelectronic Manufacturing,* Oct. 1996, Austin, TX.
[37] Stine, B., Boning, D. S., and Chung, J. E., "Analysis and decomposition of spatial variation in integrated circuit processes and devices," *IEEE Trans. On Semiconductor Manufacturing*, No.1, pp. 24-41, Feb. 1997
[38] Stine, B., et al., "Simulating the Impact of Pattern-Dependent Poly-CD Variation on Circuit Performance," *IEEE Trans. on Semiconductor Manufacturing*, Vol. 11, No. 4, November 1998.
[39] Stine, B., et al., "A methodology For Assessing the Impact of Spatial/Pattern-Dependent Interconnect Variation on Circuit Performance," *Proc. of IEDM*, p. 133, 1997.
[40] Stine, B., et al., "A Closed-Form Analytic Model for ILD Thickness Variation in CMP Processes," *CMP-MIC*, p. 266, 1997.
[41] Takeuchi, K., Tatsumi, T., and Furukawa A., "Channel Engineering for the Reduction of Random-Dopant-Placement-Induced Threshold Voltage Fluctuations," *IEDM Tech. Dig.,* Dec. 1997
[42] Van den hove, L., et al., "Optical lithography techniques for 0.25um and below: CD control issues," *Int. Symposium on VLSI Technology*, 1995.
[43] Wu, Q., Qiu, Q., and Pedram, M., "Dynamic power management of complex systems using generalized stochastic Petri nets," *Proc. DAC.*, pp. 352-356, June 2000.
[44] Yang, P., et al., "An integrated and efficient approach for MOS VLSI statistical circuit design," *IEEE Trans. on CAD*, No 1, Jan. 1986.
[45] Yu, C., et al., "Use of short-loop electrical measurements for yield improvement," *IEEE Trans. on Semiconductor Manufacturing*, vol. 8, no. 2, May 1995.

Chapter 15

Achieving 550MHz in a Standard Cell ASIC Methodology
The Texas Instruments SP4140 Disk Drive Read Channel

David Chinnery, Borivoje Nikolić, Kurt Keutzer
Department of Electrical Engineering and Computer Sciences,
University of California at Berkeley

1. INTRODUCTION

Faster custom speeds are achieved by a combination of factors that can be applied to ASICs: good architecture with well-balanced pipelines; compact logic design; timing overhead minimization; careful floorplanning, partitioning and placement. Closing the speed gap requires improving these same factors in ASICs, as far as possible. In this chapter we examine a practical example of how these factors may be improved in ASICs. In particular we show how techniques commonly found in custom design were applied to design a high-speed 550MHz disk drive read channel in an ASIC design flow.

To show how the gap between ASIC and custom can be closed, we examine an ASIC disk drive read channel chip that achieves a clock frequency of 550MHz in 0.21um CMOS [21]. The speed of 550MHz is comparable to custom design speeds. This example illustrates the importance of some of the principles outlined in the earlier chapters, such as partitioning, duplication of logic to achicvc higher throughput, and reduction of the timing overhead.

The key opportunities for closing the gap between ASIC and custom form the organizing principle of this chapter. Specifically, in Section 2 we overview the design example, and examine its microarchitecture in Section 3. In Section 4 we discuss the registers used to reduce the register overhead, and Section 5 discusses the clock tree distribution to reduce the clock skew. Section 6 compares area and performance of synthesized ASIC logic versus custom logic. Section 7 briefly mentions how the uncertainty in operating conditions was reduced. Finally, Section 8 reflects our conclusions.

While process can account for a large difference between ASIC and custom designs, it was not possible to reduce the impact of process variation in this chip. Disk drive read channels sell for a small price, which cannot accommodate the costs of additional lower yield at higher clock frequency, or testing for speed-binning.

2. A DESIGN BRIDGING THE SPEED GAP BETWEEN ASIC AND CUSTOM

Disk drive read channels are a high-speed signal processing application. Data rates in current high-performance commercial products are in the range of 500-1200Mb/s [1][10][16] and demand increasingly high speeds. For example, Marvell's 88C5500 has a throughput of 1.2Gbit/s with 0.18um technology and 3.3V supply [10].

We examine a competitive ASIC disk drive read channel design, the Texas Instruments SP4140 in 0.21um CMOS (0.18um effective channel length) with 1.9V supply [21]. It is based on EPR4 equalization [20]. The Viterbi detector operates at 550MHz, and the rest of the digital logic operates at 275MHz. The disk drive read channel dissipates at most 1.7W at full speed, and has 525Mb/s user data rate. This speed is comparable to custom-designed read channels in similar technology [12].

Good synthesis results require a rich library with sufficiently many drive strengths, as discussed in Section 3.5 of Chapter 4. This ASIC used TI's standard cell library, with 3 to 4 gate sizes for standard cells and 5 buffer/inverter sizes. To reduce timing overhead, some custom designed cells were characterized for an ASIC flow and used in the SP4140, (e.g. the memory elements on the critical paths, such as the SAFF), and their drive strength was matched to their typical load.

The SP4140 was an entirely new design taken from application concept, including new algorithms and architecture, to circuit realization, in a new process, in nine months. We concentrate on techniques to speed up bottlenecks (the Viterbi detector and adaptive equalizer in the read path) in the chip's digital portion.

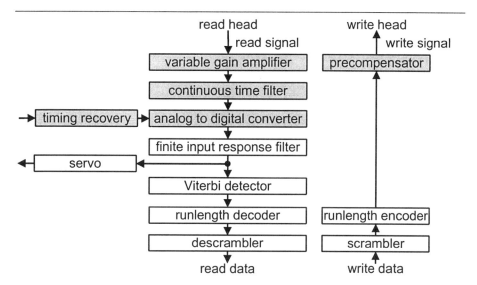

Figure 1. A block diagram for the functional blocks in a typical disk drive read channel. The shaded functional blocks are analog, and the others are digital.

2.1 Read and Write Data

The diagram in Figure 1 represents most of today's read channels. The write data is scrambled, runlength encoded, pre-compensated for magnetic channel nonlinearities, and fed to the write head. The read signal is read by the read head, preamplified, and processed by the read channel.

On the read side, the signal from the preamplifier is conditioned by the variable-gain amplifier (VGA) and continuous-time (CT) filter, before analog-to-digital conversion (ADC). Besides anti-aliasing, the continuous time filter partially equalizes the data. After the ADC, the data is processed digitally. The key blocks are the digital adaptive equalizer, the Viterbi detector, the runlength decoder and the descrambler. Then the data is converted from serial bit data to byte data, and can then be processed at a lower speed. The timing recovery and servo blocks use equalized and detected data.

2.2 Digital Portion Speed Bottlenecks

Due to increasing storage densities, to limit noise enhancement read channels use partial-response equalization, which is often done by finite impulse response (FIR) filtering. Viterbi detection resolves the remaining inter-symbol interference. The FIR filter performs partial-response equalization, with least-mean squares (LMS) algorithm adapted taps. The FIR filter critical path has a slow multiply-accumulate operation, which is not recursive, so pipelining and parallelization can achieve the desired throughput, at the expense of increased area. Whereas, the single-cycle dependency of the Viterbi algorithm prevents pipelining, and reducing the timing overhead is the only way to increase speed.

3. MICROARCHITECTURE: PIPELINING AND LOGIC DESIGN

Frequently, microarchitectural transformations reduce the critical path in signal processing datapaths. Different transformations are applicable to structures with and without cycle dependency (e.g. FIR filter and Viterbi detector respectively) [11].

Microarchitectural exploration is much easier using an HDL description, making ASIC design iteration an order of magnitude faster. Custom layouts have to be redone by hand to explore alternative structures [5], whereas, high-level HDL can be quickly rewritten and then ASIC tools produce the corresponding layout for evaluation. As discussed in Chapters 1 and 2, microarchitecture offers the greatest potential for speed improvement, and we will dedicate most of this chapter to detailing the microarchitectural improvements that netted the principal speed gains.

3.1 FIR filter

Several transformations can speed up the multiply-accumulate operation in the FIR filter critical path. LMS update of coefficients adds feedback recursion to the FIR implementation, but employing delayed or semi-static LMS allows the critical path to be pipelined (LMS coefficients are updated before the read cycle with the training sequence). The SP4140 uses several techniques to shorten the critical path [16]. The FIR equation is:

(1) $$y[n] = \sum_{k=0}^{n} h[k]x[n-k]$$

Direct implementation of the FIR equation gives the direct-form FIR [8], shown in Figure 2(a). Some possible transformations for reducing the critical path are shown in Figure 2(b) and (c). Inherent pipelining can be achieved by transposing the data flow graph, giving a transpose-form FIR, shown in Figure 2(b). To further reduce the delay, the FIR can be interleaved and computed in parallel [16]. For m parallel paths, the area increases linearly with m, and the multiply-add is performed at $1/m$ of the data rate. Figure 2(c) shows a two-path parallel transpose type FIR, performing multiply-add at half the data rate.

Booth recoding can speed up multiplication, reducing the multiply-accumulate delay [22]. With these architectural improvements, the SP4140 FIR achieves a 275MHz clock frequency, and 525Mb/s throughput (encoded data is read at 550Mb/s, and the throughput is 525Mb/s after redundancy removal).

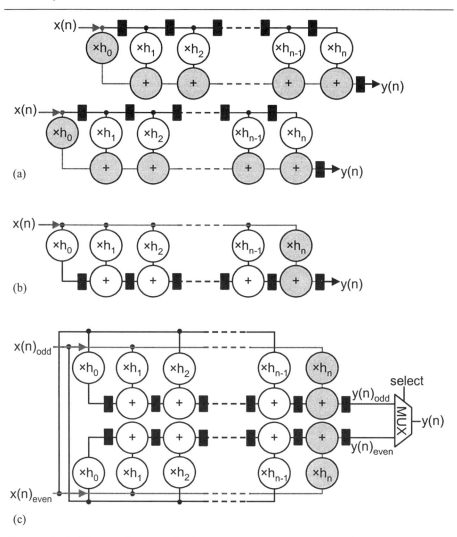

Figure 2. Different forms of a finite input response (FIR) filter. The critical paths are shown in grey. Registers are indicated by black rectangles. (a) The direct form of the FIR filter. (b) The transpose form of the FIR filter. (c) Two-path parallel transpose FIR filter.

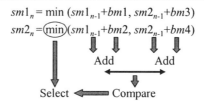

Figure 3. Illustration of the add-compare-select recursive cycle dependence in the Viterbi algorithm for a 2-state trellis.

3.2 Viterbi detector

The Viterbi algorithm has tight, single-cycle recursion. Figure 3 illustrates the 2-state Viterbi algorithm. The branch metric (*bm*) is the cost of traversing along a specific branch. State metrics (*sm*) accumulate the minimum cost of 'arriving' into a specific state.

The Viterbi algorithm is commonly expressed in terms of a trellis diagram, which is a time-indexed version of a state diagram. The simplest 2-state trellis is shown in Figure 4. Maximizing probabilities of paths through a trellis of state transitions (branches) determines the most likely sequence, for an input digital stream with inter-symbol interference. The path finally taken through the trellis is a survivor sequence, the most likely sequence of recorded data.

The Viterbi detector is a processor that implements the Viterbi algorithm, and consists of three major blocks: the central part is the add-compare-select unit (ACS); the branch metric calculation unit; and the survivor path decoding unit.

Efficient design of the ACS, which is a nonlinear feedback loop, is crucial to achieve a high throughput to circuit area ratio. The ACS calculates the sums of the state metrics (*sm*) with corresponding branch metrics (*bm*) and selects the maxima (or minima) to be the new state metrics. The throughput depends highly on the ACS addition and comparison implementations. The comparison is frequently done via subtraction, and the carry profile inside the adders and subtractors determines the speed.

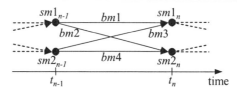

Figure 4. The two-state trellis for the Viterbi algorithm.

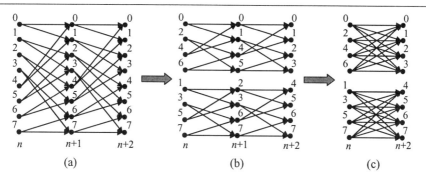

Figure 5. (a) shows two time steps in the 8-state Viterbi detector. (b) shows the decomposition of (a) into two steps of two 4-state Viterbi detectors. (c) converts this to a one-step lookahead Viterbi detector [2]. The time step is shown at the bottom of the diagrams.

Architectural transformations, like loop unrolling and retiming can be applied to an ACS recursion to reduce the critical path. The SP4140 had an 8-state Viterbi detector. A simple implementation of the 8-state Viterbi detector is shown by the 8-state trellis in Figure 5(a). The speed can be increased by using a one-step lookahead Viterbi detector, as shown in Figure 5(c).

Applying a one-step look-ahead to the ACS theoretically roughly doubles the throughput. However, in the deep submicron, this speed gain is reduced only to 40% because of increased wiring overhead. Also, the area more than doubles, increasing by a factor of ×2.7. Figure 5(c) shows the transformed algorithm, using a four-way ACS operation.

In general, a Viterbi detector performs the following operations:

(2) $p_{i,k}^{n-1} = sm_i^{n-1} + bm_{i,k}$ and $p_{j,k}^{n-1} = sm_j^{n-1} + bm_{j,k}$ (add)

(3) $sm_k^n = \min(p_{i,k}^{n-1}, p_{j,k}^{n-1})$ (compare and select)

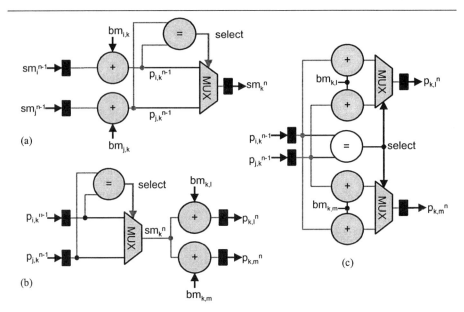

Figure 6. Critical paths are shown in grey. The comparator is indicated by '='. (a) The add-compare-select implementation for the Viterbi. (b) The transformation to compare-select-add, by retiming the adders to before the registers. (c) Retiming the compare-select-add by pushing the adders to before the multiplexer, doubling the area, but removing the comparator (subtract) from the critical path [3].

Implementing Equations (2) and (3) directly is shown in Figure 6(a). As indicated in Figure 6(a), the add, compare, and select operations are all in the critical path.

Section 1.7 of Chapter 2 discussed retiming of registers to balance pipeline stages. It is also possible to *retime combinational logic* to increase the performance. Transforming and retiming the ACS to perform compare-select-add (CSA) [3][9] removes the comparison from the critical path, as shown in Figure 6(c).

Further speed improvements are possible using a redundant number system and carry-save addition [4]. This allows deeper bit-level pipelining of the ACS, increasing the speed. A practical realization with a dynamic pipeline and latches was shown in [23]. The SP4140 Viterbi detector runs at a clock frequency of 550MHz, with a user throughput of 525Mb/s (recording code rate reduces the user data rate).

4. REGISTER DESIGN

With deep pipelining achieved by architectural transformations, the timing overhead fraction of cycle time increases. Reducing the impact of clock skew and better timing element design can significantly improve ASIC speeds. Typical high-performance custom designs keep the timing overhead down to 20% to 30% [6][7]. In a 550MHz design, this is as little as 0.4ns, whereas in common ASIC methodology the timing overhead would be about 1ns in 0.18um L_{eff} technology (see Table 2 of Chapter 3 for a comparison of ASIC and custom timing overhead).

ASIC designs typically use flip-flops, which present hard boundaries between the pipeline stages. As a result the pipeline stages must be well balanced as there is no slack passing. Also, the timing budget has to include clock skew. Level-sensitive latches allow slack passing and are less sensitive to clock skew. Latches are well supported by the synthesis tools [19], but are rarely used other than in custom designs. We believe that with latches and good clocking methodology, the speed impact of timing overhead on ASIC designs can be reduced from about 40% worse than custom to about 10% worse.

4.1 Latch Slack Passing and Skew Tolerance

Time (slack) not used by the combinational logic delay in one pipeline stage is automatically passed to the next stage in latch based designs. Likewise, a logic stage can borrow time from the succeeding stage to complete the required function [14]. Chapter 3 discusses slack passing and time borrowing.

Sections 1.3.4 and 1.3.5 of Chapter 3 discussed the clock period with latches. If latches are used, the clock skew and register setup time have less affect on the clock period. The clock skew and setup do not affect the minimum cycle time if the longest combinational logic path always arrives after the latch becomes transparent, and before the setup time plus the clock skew and jitter. Providing the latch inputs arrive while the latch is transparent, the minimum cycle time is bounded by:

(4) $\quad T_{latches} \geq 2t_{DQ} + 2t_{comb,average} + t_j$

where $t_{comb,average}$ is the average delay of combinational logic in each pipeline stage between latches (half the delay of the combinational logic between flip-flops in Equation (5)); t_{DQ} is the D-to-Q propagation through the latches; and t_j is the edge jitter. This assumes that the registers are latches that are active on opposite clock phases. (See Chapter 3 for more careful analysis.)

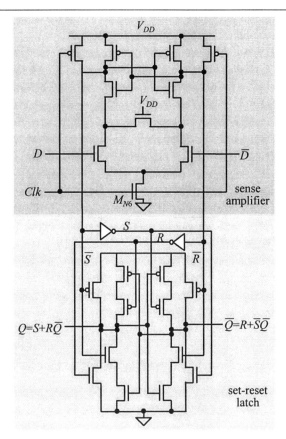

Figure 7. The modified sense-amplifier-based flip-flop (SAFF). The sizing of the M_{N6} transistor controls the skew tolerance of the SAFF.

4.2 Latch-Based FIR Filter Design

To reduce the timing overhead, the SP4140 FIR filter used latches and time borrowing [16]. The parallel transpose-type FIR architecture has two critical timing paths: from the ADC output through the multiply-accumulate; and from the previous latch output through addition to the next latch stage. In this implementation, the third path, for the coefficient update, is not an issue since the coefficients are semi-static. As the architecture is split into two paths, the timing critical multiply-accumulate operation is implemented at half the data rate.

This implementation of the FIR filter is based on one-hot Booth encoding of 6-bit data. Encoded data is distributed to all the taps of the filter. Each tap coefficient is pre-multiplied for $-4C, -3C, -2C, -C, 0, C, 2C, 4C, 8C$, and a correct partial product is selected (using a custom 9:1 multiplexer) by the encoded data. This significantly simplifies the multiplication. The use of latches naturally allows time borrowing between the FIR taps. Also, since the equalizer is the first block that follows the ADC, its clock tree insertion delay can be increased to absorb the additional delay required for data encoding.

4.3 Viterbi Detector

As the Viterbi detector has tight recursion, slack passing cannot be used. Instead, faster flip-flops can reduce the clocking overhead. In an edge-triggered system, the cycle time T has to meet the following relationship:

(5) $T_{flip-flops} \geq t_{CQ} + t_{comb} + t_{su} + t_{sk} + t_j\}$

In the critical path, the flip-flop delay is the sum of setup time and the clock-to-output delay, $t_{su} + t_{CQ}$. The clock skew is t_{sk}, the combinational delay is t_{comb}, and the hold time is t_h.

Pulsed flip-flops have less total delay, latency, than (master-slave) D-type flip-flops. Examples of pulsed flip-flops are the modified sense-amplifier-based flip-flop (SAFF) [13], Figure 7, and the hybrid latch-flip-flop (HLFF) [15], Figure 8. Similar pulsed flip-flops are used in custom designs such as the DEC Alpha 21264 and the StrongArm.

The first stage of a pulsed flip-flop is a pulse generator, and the second stage is a latch to capture the pulse. The HLFF generates a negative pulse when $D = 1$, which is captured by the D-type latch. The SAFF generates a negative pulse on either $\overline{S}$ or $\overline{R}$, which triggers the SR latch.

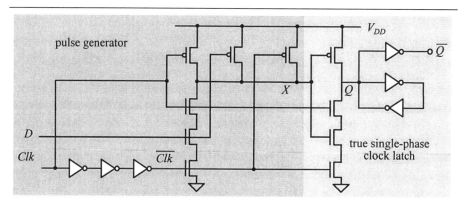

Figure 8. Hybrid latch-flip-flop (HLFF).

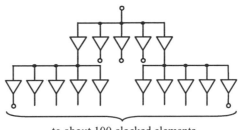

to about 100 clocked elements

Figure 9. Prescribed clock tree.

Some pulsed latches exhibit a soft edge property, which can be accounted for during cell characterization for clock skew tolerance [17]. However, characterization must include the long hold time of pulse triggered latches. For example, synthesis inserts buffers in the scan chain when clock skew is comparable to $t_{CQ} - t_h$, significantly increasing the area. The HLFF has a transparency period of about three inverter delays, while the SAFF has very small skew tolerance, controlled by sizing M_{N6}.

The SP4140 Viterbi detector uses a flip-flop derived from the SAFF [13], characterized as a standard cell, where needed in the ACS. An advantage of the SAFF is its differential output structure, doubling the drive strength *if both outputs are used in synthesis*.

5. CLOCK TREE INSERTION AND CLOCK DISTRIBUTION

Partitioning an ASIC design into blocks of 100,000 gates or less can improve synthesis results and help convergence, by limiting the maximum wire length [18]. The read channel presents a natural opportunity for design partitioning. All of the timing critical signal processing blocks are about 10,000 to 30,000 gates, with layout areas of 1 to 2 mm^2 in 0.25um CMOS. Block partitioning the design requires gated clock trees to be inserted in the blocks. Also, limiting clock distribution over a smaller area minimizes clock skew. However, the local clock trees have to be merged into a global clock tree with added clock gating, which is not generally well supported in standard ASIC methodologies.

The local clock trees in the SP4140 are designed for equal clock rise and fall times, and minimum skew, by buffer sizing and placement. Fixing the fan-out at each clock tree level controls the insertion delay, and 'prescribed' clock trees control the insertion delay to allow later matching. For example, for a given total flip-flop/latch load and block size the total clock load is computed.

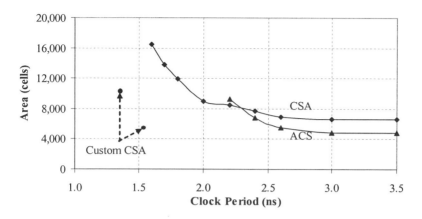

Figure 10. Area – delay comparison of synthesized and custom ACS and CSA Viterbi detectors.

By prescribing the size and number of buffers in the last stage the clock slope is met. Based on the post layout extraction data, the clock tree can be trimmed (by shorting or leaving open clock buffer outputs) to match the insertion delays, as illustrated in Figure 9. This reduces the clock skew to 60ps.

6. CUSTOM LOGIC VERSUS SYNTHESIS

ASIC designs can reach high speeds, but the area and power consumption are more than in custom designs (as discussed in Chapter 4, Sections 4 and 5). Figure 10 summarizes the basic trade-off between the area and speed in synthesized designs. ACS and CSA based Viterbi detectors with fixed microarchitecture were synthesized for different clock cycles. For longer cycles (3ns for ACS), synthesis easily achieves the speed goal. To achieve shorter cycle times, synthesis increases gate sizes to drive interconnect and loads more strongly. Interestingly, the ACS is smaller for lower speeds, but the two ASIC curves intersect at a period of 2.3ns.

To obtain some data points on the implementation efficiency of synthesized combinational logic, a functionally equivalent Viterbi detector ACS array was implemented in custom logic. The design was based on complementary and pass transistor logic, and supported the same clocking style.

Even though the adders and comparators were implemented using differential logic, the custom ACS was roughly half the size at the same speed as the synthesized version, because the custom logic used much smaller transistors and had less wiring capacitance.

When the custom design was doubled in size to the synthesized area, it ran only 20% faster than the synthesized design. However, the flip-flops were not changed, and the design was not re-optimized (wiring or placement) for the new gates.

7. REDUCING UNCERTAINTY

As discussed in Chapter 13, if the design uncertainty is reduced, performance can be improved. The disk drive read channel has an on-chip voltage regulator that gives better control over supply voltage and allows tighter worst and best-case voltage corners. This does require re-characterization of the library for non-standard corners, but results in a 5-10% speed increase.

8. SUMMARY AND CONCLUSIONS

We have examined techniques used to achieve high speeds in a 550MHz chip with an ASIC design methodology. We have identified several design techniques, common to custom designs that were used to improve the performance of a disk drive read channel designed in an ASIC methodology. The main techniques used to increase the speed were crafting the microarchitecture to increase performance, and reducing the timing overhead.

Having identified the FIR filter and Viterbi detector speed bottlenecks, architectural transformations and alternative clocking styles were used to increase their speed. The FIR filter could be pipelined, and computation in parallel doubles the speed at the price of doubling the area. Architectural transformation of the add-compare-select (ACS) in the Viterbi detector to compare-select-add (CSA) reduced the critical path length. Pipelining wasn't possible because of the recursive nature of the Viterbi algorithm, but reducing the clocking overhead can increase the speed further.

This design was amenable to achieving high clock speeds, because the datapath was not wide. Wide datapaths have larger adders and other functional elements, with more levels of logic. Having only a few levels of logic between registers reduces the combinational delay per pipeline stage.

Reducing the clock skew, and using level-sensitive latches or high speed pulsed flip-flops (instead of D-type flip-flops) reduced the timing overhead. The timing overhead factor improves from about ×1.40 worse than custom to ×1.10 worse.

We have shown that ASIC designs can be brought to within custom speeds with a proper design methodology orchestration, and attention to key design factors. Nevertheless, compared to custom implementations, ASICs will still be larger at the same speed, or slower for the same area. This was illustrated comparing custom CSA implementations to CSA and ACS ASIC versions.

9. ACKNOWLEDGMENTS

We would like to acknowledge the SP4140 design team, especially Kiyoshi Fukahori, Michael Leung, James Chiu, Bogdan Staszewski, Vivian Jia, and David Gruetter. James Chiu provided the area-delay comparison data.

10. REFERENCES

[1] Altekar, S., et al. "A 700 Mb/s BiCMOS Read Channel Integrated Circuit," *IEEE International Solid-State Circuits Conference*, Digest of Technical Papers, San Francisco CA, February 2000, pp. 184-185, 445.
[2] Black, P., and Meng, T. "A 140 MB/s 32-state radix-4 Viterbi decoder," *IEEE Journal of Solid-State Circuits*, vol. 27, no. 12, December 1992, pp. 1877-1885.
[3] Fettweis, G., et al. "Reduced-complexity Viterbi detector architectures for partial response signaling," *IEEE Global Telecommunications Conference*, Singapore, Technical Program Conference Record, vol. 1, November 1995, pp. 559-563.
[4] Fettweis, G., and Meyer, H. "High-speed parallel Viterbi decoding algorithm and VLSI architecture," *IEEE Communications Magazine*, vol. 29, no. 8, May 1991, pp. 46-55.
[5] Fey, C. F., and Paraskevopoulos, D. E. "Studies in LSI Technology Economics IV: Models for gate design productivity," *IEEE Journal of Solid-State Circuits*, vol. SC-24, no. 4, August 1989, pp. 1085-1091.
[6] Gronowski, P., et al. "High-Performance Microprocessor Design," *IEEE Journal of Solid-State Circuits*, vol. 33, no. 5, May 1998, pp. 676-686.
[7] Harris, D., and Horowitz, M. "Skew-Tolerant Domino Circuits," *IEEE Journal of Solid-State Circuits*, vol. 32, no. 11, November 1997, pp. 1702-1711.
[8] Jain, R., Yang, P.T., and Yoshino, T. "FIRGEN: a computer-aided design system for high performance FIR filter integrated circuits," *IEEE Transactions on Signal Processing*, vol. 39, no. 7, July 1991, pp. 1655-1668.
[9] Lee, I., and Sonntag, J.L. "A new architecture for the fast Viterbi algorithm," *IEEE Global Telecommunications Conference*, San Francisco CA, Technical Program Conference Record, vol. 3, November 2000, pp. 1664-1668.
[10] Marvell, Marvell Introduces HighPhyTM, the Industry's First Read Channel PHY to Exceed Gigahertz Speeds, December 2000. http://www.marvell.com/news/dec4_00.htm
[11] Messerschmitt, D. G. "Breaking the recursive bottleneck," in Skwirzynski, J.K. (ed.) *Performance Limits in Communication Theory and Practice*, Kluwer, 1988, pp. 3-19.
[12] Nazari, N. "A 500 Mb/s disk drive read channel in 0.25 μm CMOS incorporating programmable noise predictive Viterbi detection and trellis coding," *IEEE International Solid-State Circuits Conference*, Digest of Technical Papers, San Francisco CA, February 2000, pp. 78-79.
[13] Nikolić, B. et al. "Sense amplifier-based flip-flop," *IEEE Journal of Solid-State Circuits*, vol. 35, June 2000, pp. 876-884.

[14] Partovi, H., "Clocked storage elements," in Chandrakasan, A., Bowhill, W.J., and Fox, F. (eds.). *Design of High-Performance Microprocessor Circuits. IEEE Press*, Piscataway NJ, 2000, pp. 207-234.

[15] Partovi, H., et al. "Flow-through latch and edge-triggered flip-flop hybrid elements," *IEEE International Solid-State Circuits Conference*, Digest of Technical Papers, San Francisco CA, February 1996, pp. 138-139.

[16] Staszewski, R.B., Muhammad, K., and Balsara, P. "A 550-MSample/s 8-Tap FIR digital filter for magnetic recording read channels," *IEEE Journal of Solid-State Circuits*, vol. 35, no. 8, Aug. 2000, pp. 1205–1210.

[17] Stojanovic, V., and Oklobdzija, V.G. "Comparative analysis of master-slave latches and flip-flops for high-performance and low-power systems," *IEEE Journal of Solid-State Circuits*, vol. 34, no. 4, April 1999, pp. 536-548.

[18] Sylvester, D.; Keutzer, K. "Getting to the bottom of deep submicron," *Proceedings of the International Conference on Computer Aided Design*, San Jose CA, November 1998, pp. 203-11.

[19] Synopsys, *Synopsys Design Compiler*, Reference Manual, Synopsys.

[20] Thapar, H. K. and Patel, A.M. "A Class of Partial Response Systems for Increasing Storage Density in Magnetic Recording," *IEEE Transactions on Magnetics*, vol. MAG-23, no. 5 part 2, September 1987, pp. 3666-3678.

[21] Texas Instruments, Texas Instruments SP4140 CMOS Digital Read Channel, 1999. http://www.ti.com/sc/docs/storage/products/sp4140/ index.htm

[22] Weste, N., and Eshraghian, K., *Principles of CMOS VLSI Design*, 2nd Ed. Addison-Wesley, Reading MA, 1992, pp. 547-554.

[23] Yeung, A.K., and Rabaey, J.M. "A 210 Mb/s radix-4 bit-level pipelined Viterbi decoder," *IEEE International Solid-State Circuits Conference*, Digest of Technical Papers, San Francisco CA, February 1995, pp. 88-89, 344.

Chapter 16

The iCORE™ 520MHz Synthesizable CPU Core

Nick Richardson, Lun Bin Huang, Razak Hossain, Julian Lewis, Tommy Zounes, Naresh Soni
STMicroelectronics Incorporated,
Advanced Designs, Central R&D,
4690 Executive Drive, PO Box 919028,
San Diego, CA 92121, USA
{nick.richardson, lun-bin.huang, razak.hossain, julian.lewis, tommy.zounes, naresh.soni}@st.com

ABSTRACT

This chapter describes a new implementation of the ST20-C2 CPU architecture. The design involves an eight-stage pipeline with hardware support to execute up to three instructions in a cycle. Branch prediction is based on a 2-bit predictor scheme with a 1024-entry Branch History Table and a 64 entry Branch Target Buffer and a 4-entry Return Stack. The implementation of all blocks in the processor was based on synthesized logic generation and automatic place and route. The full design of the CPU from micro-architectural investigations to silicon testing required approximately 8-man years. The CPU core, without the caches, has an area of approximately 1.8mm^2 in a 6-metal 0.18um CMOS process. The design operates up to 520MHz at 1.8V, among the highest reported speeds for a synthesized CPU core [2].

1. INTRODUCTION

Ever-increasing levels of CPU performance demanded by modern embedded applications, and product design cycles that have often been reduced to only a few months, have made it important to produce synthesizable processor cores capable of execution speeds typically only achievable by complex custom solutions. In many cases it is unfeasible and prohibitively expensive to use lengthy custom design flows. The iCORE™

project was born out of the need to demonstrate a solution to this problem. The goal was to create a very high performance version of STMicroelectronics' ST20-C2 embedded CPU architecture, but it also had to be shown that the design could be quickly and easily portable as a soft-core across existing and future technologies. This ruled out the use of extensive custom circuits, and led to the adoption of a methodology close to a traditional ASIC design flow, but one tuned to the aggressive performance goals demanded by the project.

The ST20-C2 architecture is a superset of the acclaimed Inmos transputer, which was first introduced in 1984 (Inmos was acquired by STMicroelectronics in 1989). The ST20-C2 added an exception mechanism (interrupts and traps), extensive software debug capability and improved support for embedded systems. It is now used in high-volume consumer applications such as chips for set-top boxes.

The ST20-C2's basic instruction-set specifies a set of simple, RISC-like operations, making it a good candidate for relatively easy high-frequency implementation. However, it extends conventional RISC technology by having a variable-length instruction word to promote code compactness, and uses some instructions that specify complex operations to control hardware-implemented kernel functions such as task scheduling and inter-process communications [3]. Also, rather than being based on a RISC-like load/store architecture with a large number of machine registers, the ST20-C2 employs a three-deep hardware evaluation stack, also known as the register-stack, on which the majority of instructions operate. This scheme permits excellent code density, ideal for embedded applications, but must be carefully matched with an efficient local memory or cache system in order to minimize the frequency of memory references outside the CPU core.

These characteristics of the architecture required special consideration to devise a design capable of maintaining high instruction throughput on a per-clock-cycle basis (measured in Instructions Per Cycle or IPC) without compromising the CPU's frequency of operation.

For the project to be considered successful, however, we had to demonstrate optimization in all the following areas, in approximate order of importance:
- Frequency
- Execution efficiency (IPC)
- Portability
- Core size
- Power consumption

2. OPTIMIZING THE MICROARCHITECTURE

To achieve good instruction-execution efficiency without excessive design complexity, previous implementations of the ST20-C2 architecture employed relatively short pipelines to reduce both branch-delays and operand/result feedback penalties, thus producing good IPC counts over a wide range of applications. In the case of iCORE, however, the aggressive frequency target dictated the use of a longer pipeline in order to minimize the stage-to-stage combinational delays. The problem was how to add the extra pipeline stages without decreasing instruction execution efficiency, since there would be little point in increasing raw clock frequency at the expense of IPC. Following analysis using a C-based performance model, a relatively conventional pipeline structure was chosen, but one that had some important variations targeted at optimizing instruction flow for the unique ST20 C2 architecture.

2.1 The Pipeline

The pipeline micro-architecture is shown in Figure 1. It comprises four separate units of two pipeline stages each. The **IFU** (Instruction Fetch Unit) comprises the IF1 and IF2 pipeline stages and is responsible for fetching instructions from the instruction cache and performing branch predictions. The **IDU** (Instruction Decode Unit) comprises the ID1 and ID2 pipeline stages and is responsible for decoding instructions, generating operand addresses, renaming operand-stack registers and checking operand/result dependencies, as well as maintaining the program counter, known in this architecture as the Instruction Pointer (IPTR), and a Local Workspace Pointer (WPTR). The **OFU** (Operand Fetch Unit) comprises the OF1 and OF2 pipeline stages and is responsible for fetching operands from the data cache, detecting load/ store dependencies, and aligning and merging data supplied by the cache and/or buffered stores and data-forwarding buses. Finally, the **EXU** (Execute Unit) comprises the EXE and WBK pipeline stages and performs all arithmetic operations other than address generation, and stages results for writing back into the Register File Unit (RFU) or memory.

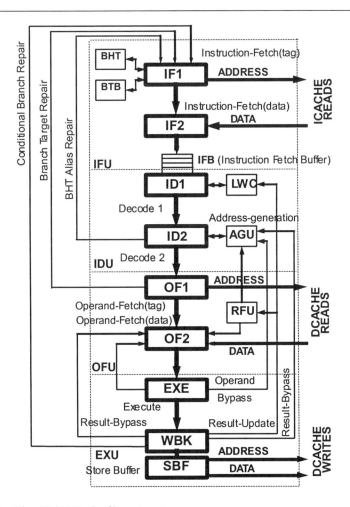

Figure 1. The iCORE pipeline structure.

The RFU contains the 3-deep register-stack, conceptually organized as a FIFO, comprising the A, B, and C registers (in top-to-bottom order). In practice, it is implemented using a high-speed multi-port register file. The memory-write interface is coupled with a store buffer (SBF) that temporarily holds data while waiting for access to the data cache.

The instruction-fetch portion of the pipeline (the IFU) is coupled to the execution portion of the pipeline (the IDU, OFU, and EXU) via a 12-byte Instruction Fetch Buffer (IFB).

Early on in the analysis of ST20-C2 program behavior it became apparent that the micro-architecture should not only be tuned for high frequency, but also for very efficient memory access, to get best value from the memory bandwidth available in its target environment: typically a highly integrated system on a chip (SoC) with multiple processors sharing access to off-chip memory. This was done by including two in-line operand caches in the pipeline, both capable of being accessed in a single pipeline throw.

To understand the function of the operand caches, a brief explanation of the ST20-C2's two addressing modes is needed. One mode is used for addressing local variables, which are accessed using an offset relative to an address pointer held in a register known as the Local Workspace Pointer (WPTR). WPTR holds the start-address of a program's procedure stack (also known as its local workspace), and so is loaded with a new value on each procedure call and return. The other mode is for addressing non-local variables anywhere in the 4 GB memory address space, using a full 32-bit address that is pre-loaded into the top location of the register-stack (the A-register).

The first operand cache is for local variables. It is called the Local Workspace Cache (LWC) and accelerates references to the first 16 words in the current local workspace, as pointed to by WPTR. The size of the LWC was determined by the results of running a number of benchmark programs on the C-based statistical performance model. It is placed early in the pipeline (in the ID1 stage) so that local variables used for non-local address calculations can be supplied to the Address Generation Unit (AGU) in the next pipeline stage (ID2) for early calculation of the full non-local address.

The second operand cache is an 8 KB in-line data cache that accelerates references to non-local variables, as well as local variables that miss the LWC. It uses non-local addresses calculated by the AGU in the ID2 stage, and supplies operands to the Execute Unit, thus occupying the two pipeline stages before the EXU. For critical path optimization, its tag RAM and data RAM are placed in two different pipeline stages. Accessing the tag RAM before the data RAM eliminates the cache's set-associative output multiplexing that would have been needed if the tag and data RAMs were placed in the same stage, and also allows cache reads and writes to share the same control flow, thus simplifying the design.

Together, these two highly integrated operand caches are complementary to the ST20-C2 architecture's A, B, and C registers, effectively acting like a very large register file.

2.2 Instruction Folding

An important effect of the in-line operand caches is to increase the opportunities for instruction folding. This is an instruction decoding technique that combines two or more machine instructions into a single-throw pipeline operation. Since the ST20-C2 has a stack-based architecture, most of its instructions specify very simple operations such as loading and storing operands to and from the top of the 3-deep register-stack, or performing arithmetic operations on operands already loaded onto the register-stack, but for simplicity, memory and ALU operations are never combined in the same instruction. Without such compound instructions that allow memory and arithmetic operations to occur as single-slot pipeline operations, it would not be possible to take advantage of iCORE's in-line memory structure. With folding, however, up to three successive instructions can be merged into one operation that occupies a single execution slot and which fully uses iCORE's pipeline resources.

As an example, take the following very common three-instruction sequence:

ldl <n>
ldnl <m>
add

The first instruction, **load local** (*ldl*) loads from memory a variable that is addressed by an offset <n> from the current value of the Local Workspace Pointer (WPTR) and pushes it onto the top of the register-stack. The second instruction, **load non-local** (*ldnl*) loads from memory a variable that is addressed by an offset <m> from the value on top of the register-stack (in this case the value that was just placed there by the *ldl* instruction). Finally the *add* instruction adds the top two values on the register-stack (A, B) and pushes the result back onto it.

Without folding, this instruction sequence would occupy three distinct pipeline slots, and could be subject to stalling due to data-dependency between the *ldl* and the *ldnl* instructions. However, with folding and iCORE's in-line memory structure they can be executed in a single pipeline slot, even though they require two memory reads. To understand how this is done, consider the stage- by-stage execution:

IF1: Instruction fetch (instruction cache tag access)

IF2: Instruction fetch (instruction cache data RAM access, load IFB)

ID1: execute the *ldl* by reading local variable from LWC at offset <n>; issue *ldl*, *ldnl*, *add* from IFB as one operation if *ldl* hits the LWC

ID2: Using the AGU, generate address to be used by the *ldnl* instruction by adding the local operand obtained by the *ldl* in ID1 to the offset <m>.

OF1: Fetch data requested by *ldnl* from data cache (tag RAM)

OF2: Fetch data requested by *ldnl* from the data cache (data RAM); fetch second data operand from register file unit (pop top of stack)

EXE: Add the data obtained by the *ldnl* in OF2 to the second operand from register-stack.

WBK: Push result onto the top of the register-stack.

Note how the three folded instructions are executed separately by different pipeline stages (the *ldl* in ID1, the *ldnl* in ID2, OF1, OF2, and the *add* in EXE, WBK), but because only one pipeline stage at a time is occupied, the three instructions are apparently executed in a single clock cycle.

There are many such folding sequences supported by iCORE, but they all follow the same principle as described above.

2.3 Data Forwarding and Register Score-boarding

In a deeply pipelined design, it is important to mitigate the effects of delays caused by pipeline latencies. One significant source of latency-based performance degradation is caused by data dependency between successive instructions. If the dependency is a true one (such as when an arithmetic operation uses the result of another instruction immediately preceding it) then there is no way of eliminating it, although micro-architecture techniques can be used to reduce the pipeline stalls it causes. This is generally done by providing data forwarding paths between potentially interdependent pipeline stages. In iCORE this includes data bypass paths from WBK to OF2, EXE to OF2, WBK to ID2, and EXE to ID2, as shown in Figure 1.

The paths from EXE and WBK to OF2 allow an instruction either immediately following an arithmetic operation, or one separated from it by two clock cycles, to use its result without penalty. If they are separated by more than two clocks, then the result can be read from the register-stack without penalty.

Similarly, the path from EXE to ID2 allows the AGU to use an operand fetched from the data cache by one instruction as an operand address for another following it as soon as possible. This commonly occurs when either an *ldnl* instruction, or an *ldl* instruction that misses the LWC, loads an address from memory that is used by a subsequent *ldnl* instruction (such as when following a pointer chain). Using the forwarding path under these circumstances results in a worst-case pipeline stall penalty of two clock cycles. If no forwarding paths were provided, this penalty would be four clock cycles, to allow time for the operand address to cycle through the register-stack.

The path from WBK to ID2 operates in a similar fashion, but is used when one instruction's result that is calculated by the Execution Unit (rather than just fetched from the data cache) is used as an AGU input by a following one. This reduces the penalty for this type of addressing to a maximum of three clocks.

To activate the forwarding paths at the appropriate times, register dependency-checking logic (also known as register score-boarding) must be used. In iCORE, this logic resides in the ID2 stage. Its operation is quite straightforward, and works on the basis of comparing the name of a register required as a source operand by one instruction with the names of all registers about to be modified by instructions further ahead in the pipeline. When a source/destination register-name match is detected, the newer instruction is stalled in ID2 if the older instruction cannot supply its result in time (typically when an ALU result is used as an operand address, or when the result of a multi-cycle ALU operation is used by the next instruction). Then after inserting any necessary stall cycles, the score-boarding logic releases the instruction from ID2, and activates the appropriate forwarding path to provide the source operands for the instruction at the earliest possible time.

For the ST20-C2 architecture, register dependency checking must handle operands constantly changing position between the A, B, and C registers of the three-deep register-stack, since most instructions perform implicit or explicit pushes or pops of the stack. The effect is that almost every instruction appears to have data dependencies on every older instruction in the pipeline, since nearly all of them cause new values to be loaded into A, B, and C, even though most of those dependencies could be considered "false", in the sense that they are caused simply by the movement of data from one register to another rather than by the creation of new results. Nevertheless, they would cause considerable performance degradation if newer instructions were stalled based on those false dependencies.

iCORE solves this problem by a simple renaming mechanism that maps the conceptual A, B, and C registers of the ST20-C2's architecture onto fixed hardware registers, named $R0$, $R1$, and $R2$. The key is that the same hardware register remains allocated to a particular instruction's result throughout its lifetime in the pipeline despite the result apparently being moved between A, B, and C by subsequent instructions pushing and popping the architectural register-stack. This is accomplished by a mapping table for each pipeline stage that indicates which architectural register (A, B, or C) is mapped onto which fixed hardware register ($R0$, $R1$, $R2$) for the instruction in that pipeline stage. Operations that only move operands between registers do so by simply renaming the $R0$, $R1$, and $R2$ registers to the new A, B, or C mapping, rather than actually moving data between them. The mapping is performed in ID2 and is based on the effect of the current instruction in ID2 on the existing mapping up to but not including that instruction.

Once the ID2 mapping is done, then the dependency checking is performed based on $R0$, $R1$, and $R2$ source and destination values. Since results are never physically moved directly between these registers, false dependencies are eliminated.

2.4 Branch Prediction

iCORE's relatively long pipeline gives rise to the danger of branches causing significant performance degradation compared to previous ST20-C2 implementations. After performance simulations showed that the effect of the longer pipeline on some benchmarks was significant, a low-cost branch prediction scheme was incorporated into iCORE's micro-architecture.

In a machine without branch prediction, the source of most branch penalties is two-fold:

First, the target address of a taken branch or other control-transfer instruction must be determined. This cannot be done until after the branch has been decoded, since it must first be identified and then (usually) an arithmetic operation performed, such as adding an offset to the current Instruction Pointer value, to obtain the address of the next valid instruction in the program flow. By the time its target address can be applied to the instruction cache to make it start fetching instructions from the new destination, several instructions have been fetched from the wrong address. These must be cancelled, creating a penalty of several clock cycles.

Secondly, conditional branches, i.e. those that are dependent on the results of previous instructions to determine their direction, usually must progress all the way to the end of the pipeline before the condition on which their action depends is resolved. Only then can instruction fetches from a new target address proceed, causing an even greater branch penalty than that incurred by the target address calculation alone.

iCORE implements a branch prediction mechanism that reduces the penalties in both these cases. A two-bit predictor scheme is used to predict branch and subroutine-call instruction behavior (taken or not taken), while a Branch Target Buffer is used to predict target addresses for taken branch and call instructions. Also, a 4-deep return stack is used to predict the special case of *ret* instruction return addresses.

The prediction scheme used by iCORE is shown in Figure 2. It employs a 1024-entry Branch History Table (BHT), each entry of which contains a two-bit predictor that implements a branch-hysteresis algorithm (weakly/strongly taken/not-taken). The BHT keeps the most current branching history of a set of instruction cache words. Predicted fetch-addresses (i.e. the target addresses of predicted-taken branches) are kept in a 64-entry Branch Target Buffer (BTB).

Whenever a new 4-byte word is fetched from the instruction cache, the instruction address that is used to index the cache also indexes the BHT and BTB. The retrieved information from the BHT is used to predict whether there is a taken branch at the current instruction address. If a taken branch is predicted, the address retrieved from the BTB is loaded into the Instruction- Fetch Pointer and used to fetch the next instruction word.

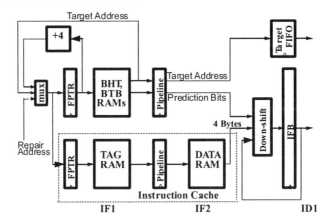

Figure 2. Branch prediction logic in the iCORE.

Otherwise, the sequential value of the Instruction-Fetch Pointer is used. All this takes place in the very first pipeline stage, ID1, so that new target addresses can be applied immediately to the instruction cache on predicting a taken branch.

The effect is that predicted-taken branches can immediately change the instruction address applied to the instruction cache without any delay cycles, and the performance impact of successfully predicted branches is negligible.

The predicted direction and target are fed forward through IF2 and loaded into the IFB in the case of the prediction bits, and into a 4-deep Target Buffer in the case of the predicted target. They are used in later pipeline stages to validate the predictions and update the IPTR.

As branches are predicted, mispredictions may occur. One type of misprediction is caused by the cost-saving measures of using only the low-order bits of the instruction-fetch address to index BHT and BTB entries while ignoring the upper address bits, and by providing only one BHT/BTB entry per 4-byte cache word, even though there could be more than one branch in a word. These factors can cause a single BHT or BTB entry to be shared by many fetched instructions, even though the entry is likely to be correct only for the one instruction that was originally used to update it; the instructions at other fetch addresses that share the entry may get wrong predictions or wrong target addresses, or both.

Other sources of misprediction include mispredicted direction of conditional branches, and incorrect return addresses. Mispredictions are detected and resolved in later pipeline stages. The instruction decoder in the ID1 stage can detect many mispredictions caused by BHT aliasing. For instance, if a *taken* condition is predicted for an instruction that is not a branch, or an *untaken* condition is predicted for an unconditional branch, the prediction is clearly in error and a repair operation can start immediately with only a small penalty. To check predicted targets, an adder in the OF1 stage calculates the correct target address of a taken branch and compares it with the predicted one (which is fed forward from the BTB). If there is a mismatch due to a missing aliased BTB entry, then recovery can start from OF1. Finally, a conditional branch must wait until it reaches the EXE stage before the condition on which its action was determined can be checked, and thus has the longest repair time when incorrectly predicted.

The steps needed to repair a branch depend on the type of misprediction. In the case where BHT aliasing causes a non-branch instruction to be taken, then all instructions fetched after that instruction are flushed from the pipeline, and instruction fetching is restarted from the *source* address of that instruction to ensure that it is correctly fetched, while the BHT is updated to indicate *not taken.* Conversely, in the case where a target address of a taken branch is incorrectly predicted, then instructions from the bad target are flushed from the pipeline and instruction fetching is restarted from the correct *target* address computed in the OF1 stage, while the BTB is updated with the new target address.

Finally, repairs needed to correct a mispredicted *conditional* branch flush all subsequent instructions from the pipeline and begin fetching again either from the target address of the branch or the next-sequential address of the branch, depending on whether the branch is taken or not, and update the BHT and BTB appropriately.

For branches caused by returns from subroutines, conventional branch prediction is not effective, since their target address constantly changes depending on where their subroutine was called from. Thus in the iCORE a 4-entry return-stack is used to predict the target of the *ret* (return) instruction. The return-stack saves the return addresses for the most current *call* instructions. When a *ret* instruction is encountered in the ID2 stage, the top of the return-stack is popped and the value is used as the predicted branch address for the *ret* instruction, incurring only a small penalty.

The *ret* predictions are validated in the EXE pipeline stage, because in the ST20-C2 architecture, the actual return addresses are obtained from a memory location in the local workspace, and typically are available only after accessing the data cache.

2.5 Cache Subsystem

Features of the cache controller's design were focused on the need for simplicity (thereby enabling high frequency) and the requirement for very tight coupling to the CPU pipeline. These requirements resulted in a two-stage pipelined approach, the first stage of which handled the tag access and tag comparison, and the second stage that handled the data access and data alignment. This gave a very good balance of delays, and had the advantage of saving both power and delay by eliminating the need for physically separate data RAMs for each associative cache-bank (Way), since the data RAM's Way-number could be pre-determined in the tag cycle and then simply encoded into extra data RAM address bits.

To further reduce delays, integrated pipeline input registers were built into both the tag and data RAMs to eliminate input wire delays from their access times. This eliminated core-to-cache wire delays from the critical SRAM access time, as well as allowing a more predictable start-time for the SRAMs' internal precharge logic. These two factors improved the overall cycle time by several percent.

A simplified block diagram of the IFU and OFU cache controllers is shown in Figure 3. One controller occupies the IF1 and IF2 pipeline stages in the IFU and the other occupies the OF1 and OF2 pipeline stages in the OFU. The same cache controller design is used in each case, so its main pipeline stages corresponding to IF1/IF2 in one case and OF1/OF2 in the other are generically known as CH1 and CH2.

Before the CH1 phase, the cache controller receives the request (read or write), address, and data from the CPU and loads them into its input pipeline registers.

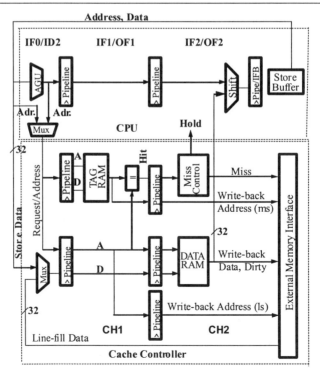

Figure 3. The cache controller.

In the CH1 phase, the tag RAMs are accessed and the tag comparison is performed. The first version of iCORE has 8 KB caches and a 16-byte line size, so 512 tag entries are required. At the end of CH1, the result of the tag comparison (hit/miss), and the CH1 address and data fields from the original CPU request are forwarded to the CH2-phase pipeline input registers.

In CH2, the data RAMs are accessed and the data is returned to the CPU's Instruction Fetch Buffer (IFB) in the case of the instruction cache, or the data aligner on the inputs to the EXE stage in the case of the data cache.

As well as returning data on a cache hit, the CH2 stage also initiates a cache-miss sequence if the requested data is not in the cache. It does this by asserting the *hold* signal, which freezes the CPU pipeline, and by then signalling the main-memory controller to start fetching the missing line. Once the missing line has been retrieved, its data is loaded into the data RAM, its upper address bits are loaded into the tag RAM, and *hold* is deasserted. The CPU is then required to restore the original state of the cache controller's internal pipeline registers by repeating the read or write requests that were in progress at the time the cache miss occurred, since this simplifies the cache design to reduce the complexity of its critical paths. That done, normal operation is resumed.

The cache controller implements a write-back policy, which means writes which hit the cache write only to the cache, and not main memory (as opposed to a write- through cache which would write to both). Thus the cache can contain a more up-to-date version of a line than main memory, necessitating that the line be written back to memory when it is replaced by a line from a different address. To facilitate this, a *dirty* bit is kept with each line to indicate when it has been modified and hence requires writing-back to main memory on its replacement. The *dirty* bits are kept in a special field in the data RAM, and there is only one per four word locations (since each word is 4 bytes, and there are 16 bytes in a line). The *dirty* bit for a given line is set when a write to that line hits the cache.

2.6 Write Buffering and Store/Load Bypass

Writes to the data cache presented a potential performance problem, since write requests are generated by the WBK stage at the end of the CPU pipeline, and so in any given clock-cycle could clash with newer read requests generated by the AGU at the end of the ID2 stage. Since the cache is single-ported, this would cause the CPU pipeline to stall, even though on average the cache controller is capable of dealing with the full read/write bandwidth. Performance simulations showed that most collisions could be avoided by the addition of a special write-buffering pipeline stage, SBF, which is used to store blocked memory write requests from the WBK stage, and which is coupled with the ability to generate write requests from either the WBK stage if the SBF stage is empty and the data cache is not busy, or from the SBF stage if it is not empty and the data cache is not busy. This gives write requests more opportunities to find unused data cache request slots. Additional write buffers would further increase the opportunities for collision avoidance, but were found to give much smaller returns than the addition of the first one, so were not implemented in the demonstration version of iCORE.

Finally, due to the high memory utilization of ST20-C2 programs, it was found very beneficial to add a store/load bypass mechanism, whereby data from queued memory writes in the EXE, WBK and SBF stages could be supplied directly to read requests from the same address in OF2, in replacement of stale data obtained from the data cache. This avoids having to stall memory reads from data cache locations that need to be updated with outstanding writes to the same location.

All data alignments and merges with data from the cache and write buffers are catered for, so the pipeline never needs to be stalled for this condition.

2.7 Complex Instruction Support

Some of the more complex ST20-C2 instructions are designed to optimize operating system functions such as hardware process scheduling and inter-process communication, to give extremely fast real-time response. One of the iCORE project goals was to ensure that the new micro-architecture could support these instructions efficiently.

The first prototype version of iCORE was designed to demonstrate execution speed of the regular instructions, so in the interests of expediency it did not support the complex instructions. However, for future versions of the CPU, a configurable solution was selected in which the regular instructions are implemented by hardware decoders and state machines (as in the first version), but the complex instructions are supported by a microcode engine and microinstruction ROM. This interfaces to the main instruction decoder in the ID2 stage, and supplies sequences of regular (non-complex) instructions to forward pipeline stages for execution of the complex instruction-driven algorithms, while stalling backward pipeline stages until the sequence is complete.

For systems that require full backward compatibility with the complex instructions but do not require their full-speed execution, the microcode ROM can be eliminated and replaced with a mechanism that on encountering a complex instruction in an early pipeline stage, generates a very fast instruction-trap into a specially protected region of main memory that contains pre-loaded emulation code that implements the instruction's functionality. A minimal amount of hardware support is provided (e.g. extra hardware registers and status bits for intermediate results, instruction cache lock-down lines for critical sections) to ensure that the instruction emulation runs at reasonable speed. A preliminary paper analysis showed that this approach provides satisfactory performance for the majority of iCORE's applications at very low cost.

2.8 Effects of Micro-Architecture Features on IPC

Most of the micro-architectural enhancements' effect on IPC were first studied through the use of a C-based statistical performance model, which "executes" various benchmark program traces produced by an ST20-C2 Instruction Set Simulator (ISS). They include basic instruction folding, the Local Workspace Cache and its alternative designs, enhanced instruction folding enabled by the use of LWC, and branch prediction schemes and designs. The architecture of some of the features was fine-tuned during RTL implementation, as more accurate performance and area trade-off analyses were made possible. Examples of some of the more noticeable implementation-level enhancements include modification of branch predictor state transition algorithm to handle the case where multiple predicted branch instructions reside in same cache-word, and the addition of a dynamic LWC disable feature to minimize the time when it cannot be used during coherency updates. A summary of the effects of these micro-architecture features on IPC is listed in Table 1.

3. OPTIMIZING THE IMPLEMENTATION

Implementation of the design was based on a synthesis strategy for the control and datapath logic. Only the RAMs used custom design methods.

Features	IPC Effect	Overall IPC Improvements
Add basic fold	10%	10%
Add LWC	10%	20%
Enhance instruction folding	7%	27%
Add LWC rotating index	8%	35%
Add branch prediction	5%	40%
Enhance branch prediction transition algorithm	1%	41%
Add Return Stack	3%	44%
Add LWC dynamic disable	4%	48%

Table 1. IPC enhancements on Dhrystone 2.1.

3.1 Pipeline Balancing

Since the pipeline had to be carefully delay-balanced to achieve the target frequency, the use of 2-phase transparent latches between pipeline stages was considered. This would have allowed flexibility for transferring delays between pipeline stages, totaling up to 50% of the clock period, cumulative across the pipeline. Attractive as this benefit might have been, the approach was rejected for the traditional reasons of increased difficulty of applying design automation tools to latch-based systems, and of splitting synchronous pipeline stages into phase 1 and phase 2 latched segments. Additionally, a scan-based/ATPG test methodology would have been significantly harder to implement for a non-registered design, and the use of tightly-coupled memories requiring a stable address at a precise delay from a clock-edge would have been problematic.

Instead, the less-glamorous approach of manual timing estimation and code-modification for pipeline balancing was used, resulting in two to three moderate RTL iterations attributable to this process. The key to controlling the overall amount of effort put into this activity was to continually estimate the type of logic and circuit structures that would be needed during the micro-architecture development phase, thereby gaining visibility into potential problems before investing time in the detailed RTL implementation.

3.2 Synthesis Strategy

The synthesis process was divided into two steps. The first step employed a bottom-up approach using Design Compiler in which all top-level blocks were synthesized using estimated constraints. Blocks were defined as logic components generally comprising a single pipeline stage. The execute unit of the processor included a large multiplier, for which Module Compiler was used as it was found to provide better results than Design Compiler. As the blocks were merged into larger modules, which were finally merged into the full processor, several incremental compilations of the design were run to fine tune the performance across the different design boundaries.

The only region of the design where synthesis provided unacceptable path delays was in the OF2 and the IF2 stages. In both cases the problems involved the use of complex multiplexer functions. Custom multiplexer gates were designed for these cases and logic was manually inserted with a *dont_touch* attribute to achieve the desired results. The custom multiplexers took about a week to design, and were characterized for the standard cell flow in .lib format. Other complex multiplexer problems were resolved by re-writing the HDL to guide the synthesis tool towards the preferred implementation.

The second step of the synthesis strategy employed a top-down approach, with cell placement using Synopsys' Physical Compiler, which optimized gates and drivers based on actual cell placement. Physical Compiler proved to be effective in eliminating the design iterations often encountered in an ASIC flow due to the discrepancy between the gate load assumed by the wire-load model and that produced by the final routed netlist. The delays based on the placement came within 10% of those obtained after final place and route. Physical Compiler can be run with either an "RTL to placed gates", or a "gates to placed gates" flow. The "gates to placed gates" flow was found to provide better results for this design.

3.3 RAM Design

The processor core incorporates three small RAMS: The 2-bit 1024-entry Branch History Table (BHT); the 34-bit 64-entry Branch Target Buffer (BTB); and the 32- bit 16-entry Local Workspace Cache (LWC).

Of the three small RAMs, the BTB and LWC employ asynchronous reads and synchronous writes. The BHT has synchronous reads and writes. To improve access times, the read bit-lines are precharged during the low phase of the clock, and the read access occurs during the high phase. An inverter type circuit is used as for the sense amplifiers.

In addition, the test chip was built with in-line 8 KB instruction and data caches that use 1Kx32 synchronous SRAMs for their data RAMs, and 256x21 synchronous SRAMs with built-in comparators for their tag RAMs.

3.4 Standard Cell Library

A higher performance standard cell library (H12) was developed for iCORE to improve circuit speed compared to a supplied generic standard library. Speed improvement of more than 20% was observed in simulation and on silicon when the H12 library was used. The cells of the H12 library were manually laid out. The library design time was about 25 man-months.

The conventional CMOS static designs used in the generic library were used in the H12 library, but with the following differences:
- Reduced P/N ratio
- Larger logic gates (vs. buffered)
- More efficient layout
- Increased cell height

The P/N ratio for each cell was determined by the best speed obtained when the gate was driving a light load. It was observed that when the gate-load increased, the P/N ratio needed to be increased to get optimal speed.

However, the physical-synthesis tool minimizes the fan-out and line loading on the most critical paths, so the H12 library was optimized for the typical output loading of those paths.

Most cells in the H12 library had drive strengths of 0.5X, 1X, 2X and 4X. Some cells also had drive strengths of 3X, 8X, 12X and 16X. There were about 500 cells total, which includes both polarities for all the combinational cells. In comparison, the generic library had 600 cells, with a similar range of drive strengths.

Delays were also reduced by creating larger, single level cells for high-drive gates. The original library added an extra level of buffering to its high-drive logic gates to keep their input capacitance low and their layout smaller, but at the expense of increasing their delay. Eliminating that buffer by increasing the cell size also increased the input capacitance, but still the net delay was significantly reduced. These cells were only used in the most critical paths, so their size did not have much impact on the overall layout size.

The H12 cells generally had larger transistors that required the layouts to be taller. However, by using more efficient layout techniques, many of the cells widths were reduced compared to the generic library. Also, the larger height of the cells allowed extra metal tracks to be used by the router that in turn increased utilization of the cell placement.

4. PHYSICAL DESIGN STRATEGY

Physical Compiler was used to generate the placement for the cells. The data cache and instruction cache locations were frozen during the placement process and the floor-planning was data-flow driven. Clock lines were shielded to reduce delay uncertainty by using Wroute in Silicon Ensemble by Cadence. To minimize the power-drops inside the core, a dense metal5/Metal6 power grid mesh was designed.

The balanced clock tree was generated with CTGen. The first design gave a maximum simulated skew across the core of about 120ps across all process and environmental conditions. Typically this would be significantly less. Then another technique to keep skew under control between process corners was applied, which was to use higher than normal rise and fall times. The drawback was that this resulted in a slightly larger clock tree, and higher power consumption. However, the final skew was reduced to about 80ps.

The final routed design was formally compared to the provided netlist using Formality. The Arcadia extracted netlist delay was calculated using PrimeTime.

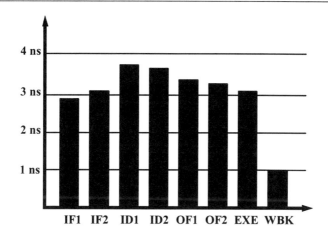

Figure 4. Static timing analysis of the worst-case pipeline paths.

5. RESULTS

The iCORE processor was fabricated in STMicroelectronics' 0.18m HCMOS8D process. This is a multi-threshold process, and threshold voltages of 0.47V and 0.57V were used. The majority of the cells used were low threshold (high speed) for the prototype. A GDSII plot of the test-chip is shown in Figure 5. Excluding the memories, the entire chip is seen to have the distinctive random layout of synthesized logic. The large memories on the right and on the top and bottom of the plot are the data and instruction RAMs. The other small memories seen in the plot are the Local Workspace Cache, the Branch History Table and Branch Target Buffer.

Timing simulation indicated that a good balance of delays between the pipeline stages was achieved, as shown in Figure 4. Power consumption of the CPU core itself, as measured by PrimePower on the synthesized part under worst-case power consumption conditions (fastest process, 1.95V, -40 degrees C), averaged about 147mW.

Silicon testing on several devices over two or three wafers showed functional performance of the design from 475MHz at 1.7V to 612MHz at 2.2V and 25°C ambient temperature. The clock frequency was 520MHz under typical operating conditions of 25°C ambient temperature and 1.8V. The wafers used for conducting the frequency tests were very close to typical process speed.

The measured clock frequencies on these near-typical devices were in-line with the expected improvements from the worst-case process and worst-case operating conditions, post-layout simulated frequency estimate. This was about 275MHz at 125°C, 1.5V, with the longest paths being in the instruction decoder stages. The pre-place-and-route simulated timing measurements varied from this by only a few percent, a good testimony to the Physical Compiler methodology.

The area of the CPU core without caches was approximately $1.8mm^2$ and, with the 8 KB instruction and data caches, approximately $3.7mm^2$.

Analysis of the Dhrystone 2.1 benchmark showed that iCORE achieved an IPC (Instructions Per Cycle) count of about 0.7, which met the goal of being the same or greater than that of previous ST20-C2 implementations, indicating that the various pipeline optimizations were functioning correctly.

The total design time was around 100 man-months, excluding the H12 library work. The design took about 15 months from concept to tape-out.

6. CONCLUSIONS

The performance gap between custom and synthesized embedded cores can be closed by using deep, well-balanced pipelines, coupled with careful partitioning, and mechanisms that compensate for increased pipeline latencies. This enables the use of simple and well-structured logic functions in each pipeline stage that are amenable to highly optimal synthesis [1]. The performance gains are consolidated by use of placement-driven synthesis, and careful clock-tree design.

This chapter explored the use of advanced logic design to implement a high performance CPU core using an ASIC methodology. By avoiding custom circuit design, we have been able to reduce the design time and high costs traditionally associated with it, while still achieving excellent performance.

The final chip has been demonstrated to be functional in silicon.

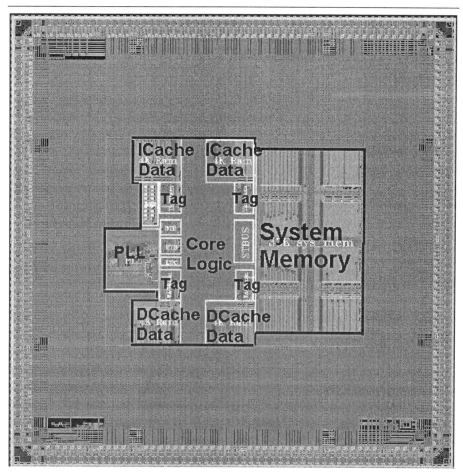

Figure 5. A GDSII plot of the iCORE test-chip.

7. REFERENCES

[1] Chinnery D. G., and Keutzer, K., "Closing the Gap Between ASIC and Custom: An ASIC Perspective," *Proceedings of the 38th Design Automation Conference*, Las Vegas, NV, June 2001, pp. 420-425.
[2] Snyder, C. D., "Synthesizable Core Makeover: Is Lexra's Seven-Stage Pipelined Core the Speed King?" *Cahner's Microprocessor Report*, July 2, 2001.
[3] STMicroelectronics, "ST20-C2 Instruction Set Reference Manual," *STMicroelectronics, 72-TRN-273-02,* November 1997.

Chapter 17

Creating Synthesizable ARM Processors with Near Custom Performance
Migrating from full-custom design techniques

David Flynn
Fellow, ARM R&D,
Cambridge, UK
David.Flynn@arm.com

Michael Keating
Synopsys,
700 E. Middlefield Road,
Mountain View, CA 49043, USA

1. INTRODUCTION

In 1997, ARM and Synopsys began a collaboration to create a synthesizable ARM7 core. After completing that project successfully, the two companies continued working together to develop more advanced synthesizable processors and to refine the process for designing, delivering, and integrating soft processor cores. Through these efforts, both teams learned a great deal about the challenges of delivering processor cores to a large, heterogeneous customer base and the appropriate technologies for meeting these challenges.

This chapter describes several of the key projects and the lessons learned from them. We start by describing, in detail, the process of converting the full-custom version of the ARM7 into a synthesizable processor. Then we describe two follow-on projects: the conversion of the ARM9 into a synthesizable core, and the extension of the ARM9 to a synthesizable ARM9E. Finally, we summarize how these projects have influenced the development process ARM uses today.

Parameter	0.35um technology	0.25um technology
Speed (MHz)	77	85
Area (3LM, mm^2)	2.10	1.00
Power (mW/MHz)	1.10	0.31

Table 1. Speed, area, and power for the ARM7TDMI.

2. THE ARM7TDMI EMBEDDED PROCESSOR

2.1 Flagship Product in 1997

Designers looking to license a CPU are typically concerned first about performance, with area and power consumption as secondary but still very important concerns. By 1997, the ARM7TDMI had established a reputation for effectively addressing these needs across a large number of semiconductor processes. At that point, the ARM7TDMI was clearly the flagship product for ARM.

The ARM7TDMI was implemented using full-custom techniques in a process-portable manner that, by late 1997, had been ported to over 40 semiconductor processes (0.35 and 0.25 micron technologies at that time).

Key to the success of the ARM7TDMI was the fact that each new port of the processor started from an implementation that exhibited an excellent balance between speed, power, and area. In addition, the ARM porting team had a very good idea of process spread across similar design rules. A pre-fabrication characterization flow that provided a highly accurate prediction of the performance of the actual silicon was developed using the Synopsys TimeMill and PathMill tools.

2.2 A Flexible Architecture

The ARM7 family implements a micro-architecture for the ARM Instruction Set Architecture with a unified 32-bit instruction and data bus interface and a three-stage pipeline design. It is optimized for general-purpose embedded microprocessor system designs.

The "TDMI" suffix indicates additional hardware features for the basic CPU core:
- 'T' - Support for the compressed 16-bit Thumb instruction set in addition to the 32-bit native ARM instruction set
- 'D' - Scan-based embedded debug support to allow the CPU to be stopped and restarted, and a scan chain on the data bus to allow

instructions to be scanned into the CPU pipeline and load and store data to be intercepted
- 'M' - Enhanced multiplier for 64-bit result multiplication operations over and above the basic 32-bit results from ARM7 instructions
- 'I' - In-circuit debug emulation support logic that implements two watchpoint/breakpoint units, serial communications, and debug control scan chains for the external debug agent

Although the T, D, M, and I extensions were architecturally modular, every licensee that purchased the Thumb-enhanced CPU received the complete ARM7TDMI. While alternative cores with and without debug (D and I) extensions sounded attractive in theory, the advantage of qualifying and supporting a single fully featured hard macro core appeared to outweigh the need for flexibility in area.

2.3 Vital Statistics

Representative results under worst-case conditions (120°C, voltage 90% of nominal, slow-corner silicon with nominal 3.3V for 0.35um and 2.5V for 0.25um) are shown in Table 1.

2.4 Micro-architecture

The basic ARM7 architecture is a three-stage pipeline:
1. Fetch - Instruction read from memory
2. Decode - ARM instruction decode with optional Thumb decompression
3. Execute - Controls the datapath and sequence for multi-cycle operation

The micro-architecture balances [1]:
- The memory access cycle time, which dominates the Fetch timing
- Instruction decoding cycle time, which dominates the Decode timing
- Datapath sequencing cycle times, which dominates the Execute timing

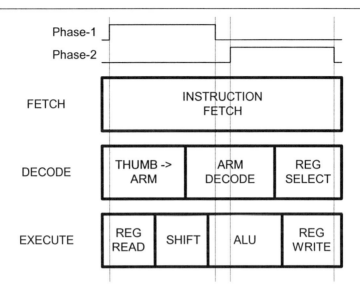

Figure 1. The ARM7TDMI pipeline.

Figure 1 depicts the concurrent pipeline stages, and the clock phases are described below.

2.5 Datapath

The Execute stage controls the datapath. Within a single cycle:
- Two source registers are read onto internal A and B buses.
- The B operand is optionally barrel-shifted left or right up to 32 bits with appropriate sign extension.
- The Arithmetic Logic Unit is then presented with operand A, and the result is computed and broadcast on the ALU bus.
- The result is optionally written back to the appropriate destination register.
- Condition code flags (including zero detect on the ALU result) are computed in time to update the Program Status Register at the end of the same cycle.

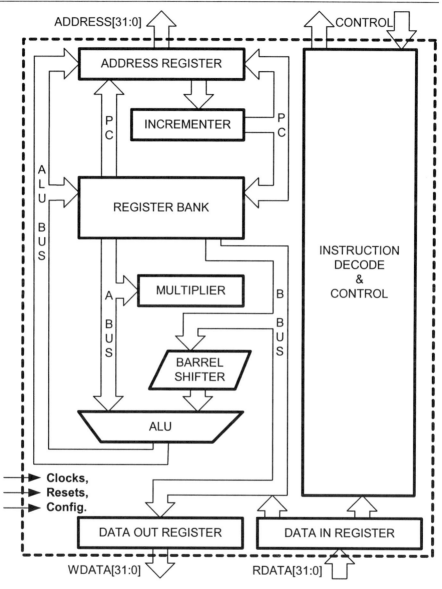

Figure 2. The ARM7TDMI datapath.

In parallel with this complex Execute cycle, sequential addresses may be generated using an incrementer, so only branch target and load/store target addresses use the full ALU. The unified instruction and data interface provides the instruction opcode during the first cycle of execution. Because it is unified, any data read or write operation creates a pipeline stall. Figure 2 illustrates the basic datapath.

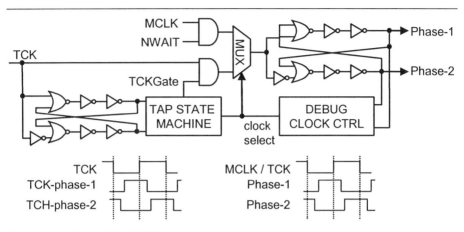

Figure 3. The ARM7TDMI clocking structure.

2.6 Implementation - Clocking

Figure 3 shows the clocking scheme. There are two clock inputs to the ARM7TDMI design, a main clock, MCLK, and a test and debug clock, TCK. Normally, these two clocks drive separate clock domains in the core: the main clock drives the core itself, and the test and debug clock drives a set of scan registers. In certain debug modes, the test and debug clock is switched to drive the core itself, in order to scan out register contents as part of debug.

These two clocks can be entirely asynchronous to one another; the synchronization and safe clock switching required for correct operation is handled internally.

Internally, the full-custom core uses transparent latches and a two-phase, non-overlapping clock scheme. The design is fully static, and the clocks may be phase-stretched appropriately, subject to minimum low-phase and high-phase constraints. To support wait states for slow memory, a wait state input is provided to gate MCLK. Clocking to the entire core is suspended during waits, reducing power. This clock gating circuit requires careful design to assure correct, glitch-free operation.

The timing of the memory interface design is kept as simple as possible. However, timing arcs originate from both edges of the clocks, and the timing models do have to factor in mode-dependent timing for both memory and scan clocks (MCLK, TCK) for synthesis and static timing analysis.

Finally, from a design perspective, the internal high-phase and low-phase critical paths determine the required external clock duty cycle and the maximum clock frequency. Pre-fabrication characterization measures these paths, and this data is used to define the minimum phase widths for the low and the high clock times, and the minimum clock period. Distributing a clock signal at the chip level to meet exactly these conditions is a serious challenge.

Figure 3 also shows the non-overlapping clock phase generator that is implemented within the core. The TAP state machine and scan chains are clocked by the TCK serial clock and internally also use two-phase, non-overlapping clocks ("TAP STATE MACHINE" in the diagram). The primary internal non-overlapping clock phases to the microprocessor are multiplexed between MCLK (gated with NWAIT) and TCK pulses gated in specific TAP controller states. The multiplexer source selection ("DEBUG CLOCK CTRL" in the diagram) handles the safe switching between the system and debug clock sources.

2.7 Implementation - Bus Interface

The primary external interface for the core is the memory bus, which is used for memory accesses and for reading and writing registers in the rest of the chip.

Basic operation is illustrated in Figure 4. The memory interface timing is derived from the *falling* edge of MCLK. The address and control signals are generated in the cycle before the data access. The single NWAIT clock qualifier must be driven low to add a wait state and high to complete the access. It is used to acknowledge both the data transfer completion and the acceptance of the next address and control information into the memory controller that is implemented outside the core.

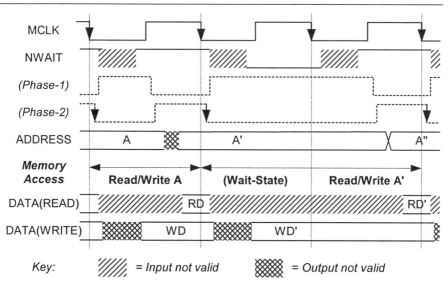

Figure 4. ARM7TDMI memory access.

The memory interface is half-duplex and is used to transfer read or write data where the direction is signified by the pipelined read/write control signal. The ARM7TDMI, in fact, presents a bidirectional, tristate-able bus as well as unidirectional read and write data buses—96 bits in total.

2.8 Datapath - Critical Paths

The pipeline diagram in Figure 1 shows the basic functional operations that happen within a clock cycle for the concurrent Fetch, Decode, and Execute stages. If the on-chip memory is able to support single-cycle accesses, then the Execute stage presents the greatest challenge in terms of overall cycle time.

Using full-custom datapath design and layout techniques, the block-level contributions to the overall critical path are, in decreasing order of significance:

- ALU (~45%) - Valid A and B inputs to valid result calculation (32-bit result carry-select propagation path)
- Register read (~25%) - Dual-ported 32-bit A and B register data selection from the 31 internal registers (with pre-charged bus implementations for area/speed)
- Shifter (20%) - Full-custom, cross-bar switch design
- Register write set up (~5%) - In order to be set up for reading next cycle without interlock requirements

However, phase non-overlap was a significant factor for some internal paths. Often, this non-overlap amounted to up to 15% of the cycle time *for each phase*. The transparent latch design, together with application of specialized timing analysis tools, was able to hide some of this effect by "borrowing" time from the phase before and phase after. But this required extremely precise control of the duty cycle of MCLK that typically can only be achieved by using a double-frequency PLL clock source with a divide-by-two, derated for clock distribution and buffering across the final chip layout.

2.9 Testability

High fault coverage is a major challenge for full-custom cores and a fundamental requirement for subsequent system-level integration. The ARM7TDMI was developed from earlier processor cores and benefited from a legacy of handcrafted vectors developed over many years to test the full-custom structures and control paths in the design. When extended for the debug, multiplier, and Thumb support, a test vector set with 96% fault coverage was developed.

Testing a full-custom core inside an ASIC requires either serial or block-level test access. Both JTAG serialized patterns and 32-bit parallel AMBA

Test (TIC) "canned" test vector sets were provided with the CPU to support either test approach.

2.10 Characterization

A Pre-Fabrication Characterization (PFC) methodology was developed using a mix of dynamic and static timing analysis approaches to provide sign-off quality AC characteristics for the full-custom core over a representative range of input slopes and output loading.

2.11 Qualification

Final qualification of the ARM7TDMI design ported to the specific process always required that a test chip be fabricated. Both the timing correlation with the PFC numbers and the architectural compliance, checked by running the entire processor validation suite on the hardware, were used to sign off the embedded macrocell as compliant and qualified for customer ASIC design.

3. THE NEED FOR A SYNTHESIZABLE DESIGN

As the ARM7 became increasingly popular, the ARM Design Migration team was under growing pressure to port, validate, and characterize a large number of ARM7TDMI implementations—all urgent for particular licensees with their own commercial priorities. As the queuing times increased, a number of ARM7TDMI licensees asked for synthesizable implementations to allow them to port the design themselves.

However, the specification for such a synthesizable product was difficult to resolve. Most customers wanted not only a CPU core that they could synthesize with standard tools and design flows, but also the security of having the identical interface and design-in appearance as the full-custom core.

3.1 Initial Synthesizable ARM7TDMI

ARM developed an initial synthesizable ARM7TDMI with one licensee. This version modeled the multiple clock-phase design exactly, complying with the full-custom, block-level specification. In order to make the interface look compatible, transparent latches had to be implemented at the periphery.

The resulting design achieved the target clock frequency, but at the expense of the following:
- Complex balancing of the internal clock trees
- Expert CPU knowledge required to optimize critical paths (that were both cell library and wire load dependent)

- Sign-offs of many internal multi-cycle paths

Testability was also compromised by having to split up the scan chains across the internal clock tree segments and constrain the sampling of states across these multiple clock domains.

At this point, the product was not in a form that could be widely licensed with a sustainable level of support.

4. THE ARM7S PROJECT

Late in 1997 a new project was set up between ARM and the Design Reuse Group (DRG) at Synopsys (now part of the Synopsys IP and Systems Group) to develop a cycle-compatible, soft implementation of the full-custom ARM7TDMI that ARM could deploy to multiple licensees without incurring a large support demand. The design needed to work with standard synthesis, timing, and test tools. The final design had to provide a robust, deterministic implementation that users could integrate without having expert knowledge of the CPU internals. Finally, to make the goal even more challenging, the mandate was to deliver this to the first customer within six months in Verilog HDL, with the VHDL release to follow within two months.

4.1 Challenges for a Synthesizable ARM7

The first step in the joint project was for the ARM engineering team to come up with a specification for the new design. This specification had to reflect the lessons learned from the initial synthesizable ARM7 project. It also had to reflect the best practices for synthesizable design as understood by the Synopsys design team.

The key issues in developing the specification all centered around clocking. The key clock-related issues were directly related to the choice of a dual-phase, non-overlapping clocking scheme—a standard practice for full-custom processors, but problematic for a synthesizable core. The clocking difficulties were as follows:

- The non-overlapping clock generator was difficult to design.
- The resulting clocks had a gap between the falling edge of Phase 1 and the rising edge of Phase 2. This gap was effectively dead time leading to lost performance.
- To recover some of this lost performance, designers of the full-custom implementation used time borrowing. This resulted in a design that required a great deal of expertise to produce and was not amenable to automated static timing analysis.

- This design complexity was compounded by the fact that the main clock could be switched between MCLK and TCLK, which were asynchronous to each other.
- The wait state clock gating required very careful design to ensure correct operation.

In addition to clock-related problems, bidirectional data bus support was an issue. Internal control of the tristate output enables, even with comprehensive enable and disable input controls, always required manual effort in the design flow.

Developing the synthesized ARM7 presented an opportunity to address all these issues. But one of the concerns of the design team going into the project was the potential loss of predictability and performance in the design. When moving to a synthesizable product, the implementation of register banks, barrel shifters, and arithmetic logic units changed from deterministic, handcrafted structures into larger, less deterministic structures with data flowing between them.

Accurately predicting the area and power penalties resulting from the move to a synthesizable implementation presented a serious technical and commercial challenge. The pre-fabrication characterization used on the full-custom core relied on a precise knowledge of the underlying transistor structures. The synthesizable design would rely on static timing analysis. How much performance and predictability would be lost in this conversion remained an open question.

4.2 The Specification

ARM engineers carefully reworked the specification in the light of the challenges described above. The new specification called for:
- A single clock domain with industry-standard rising-edge D-type clocking. Initially this was a contentious issue because it meant inverting the primary clock polarity. MCLK was essentially falling-edge triggered, whereas the clock in the synthesizable design, CLK, became rising-edge.
- Scan-based debug, synchronized to CLK.
- A cycle-compatible memory interface, but with all phase-dependent timing removed. This allowed reworking of the wait state interface (always a critical path on the full-custom CPU) while requiring minimum applications changes by the user.
- Asynchronous reset with two reset domains: one for the processor itself, and the second for the debug subsystem.
- Unidirectional data buses. Those (few) designers who really insisted on tristate buses would have to implement them external to the core.

- No combinational input to output paths. Ideally, the specification would have required all inputs to be registered, with no combinational logic at the periphery of the core. But this could not be achieved without sacrificing cycle compatibility with the full-custom core. So combinational paths on inputs and outputs had to be handled in the design and integration flow. The requirement for a level of expertise needed to integrate such a CPU core with appropriate timing budgeting was an acceptable compromise.
- Some work put into making the Thumb, multiplier, and debug functionality more modular and more synthesis-friendly.

4.3 Re-specifying the Memory Interface

Figure 5 shows the conversion of the memory interface to be driven by the rising edge of CLK. Verifying that the new interface was cycle accurate to the old interface was non-trivial, because the old interface relied on the different phases of MCLK. By doing some phase stretching on MCLK, the team was finally able to use the same input clocks and vectors to perform a side-by-side comparison of the old and new interfaces.

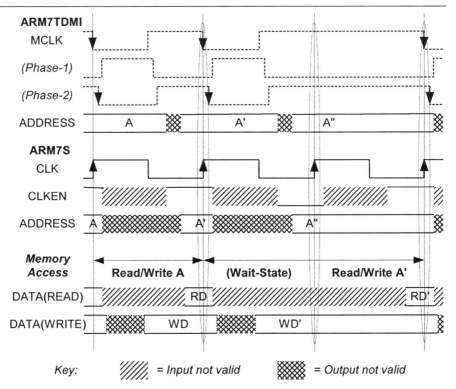

Figure 5. ARM7S cycle-based memory access.

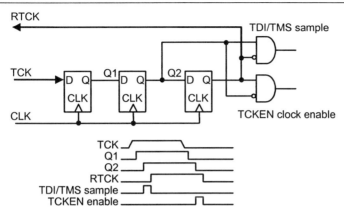

Figure 6. ARM7S debug clock domain synchronization.

4.4 Handling the Scan/Debug Clock Domain

In order to support external debug agents with asynchronous clocks, the decision was made to require the external TCK clock (typically 10MHz and often less for simple debug units) to be synchronized to CLK. The systems integrator was required to implement a basic synchronizer of the form shown in Figure 6. This circuit generates pulses to sample scan inputs TDI and TMS following the rising edge of TCK and advances the CPU debug state using TCKENABLE following the falling edge of TCK.

To support systems where the internal CLK frequency is synthesized from an on-chip PLL, and where standby slow clocking modes are supported, the concept of returning the synchronized test clock, RTCK, was introduced. This approach provided external clock edge handshaking to prevent overrunning the synchronizer period.

4.5 Project Team and Planning

A project team of seven ARM and Synopsys engineers started from the revised specification and the original block-level specifications for the full-custom core, together with the internal ARM C-based simulation environment. The project was planned around the team members as follows:
- Two RTL designers working on data path/pipeline design and control block implementation
- One RTL engineer responsible for debug
- One synthesis engineer working on scripts, synthesis, and test regression (plus some RTL)
- One dedicated verification engineer responsible for regression suites and infrastructure

- Two engineers supporting the validation, resolving specification/ implementation ambiguities and mapping transparent-latch-based micro-architectural elements to registered clock boundaries

4.6 Verification Approach

The team developed a Verilog test harness that compared the behavior of the processor under development to the ARM7TDMI reference design. The harness supported a fully synchronous comparison of the RTL under development against both a detailed Verilog model of the ARM7TDMI and a physical test chip on a logic modeler. As well as observing all ports of the CPU, the Verilog harness supported interrogation of register contents. This capability allowed complete machine state checking at the boundary of every instruction, a key for architectural compliance testing.

One of the key challenges in setting up this test harness was dealing with the different clocking schemes between the processor designs. By carefully generating the memory clock waveform as described in the cycle-based interface re-specification, and transforming the coprocessors used in the test harness from phase-based design to single-edge clocking, the primary clock domain could be tracked between the processors.

The entire ARM Architecture version 4T [2] validation suite was run on the design to verify instruction set compliance. In addition, the team used a further set of device-specific test pattern sets to verify detailed micro-architecture equivalence on a cycle-by-cycle basis.

4.7 Synthesis

After some early experiments, it became clear that two parts of the design would most affect the overall quality of results from synthesis and physical design: the register bank and the arithmetic components in the datapath.

The register bank was implemented using DesignWare dual-port memories. Experiments indicated that instantiating these memories as 4-bit slices minimized routing congestion.

Library	Area (NAND-2 equivalents)
Lib1	51688
Lib2	42281
Lib3	54810
Lib4	45495
Lib5	57430
Lib6	46187

Table 2. ARM7S area results for a range of standard cell libraries.

Experiments also indicated that selecting different architectures for the ALU and the PC incrementer allowed for different overall speed vs. area performance trade-offs. The team used inferred DesignWare arithmetic units for both the ALU and the PC incrementer. The synthesis scripts were written so that the top-level synthesis strategies would drive automatic selection of the appropriate architectures.

The final synthesis scripts used a top-down strategy:
- Read in all files, and constrain and compile from the top level.
- Perform a single pass compile, with an optional second pass if there were timing violations on the first pass.
- Perform a "test-smart" compile with the scan flops and ports inferred by a top-level script.

Table 2 shows synthesis results for a representative range of six standard cell libraries for the 0.35um technology that the first customer was targeting. The areas varied considerably due, apparently, to the varying richness of multiplexer cells in the libraries.

The performance numbers also varied considerably, with the worst library failing to meet timing at the 40MHz target, and the best meeting timing at 75MHz.

4.8 Performance Analysis

In the search for high performance, there is often a tendency to over-constrain the frequency and then suffer from long synthesis runs that never meet timing. In November 2000, some work with 0.18um technology libraries was conducted to relate area growth characteristics to increasing frequency targets. The aim was to ensure that one works below the "knee" of the curve where the combinational logic and routing overheads do not violate timing closure. Figure 7 shows the usable range, up to about 100MHz with the library in question, with the next target frequency in the series (120MHz) being beyond the timing closure range.

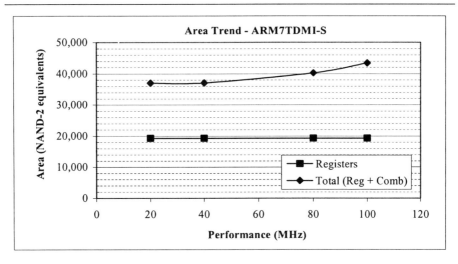

Figure 7. ARM7S target frequency effects on area.

4.9 Area Analysis

Table 3 shows area analysis results for the first layout. The A7S (the synthesizable ARM7) was initially 54% larger than the full-custom version. Analysis showed that the two key culprits were the register bank and the barrel shifter. Eliminating scan reduced the overall area of the core only a few percent. But moving from a flip-flop based register bank to one using a dual-port RAM reduced the area by about 25%. Using the ARM custom register bank reduced the area by a further 13%.

The barrel shifter was the one arithmetic component where the full-custom version was significantly better than the synthesized version. This is due to the fact that the full-custom barrel shifter was able to use pass transistors to great advantage.

Overall, the synthesizable design using a standard dual-port RAM for the register bank was about 20% larger than the full-custom design.

Area/Unit	ARM7TDMI	A7S	A7S	A7S	A7S
Post layout (mm^2)			No scan	Dual-port RAM register bank	ARM custom register bank
Shifter (mm^2)	0.028	0.066		0.085	0.085
Register bank (mm^2)	0.060	0.615		0.200	0.060
Core (mm^2)	1.050	1.620	1.585	1.250	1.110
Ratio (×ARM7TDMI)	1.000	1.540	1.510	1.200	1.060

Table 3. ARM7S area analysis for soft and full-custom implementations.

Statistic	Item
2314	D-type registers in design
99.90%	Fault coverage
1371	Test patterns (after compaction)

Table 4. Test coverage statistics for the ARM7S.

4.10 Testability

Using Test Compiler for scan insertion and automatic test generation, the design team achieved excellent test coverage. Table 4 shows the simplest case: using a single scan chain for the entire core.

Note that if the custom shifter or register bank structures were substituted, then these would require extra test support: some form of BIST for a RAM-based register port, and either a test isolation collar or specific test vectors for the custom shifter.

4.11 Project Pressures

Getting the basic pipeline working to the point of executing the first few instructions took a great deal more time and effort than first anticipated. For example, the pipeline design using transparent latches was not only very area efficient, but also exhibited flow-through characteristics to grow and shrink pipe depth. To achieve identical cycle behavior, it was necessary to add a number of forwarding paths using D-type registers and multiplexers. But once the basic pipeline was working, adding instructions proceeded much more smoothly and made up for lost time.

After four months, hitting the six-month schedule target looked possible from the processor's architectural verification progress and synthesis results. However, the debug functionality proved much more demanding because it was defined more in terms of the existing implementation than a full-detail architectural specification. Getting the final suite of debug validation tests to pass was painful because fixing the more complex tests frequently broke previously passing tests.

After intense work over the last two weeks (and weekends), thousands of lines of Verilog source, the synthesis scripts, and the validation environment had all passed acceptance criteria and been peer-reviewed and documented.

The goal of zero false or multi-cycle paths was met, ensuring the best possible chance of successful deployment by licensees of what was subsequently packaged as the ARM7TDMI-S product.

4.12 AMBA AHB

A final spin-off from the ARM7S project was the AMBA Advanced High-speed Bus (AHB). As part of the project, it was necessary to extend the memory interface to provide a full bus interface compliant with the ARM Advanced Modular Bus Architecture (AMBA) interconnect specification. The existing Advanced System Bus (ASB) had evolved from the interface to the full-custom processors and was now a latch-based design. The work done on the ARM7S resulted in a new bus specification, the AHB, which was based on a single-edge, flop-based clocking scheme. The AHB has now become the standard bus for all ARM processors and processor-based systems.

5. THE ARM9S PROJECT

After the success of the ARM7S project and a subsequent joint project to harden and prove the CPU in a test chip, the opportunity arose in late 1998 for another collaboration between ARM and Synopsys DRG to work on the ARM9 CPU family. The ARM9TDMI micro-architecture had a five-stage pipeline and separate instruction and data interfaces (Harvard architecture). Much of the control logic was available in VHDL RTL form, but again, the two-phase clock implementation needed to be mapped to single-edge register design.

5.1 The ARM9TDMI Embedded Processor

The improved performance of the ARM9 over the ARM7 family, in terms of raw clock rate, is due to a deeper pipeline with fewer logic stages per pipe stage. The Harvard architecture supports concurrent instruction and data memory access, so that single loads and stores can occur in parallel with instruction fetches. Register read operations are decoded and accessed in the cycle before they are used, and writes to registers occur in the cycle after calculation. Forwarding paths are added to the main datapath and more complex control logic is added to handle the interlocks.

The register file becomes more complex to support data access in parallel with the next instruction decode and execution. It has:
- Three read ports, with the third port providing early write data in parallel with address calculation
- Two write ports to support indexed address updates in parallel with data read updates

Figure 8 shows the basic pipeline operations with the clock phases overlaid on the pipe stages. Here, the registers are read at the end of the decode cycle and written back at the start of the final Write stage.

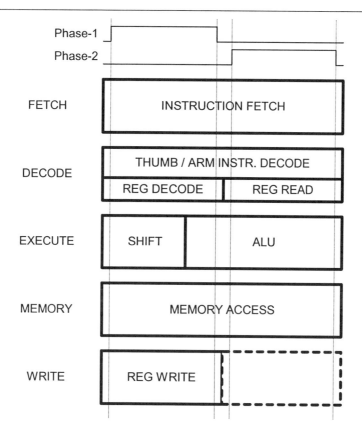

Figure 8. The ARM9TDMI pipeline.

The ARM9TDMI core, unlike the ARM7TDMI, would normally never be used standalone as a micro-controller, as the added performance is only realized if local fast memory or cache is implemented adjoining the core. The memory interface requires some expert work on the timing budgeting for the interconnect and providing coherency between data and instruction sides (to support data access to the instruction side for literal data in code and writing to the instruction side for vectors and debug access).

5.2 Project Team and Planning

The project team included three ARM7S "veterans," two other senior engineers, and one engineer from the ARM9TDMI team.

The ARM9S technical specification was developed using the experience of clock domain unification and some judicious rework of the more complex coprocessor interface to ensure clean cycle-boundary behavior.

To meet the requirement to interface to single-cycle caches and memory at high clock frequencies, a lot of time was spent negotiating the input and output timing constraints, trying to hit the right balance between synthesis performance goals and the timing requirements for real-world memories and caches.

5.3 Synthesis Strategies

A number of synthesis and layout strategies were evaluated to address the I/O constraints and internal maximum operating frequency goals:

Flow 1: DesignCompiler baseline
Synthesis: DesignCompiler-Ultra, Test Compiler
Layout: Timing-driven layout flow
ECO: Links to Layout flow with Floorplan Manager
   1. Fix max timing
   2. Connect scan paths
   3. Area recovery
   4. Fix max timing, repair buffer trees

Flow 2: Power Compiler
Synthesis: DesignCompiler-Ultra, PowerCompiler (clock gating), Test Compiler
Layout: Timing-driven layout flow
ECO: Links to Layout flow with Floorplan Manager
   1. Fix max timing
   2. Connect scan paths
   3. Area recovery
   4. Fix max timing, repair buffer trees
   5. Power optimization

Flow 3: Aggressive PowerCompiler
Synthesis: DesignCompiler-Ultra, PowerCompiler (clock gating and initial RTL optimization), Test Compiler
Layout: Timing-driven layout flow
ECO: Links to Layout flow with Floorplan Manager
   1. Fix max timing
   2. Connect scan paths
   3. Area recovery
   4. Fix max timing, repair buffer trees

Parameter	Flow 1	Flow 2	Flow 3
Speed (MHz)	95.4	81.0	90.9
Area (10^3 NAND-2 equivalents)	61.4	58.1	57.4
Power (mW/MHz)	1.59	1.00	0.99

Table 5. ARM9S post-layout speed, area, and power (5LM, 0.25um TSMC process).

5.4 Results

At high clock frequency targets (105MHz goal) the input/output constraints proved to be the significant challenge. The requirement for output signals to be valid early enough to be used before the clock edge, for real-world deployment in a cached controller, limited the achieved performance rather than register-to-register internal paths.

As Table 5 shows, Power Compiler clock gating inference provided significant improvement to power consumption and had the side effect of slightly reduced area.

6. THE ARM9S DERIVATIVE PROCESSOR CORES

The synthesizable ARM9S has become the technology backbone for enhancements to the ARM9 product family:
- ARM9E-S CPU core with enhanced DSP support
- ARM9EJ-S subsequent development with Java byte-code acceleration support
- ARM966E-S, ARM946E-S, and ARM926EJ-S system cores with parameterized local memory, caches, and virtual memory support, respectively

With the ARM9S designed for standard synthesis, static timing, and test flows, micro-architectural extensions and improvements could be developed and tested for impact on synthesis and trial layout in a matter of days, allowing prototyping orders of magnitude faster than in full custom.

With advances in programmable logic array capacity, the designs could also be synthesized with FPGA tools to allow functional testing and validation with the debug tool chain, even if only at 10s of MHz.

6.1 The ARM9E-S Embedded Processor

The ARM9E-S was developed to address a number of markets where fast, real-time software was required and a mix of DSP and general purpose processing technology was being used. Storage and audio application areas required a modest increase in multiply-accumulate performance together with packed data operations and optional ALU result saturation support.

By carefully mapping these onto the ARM9S micro-architecture, the new product was developed on a tight schedule by a small engineering team. The team was able to deliver the enhanced core to a number of disk drive customers within six months. In this case, time-to-market was the most important factor. The fact that the resulting core had 99+% fault coverage with standard test tools was also vital to the target high-volume applications.

The enhancements to the pipeline are shown in Figure 9. The Write pipe stage actually commits the write at the rising clock edge and takes half a cycle, so that this write data is available for decode register read time of the appropriate subsequent instruction in the pipeline.

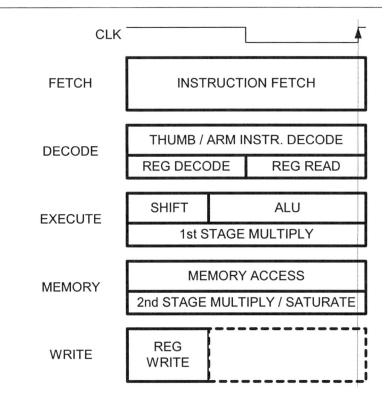

Figure 9. The ARM9E-S pipeline.

6.2 User-Parameterized ARM9E-S Cached Cores

In the years of ARM delivering full-custom cached CPU cores, there had always been tension regarding fixed cache sizes. The full-custom RAM and CAM structures were designed to portable design rules, so they could never provide as dense a RAM structure as process-specific compiled RAM with specific design rule waivers. This is because ARM's hard macros have to integrate into a system design without any top-level DRC problems or expert knowledge waivers.

Making caches sizes configurable was often a customer request, but could never be offered with full-custom cores due to the detailed characterization complexities.

As an example, the ARM920T, a low-power ARM9TDMI cached product with full virtual memory support for Windows-CE, was offered with dual 16 KB instruction and data caches. It took another 18 months before this was complemented by the ARM922T with half-size caches.

Allowing the user to generate specific RAMS and caches using a RAM compiler gave the customer flexibility and better RAM density, but at a cost in:

- Micro-architectural implementation - In particular, using standard synchronous RAM rather than CAM structures for cache tags and translation look-aside buffers results in low-associativity designs
- Power consumption - In the cache due to the simple RAM structures for cache address tags and realistic 4-way set of associative cache structures
- Timing closure - On designs with many physical RAM blocks and significant layout-dependent interconnect
- Parameterized verification suites - For the various memory sizes

Three ARM9E-S family products were developed in the following order:

1. ARM966E-S offering user-configurable, local fast memory for both instruction and data together with a unified external memory bus having write buffer support
2. ARM946E-S with user-configurable, tightly coupled memories and instruction and data caches controlled by a programmable protection unit for up to eight overlaid regions (a superset of the full-custom ARM940T product)
3. ARM926EJ-S with a Java-enabled core and virtual memory support in addition to the separate caches and deterministic local memory banks

ARM developed the 966E-S product internally and developed a real-time trace support macrocell to suit the early real-time markets that required this core.

A Synopsys Design Reuse Group team then took the prototype ARM946E-S design and developed a version that would be easier to support and improved the synthesis strategies.

ARM internally developed the Java extensions to the ARM9E-S and the more complex virtual memory ARM926EJ-S in parallel with the completion of the ARM946E-S.

7. NEXT GENERATION CORE DEVELOPMENTS

The ARM10 is the last processor family envisaged to be developed as full custom. Later generations will all start "soft" and therefore will benefit from rapid prototyping flows and standard EDA tools.

However, there still remain opportunities to reduce power and increase performance by full-custom optimization of datapaths and cache structures. For this reason, there is likely to be some optimal hardening of some CPU cores for mainstream process technologies.

7.1 Full-Custom vs. Synthesizable Design

The experience from the projects described above provides a variety of lessons about the trade-offs between full-custom and synthesizable design.

In the case of area, the ARM7S with an aggressive register file was only a few percent larger than the full-custom design, a negligible amount for all but the most extreme applications.

In the case of power, the ARM9S with an aggressive Power Compiler flow achieved power levels as low as the full-custom design, again within a few percent, and again negligible for all but the most extreme applications.

In the case of speed, the speed of the ARM9 was limited by I/O timing at the memory interface, so this is not terribly informative. The ARM7S appears to be perhaps a few 10s of percent slower than the full-custom version, with considerable variation depending on the technology library. The very best libraries achieved performance very close to that of the full-custom implementation.

These results are impressive, considering that the processors were originally architected and designed for full-custom implementation. One contributing factor was the constraints of the full-custom implementations: they were done with portable process rules and a conscious goal of achieving a balance between timing, power, and area. A processor design that pursued timing at all costs may well not have yielded results this good.

Our belief is that, for most parts of the processor design, full custom provides implementations different from, but not necessarily better than, synthesizable. We also believe that where the full-custom implementation is superior, these differences would be minimized for designs architected for a

synthesis-based implementation. Synthesis may do a better job at optimizing state machine logic, and full-custom designers may be able to do a better job on some highly symmetrical datapaths. But for designs that are intended for use in a wide variety of applications and on a wide variety of processes, the optimal strategy is clear:
- Start with an RTL design which is configurable according to the needs of the application.
- For the desired configuration and technology, harden the design using a synthesis-based flow.
- For those applications where the results do not meet the required performance, choose a few parts of the critical path that can most benefit from full custom, and re-design these by hand.

This process has been shown to produce the best value to the customer, in terms of delivering the performance needed, when it is needed, and on the technology needed.

7.2 Further Work

One key to the process outline above is the flow for hardening the core. Deterministic design flows to harden parameterizable IP quickly, but without jeopardizing CPU performance and power consumption, are essential to making synthesizable processors successful. ARM and Synopsys are now jointly working on reference implementation flows that can successfully automate hardening, characterization, modeling, and deployment to make core-based system-level design successful.

8. REFERENCES

[1] Furber, S. *ARM System-on-Chip Architecture*. Addison Wesley Longman: ISBN 0-201-67519-6.

[2] Seal, D. (ed.) *ARM Architecture Reference Manual*. Second Edition. Addison-Wesley, 2001. ISBN 0-201-73719-1.

David Chinnery

David Chinnery is a Ph.D. student at the University of California, at Berkeley. David has interned at Synopsys, Inc., examining semi-formal verification, and Tensilica, Inc., where he researched latch-based ASIC design. He received B.Sc. (Hons) and B.E. (Hons) degrees at the University of Western Australia in 1998. His research interests include design for high-performance and low power in an ASIC methodology.

Kurt Keutzer

Kurt Keutzer received his B.S. degree in Mathematics from Maharishi International University in 1978 and his M.S. and Ph.D. degrees in Computer Science from Indiana University in 1981 and 1984 respectively. In 1984 Kurt joined AT&T Bell Laboratories where he worked to apply various computer-science disciplines to practical problems in computer-aided design. In 1991 Kurt joined Synopsys, Inc. where he continued his research in a number of positions culminating in his position as Chief Technical Officer and Senior Vice-President of Research. Kurt left Synopsys in January 1998 to become Professor of Electrical Engineering and Computer Science at the University of California at Berkeley where he serves as Associate Director of the MARCO funded Gigascale Silicon Research Center.

Kurt has researched a wide number of areas related to computer-aided design of integrated circuits and his research efforts have led to three Design Automation Conference (DAC) Best Paper Awards, a Distinguished Paper Citation from the International Conference on Computer-Aided Design (ICCAD) and a Best Paper Award at the International Conference in Computer Design (ICCD). He co-authored the book entitled *Logic Synthesis*, published by McGraw-Hill in 1994 and has a book entitled *Closing the Gap Between ASIC and Custom* forthcoming from Kluwer Academic Publications.

From 1989-1995 Kurt served as an Associate Editor of IEEE Transactions on Computer-Aided Design of Integrated Circuits and Systems and he currently serves on the editorial boards of three journals: Integration – the VLSI Journal; Design Automation of Embedded Systems and Formal Methods in System Design. Kurt has served on the technical program committees of DAC, ICCAD and ICCD as well as the technical and executive committees of numerous other conferences and workshops. Kurt is a Fellow of the IEEE.

TK 7874.6 .C47 2002
Chinnery, David.
Closing the gap between AS
 & custom

DATE DUE

THE LIBRARY STORE #47-0204